Edmund Ledger

The Sun

It's Planets And Their Satellites

Edmund Ledger

The Sun
It's Planets And Their Satellites

ISBN/EAN: 9783744743693

Printed in Europe, USA, Canada, Australia, Japan

Cover: Foto ©berggeist007 / pixelio.de

More available books at **www.hansebooks.com**

THE SUN:

ITS PLANETS AND THEIR SATELLITES.

PRINTED BY HAZELL, WATSON, AND VINEY,
LONDON AND AYLESBURY.

THE SUN:

ITS PLANETS AND THEIR SATELLITES.

A Course of Lectures
UPON
THE SOLAR SYSTEM.

READ IN GRESHAM COLLEGE, LONDON,
IN THE YEARS 1881 AND 1882,
PURSUANT TO THE WILL OF SIR THOMAS GRESHAM.

BY

EDMUND LEDGER, M.A.,
Rector of Barham, Suffolk;
Late Fellow of Corpus Christi College, Cambridge.

Illustrated by 94 Woodcuts, a Chart of Mars, and eight Woodbury and Lithographic Plates.

"The Sun, a world whence other worlds drink light;
The crescent Moon, the diadem of night."

London:
EDWARD STANFORD, 55, CHARING CROSS, S.W.
1882.

[*All rights reserved.*]

"I will admire God more, and fear astrologers less."—THOMAS FULLER.

———

" By night an atheist half believes in God!"—YOUNG.

———

" Hæc ita fieri omnibus inter se concinentibus mundi partibus profecto non possent, nisi ex uno divino et continuato spiritu continerentur."—CICERO.

PREFACE.

The Lectures contained in this volume have been delivered at GRESHAM COLLEGE during the last two years. In response to various requests, and by the kind and liberal assistance of the GRESHAM COMMITTEE of the City of London, to whom the author desires to tender his grateful thanks, they are now published.

In the description of the Planets and Satellites which they contain, such information as may be found in almost every astronomical text-book is briefly recapitulated; that which is less familiar, or of recent announcement, is discussed more fully. In the case of the Sun and Moon it has only been possible to mention a few points of special interest. Those who desire fuller information may consult the larger works of Professor Young, Mr. Proctor, and the late Padre Secchi, upon the Sun; of Mr. Neison, Mr. Proctor, and others, upon the Moon. The time and space at the author's disposal have prevented him from making more than a passing reference to the cometary and meteoric members of the Solar System.

A Gresham Lectureship only provides for the employment of a limited amount of time in the duties of the office. It is extremely difficult for one who holds it to maintain a satisfactory acquaintance, amidst the press of other work, with the vast range and the startlingly rapid progress of such a subject as Astronomy, for the study of which the longest lifetime is far too short. Information as to any errors or misprints in this volume will therefore be most gratefully received.

Although the Lectures have been printed from the original manuscript used in their delivery, alterations and additions have been freely introduced. At GRESHAM COLLEGE they were copiously illustrated by means of a lime-light lantern. The illustrations now supplied are necessarily far less numerous.

In accordance with the author's usual practice in his Lectures, references for further information are in general made to such works upon Astronomy as are most easily accessible, rather than to the original authorities, which he has himself in almost every instance consulted.

In footnotes various numerical calculations are introduced. When it is possible to obtain a result by means of arithmetic it seems advisable that the process should be shown. Schoolmasters, pupil-teachers, and others have attended the Lectures, who have found such problems useful and interesting.

Some readers may perhaps think that too much space is given to the celestial phenomena which would be seen by an observer upon the various planets. Their investigation, however, affords a useful mental exercise, and enables a student to realize more easily other celestial movements, a thorough acquaintance with which is essential.

So rapid is the progress of Astronomy, that, even while the present volume has been passing through the press, advances have occurred which have rendered some of its pages almost antiquated. The only reference to Dr. Siemens' theory of the Conservation of Solar Energy is in a footnote printed very shortly after his Memoir was read to the Royal Society, and before it had been possible properly to appreciate it. In like manner it has been impossible to discuss Professor Langley's researches with regard to the atmospheric absorption of the Sun's radiations and the relation of the possible temperatures

PREFACE. vii

upon the surfaces of the various planets to the constitution of their gaseous envelopes.

In the preparation of the Lectures most existing manuals and popular works upon Astronomy were freely employed, but the references were not in every case retained, as they were not needed for *vivâ voce* use. The author has, however, endeavoured to record his indebtedness wherever it was possible. Any omission to do so is quite unintentional.

He begs to state that he owes much to Mr. Proctor's works, especially to those upon the Moon and Saturn; to various volumes and articles by Mr. Lockyer; to Mr. G. F. Chambers' "Handbook of Descriptive Astronomy;" to many pages of the *Observatory;* to Sir G. B. Airy's "Ipswich Lectures;" to Professor Newcomb's "Popular Astronomy;" to Grant's "History of Physical Astronomy;" to Sir G. Herschel's "Outlines of Astronomy;" to Professor Young's "Treatise upon the Sun;" to Dr. Ball's "Elements of Astronomy;" to the "Annuaire of the Bureau des Longitudes," and to that of the Royal Observatory of Brussels; to Sir E. Beckett's "Astronomy without Mathematics;" and to the Rev. T. W. Webb's "Celestial Objects for Common Telescopes."

He tenders his thanks, for much kind advice and information, to the Astronomer Royal; to Dr. Huggins, for very valuable assistance, particularly in connection with Lecture II.; to Dr. Hind, Superintendent of the Nautical Almanac; to Mr. Dale, Fellow of Trinity College, Cambridge; to Mr. Bentley, who has allowed two diagrams to be copied from the English edition of Guillemin's "The Heavens"; to M. Niesten, for permission to reproduce a drawing of the orbits and distribution of the Minor Planets; to Mr. N. Green, for his most liberal loan of the lithographic stone of his chart of Mars; to the Council of the Royal Astronomical Society, for the picture of the Sun's Corona in Plate IV., and for the views of Jupiter in Plate VIII.; to Mr. G. D. Hirst, who presented these last-named views to the Society; to Mr. Trouvelot, whose drawing of Saturn is shown in Plate IX.; to Mr. Nasmyth, for the plate of the lunar mountain, Copernicus; to Mr. G. F. Chambers, for ten woodcuts (six of which are from his "Handbook of Descriptive Astronomy"), and for permission to reproduce two others; to Messrs. Kegan Paul, Trench, & Co., who have kindly supplied the copies of a photograph of the Moon, by Mr. Rutherfurd; to Mr. W. H. Wesley, the Assistant Secretary of the Royal Astronomical Society, for invaluable assistance in the preparation of nearly all the woodcuts, and of Plates II., III., VII., VIII., and IX.; to Mr. W. T. Lynn, who has carefully revised the proof sheets, and whose suggestions whilst doing so have not only much contributed to the accuracy of the statements made in the volume, but have afforded important additions to its information. The kind assistance of many other friends has also been most helpful.

> "With what a perfect, world-revolving power,
> Were first th' unwieldy planets launch'd along
> Th' illimitable void! Thus to remain
> Amid the flux of many thousand years,
> Unresting, changeless, matchless, in their course:
> To night and day, with the delightful round
> Of seasons, faithful; not eccentric once:
> So pois'd, and perfect is the vast machine!"

GRESHAM COLLEGE,
October, 1882.

CONTENTS.

LECTURE I.—THE SUN.

PAGE

Determination of the Sun's distance from the Earth—By means of Transits of Venus—By the Eclipses of Jupiter's Satellites—By Observations of the Planet Mars—And by other Methods 1

LECTURE II.—THE SUN (*continued*).

Description of the Spectroscope—Its Use in Solar Observations—Prominences and Spots upon the Sun—Relation of Sun-spots to Magnetic and Meteorological Phenomena—The Corona and its Long Streamers—The Sun's Temperature 20

LECTURE III.—THE MOON.

The Moon's Distance from the Earth—Methods by which the Moon is weighed—The Moon's Orbit—Its Phases and Librations—The Extent of the Moon's Surface seen from the Earth 49

LECTURE IV.—THE MOON (*continued*).

Lunar Mountains—Their Varied Forms and Telescopic Appearance—Volcanic Action, Past or Present, upon the Moon—The Lunar Atmosphere—Photographs of the Moon—Lunar and Solar Eclipses—The Moon and Navigation—The Moon and Easter—The Moon and the Weather . 69

LECTURE V.—PTOLEMY *versus* COPERNICUS.

The Apparent Paths of the Planets seen from the Earth—The Ptolemaic Theory—Its Cycles and Epicycles—The Copernican Theory—The Orbits of the Planets—Kepler's Three Laws—Bode's So-called Law of Distances—The Phases of an Inferior Planet—The Retrogression and the Stations of the Planets 98

LIST OF ILLUSTRATIONS.

FIG.		PAGE
1.	A Well-conditioned Triangle	3
2.	An Ill-conditioned Triangle	3
3.	Transit of Venus, 1874—Ingress as seen at Rodriguez	10
4.	„ „ „ Ingress as seen at Kerguelen's Land	10
5.	„ „ „ Egress as seen in Egypt	10
6.	Distance Traversed by Light from Jupiter to the Earth	14
7.	Sun Spots seen as Depressions in the Photosphere	21
8.	Form of a Solar Prominence indicated by Bright Lines	30
9.	First Solar Prominence seen by Dr. Huggins with a widely-opened Slit	30
10.	Curves of Sun-spot Frequency and Magnetic Force	39
11.	The Moon's Path curved by the Earth and by the Sun	58
12.	The Moon's Orbit always Concave to the Sun	61
13.	Path of the Moon seen from the Earth	61
14.	False Representation of the Moon's Orbit	62
15.	The Moon Full and Half-full	64
16.	One-fourth of the Moon's Surface seen as a Semicircle	64
17.	Phases of the Moon	65
18.	Phases of the Moon	65
19.	The Great Crater of Teneriffe	75
20.	Line of Nodes of the Moon's Orbit passing through the Sun	85
21.	Eclipse of the Moon—its Maximum Duration	85
22.	The Full Moon escaping Eclipse	86
23.	The Moon nearer in the Zenith than in the Horizon	90
24.	Path of Mercury relatively to the Earth, by Cassini	100
25.	Apparent Paths of Jupiter and Saturn	102
26.	Apparent Path of Mercury in 1881 and 1882	103
27.	„ „ Venus, Feb. 1881 to Feb. 1883	104
28.	„ „ Mars, 1877-9	105
29.	Cycles and Epicycles of Venus and Mercury	107
30.	„ „ Mars, Jupiter, and Saturn	110
31.	Time occupied by Jupiter in One Loop of its Apparent Orbit	112
32.	Cycle and Epicycle, showing the Description of the Apparent Orbit of Mars	113
33.	The Copernican Theory of the Solar System	118
34.	Ellipses of Different Degrees of Ovalness	119
35.	Kepler's Second Law	122

LIST OF ILLUSTRATIONS. xiii

	PAGE
36. Greatest Apparent Distance of an Inferior Planet from the Sun	125
37. Inferior and Superior Conjunctions—Tilt of an Orbit to the Plane of the Ecliptic	126
38. The Phases of an Inferior Planet	127
39. Mercury and Venus at their Greatest Elongations from the Sun	129
40. Alternate Progression and Retrogression of Venus	132
41. Path of Venus during a Part of the Year 1884	133
42. Venus Stationary	135
43. Greatest and Least Disc of the Sun seen from Mercury	147
44. Comparative Discs of the Sun seen from Mercury, Venus, and the Earth	148
45. Mercury in Transit, Nov. 5, 1868	152
46. The Phases and Varying Apparent Diameter of Venus	159
47. Incidence of the Sun's Rays in Summer	173
48. ,, ,, ,, Winter	174
49. The Sun's Diurnal Path at Different Times in a Year	175
50. The Poles of the Heavens seen in the Horizon at the Equator	178
51. The Apparent Altitude of a Pole of the Heavens	180
52. Diurnal Path of the Sun at the Summer Solstice in Latitude $66\frac{1}{2}°$	183
53. ,, ,, ,, seen from the Equator	183
54. A Spheroid generated by the Revolution of an Ellipse	194
55. The Length of a Degree of Latitude increases with the Latitude	195
56. Foucault's Pendulum Experiment	200
57. Foucault's Pendulum at a Pole and at the Equator	201
58. The Meaning of Centrifugal Force	203
59. The Cavendish Experiment	210
60. Oppositions of Mars from 1877 to 1892	232
61. Distances of the Earth, Æthra, Mars, Hilda, and Jupiter from the Sun	269
62. The Forms of a Parabola, an Ellipse, and a Hyperbola	282
63. Orbits and Distribution of Minor Planets	290
64. The Polar Compression of Jupiter	297
65. The Centre of Gravity of Jupiter and the Sun	298
66. The Velocities of a Point upon Jupiter's Equator	301
67. Jupiter, drawn by Mr. De La Rue, October 25th, 1856	304
68. ,, showing the Great Red Spot, July 9th, 1878	306
69. ,, ,, ,, Sept. 25th, 1878	307
70. ,, March 13th, 1872	309
71. ,, April 18th, 1872	309
72. ,, Jan. 28th, 1872	310
73. ,, Jan. 30th, 1872	310
74. Configurations of the Satellites of Jupiter in 1610	322
75. The Length of Jupiter's Shadow	332
76. Relative Positions of the Three Inner Satellites of Jupiter	333
77. Occultations and Eclipses of Jupiter's Satellites	336
78. Transit of a Satellite and of its Shadow	340
79. Size of the Shadow of the 2nd Satellite upon Jupiter	342
80. Path of the Fourth Satellite of Jupiter in Space	351
81. Path of the First Satellite of Jupiter in Space	352
82. Saturn, drawn by Mr. De La Rue, March 27th and 29th, 1856	358
83. The Phases of Saturn's Rings	359
84. Saturn's Ring-plane sweeping across the Earth's Orbit	360
85. Disappearances and Reappearances of Saturn's Rings	363

LIST OF ILLUSTRATIONS.

	PAGE
86. Saturn, drawn by Mr. Wray, Dec. 26th, 1861	371
87. Saturn's Rings seen from the Planet—their Effect in producing Eclipses of the Sun	379
88. Latitudes upon Saturn from which its Rings are Visible	383
89. Transit of Titan, Dec. 9th, 1877	391
90. Representation of a Change of Inclination of 98° in a Planet's Axis	404
91. A Change of Inclination of 82° in the Opposite Direction	406
92. Illustration of the Perturbation of Uranus by Neptune	411
93. Perihelia of the Orbits of Jupiter, Saturn, Uranus, and Neptune between the Years 1876 and 1887	419
94. Comparative Sizes of the Sun and the Planets	426

PLATES AND WOODBURY PRINTS.

1. Chart of Mars, by Mr. N. E. Green	*Frontispiece*	
2. Solar Prominences	*To face page*	32
3. Sun Spots and Faculæ	,,	35
4. The Corona during a Total Solar Eclipse	,,	43
5. Photograph of the Moon, by Mr. L. M. Rutherfurd	,,	49
6. The Lunar Mountain Copernicus	,,	72
7. Drawings of Mars in 1877, by Mr. N. E. Green	,,	240
8. Drawings of Jupiter in 1878, by Mr. G. D. Hirst	,,	305
9. Saturn, drawn by Mr. L. Trouvelot	,,	369

DESCRIPTION OF THE DESIGN ON THE COVER.

THE comparative sizes of the Sun and the Planets are represented in gold as correctly as it was possible to print them upon the cloth. They are shown more accurately in a diagram at the end of Lecture XV.

The symbols above and below the group of Sun and Planets are those in general employed by astronomers for the sake of brevity. In order from left to right they represent—the SUN, the MOON, MERCURY, VENUS, the EARTH, MARS, JUPITER, SATURN, URANUS, NEPTUNE.

They are also given in two sets of five upon the back of the cover above and below the title, underneath which is a representation of Jupiter and its four Satellites.

Amongst other interpretations of the above-mentioned symbols, that of MERCURY has been supposed to represent his *herald's wand*; that of VENUS, her *looking-glass*; that of MARS, his *spear* and *shield*; that of JUPITER, his *throne*; that of SATURN, his *sickle*. Those of URANUS and NEPTUNE are respectively derived from the first *letter* of Sir William Herschel's name, and the *trident* of the great Sea-god.

ADDENDA ET CORRIGENDA.

PAGE
    3, line 14, *before* the Sun *insert* a point upon.
   13, ,, 13, *for* Leverrier *read* Le Verrier.
   13, ,, 30, *for* lights *read* light.
   34, ,, 27, *for* vertical *read* vortical.
  161, ,, 18, *before* Venus *insert* the orbit of.
  181, ,, 2 of footnote, *after* perpendicular to *insert* and in the same plane with.
  186, ,, 12, *before* monsoons *insert* south-west.
  186, ,, 14, *for* the African deserts *read* southern Asia.
  220, ,, 31, *omit* and twenty-five minutes.
  228, ,, 4, *for* equatoreal and polar *read* polar and equatoreal.

In connection with the footnote, p. 208, Lecture IX., reference may be made to *Notes and Queries* of September 30th, 1882, where the etymology of the name of the mountain called *Shehallien* by Dr. Maskelyne is discussed. It appears that Shehallion best represents the pronunciation of the name.

Since Lecture XI. was printed two additional Minor Planets have been discovered; viz., No. 230 by Dr. de Ball on September 3rd; and No. 231 by Dr. Palisa on September 10th, 1882.

The references in Lecture XIV. are to the first edition of Mr. Proctor's "Saturn."

LECTURE I.

THE SUN.

"Sire of the seasons! Monarch of the climes
And those who dwell in them! for, near or far,
Our inborn spirits have a tint of thee,
E'en as our outward aspects."

THE progress of astronomical science during the last five-and-twenty or thirty years has been so rapid as almost to approach the marvellous. This statement is true of astronomy as a whole; but it applies with especial force to the advances that have been made in the study of the Sun's physical constitution and condition. Discovery has followed upon discovery, and victory upon victory, as astronomers have stormed one outlying fortress after another of the Sun's hidden secrets, which, until lately, appeared impregnable. At the same time, the more the horizon of our knowledge widens, the less we seem to know, in contrast with the boundless field of still untrodden truth that each successive step opens to our view when we turn our thoughts to the Sun.

It is therefore not only in order to expatiate upon *past* triumphs that we begin our discussion of the Solar system by drawing the attention of our readers to that great luminary, without whose quickening rays and potent sway the Earth would, in a few brief days, or weeks, become a frozen desert, and the whole family of planets a wild band of rioters; but we have a still more cogent reason for doing so in the intense and increasing interest which must inevitably attend the study of the Sun for many a year to come. We can confidently affirm that this must be the case, when we notice the very remarkable manner in which this branch of astronomy is linked to other kindred sciences. The discoveries which have recently been effected in chemistry and physics by means of the spectroscope; investigations as to the simple or compound nature of substances hitherto classed as elementary; refined researches into the phenomena of the highest obtainable vacua; the most advanced methods of photography; and some of the most important problems of meteorology, magnetism, and electricity, are all found to be intimately connected with the study of the Sun's physical condition.

But before we discuss other points of greater fascination, and in some respects of deeper interest, it may be well, by way of preface, to explain what has been accomplished by means of astronomical observations, and by other scientific methods, towards an accurate determination of the most fundamental portion of all our knowledge of the Sun, viz., its distance from the Earth. For our greatest hindrance in all investigations of its condition is the vastness of that distance; and the importance of its more accurate determination has of late been brought into special prominence in connection with the transits of Venus of December 8th, 1874, and December 6th, 1882; the latter of which will happily (weather permitting) be partly visible at Greenwich.

To find how far the earth is from the Sun is a problem well worth solving, for almost every other astronomical measurement depends upon it. But it is in no wise an easy problem. It gives scope for the most refined investigations, and yet it is one in which, we believe, even those who have had very little scientific or mathematical learning may take an intelligent interest. Its direct solution by the ordinary methods used in a terrestrial survey is altogether impossible. A surveyor measures the distance between two stations, A and B, as in Fig. I., and observes with his theodolite the directions in which a distant object, S, is seen from them. He is then, in general, able, by means of certain trigonometrical formulæ, to calculate the distance of that object from either of his stations. But, in order to use his formulæ with success, it is essential that the triangle formed by lines joining the stations with each other and with the distant object, be what is termed a *well-*

conditioned triangle. This involves that the distance of the places of observation from one another must *not* be, as they are in Fig. II., very small compared with their distances from the object. In such a case the formulæ fail to give any accurate or useful result.

Now, if we attempt to apply the above method to the determination of the Sun's distance, it is easy to see that we

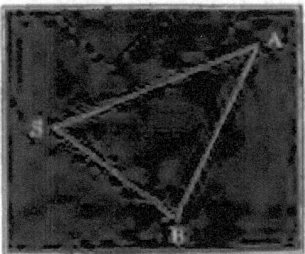

Fig. I.—A well-conditioned triangle.

shall not only have an ill-conditioned, but an *exceedingly* ill-conditioned triangle to deal with.

The utmost distance, AB, measured in a straight line, between any two stations that we can select upon the Earth, is equal to the length of its diameter, which is rather under 8,000 miles. And this is less than the $\frac{1}{11000}$th part of the

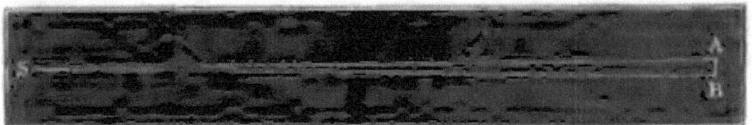

Fig. II.—An ill-conditioned triangle.

distance, AS or BS, from any such station to the Sun, or very nearly in the ratio of one inch to a thousand feet.

It may, however, be remarked that if the above-described surveying method could be used, the trigonometrical formulæ involved would equally well determine, *either* the long sides of such a triangle, corresponding to the Sun's distance, or the very small angle between them. This angle (ASB in Fig. II.) is evidently that which is subtended at the Sun by the straight line joining the two points of observation; or, in the most favourable case possible, by the length of the Earth's diameter.

In agreement with a nomenclature used in other astronomical calculations, one-half of this angle, or one such as the Earth's radius of 4000 miles would subtend at the Sun, is called the *Solar Parallax*. It will, therefore, be understood, that, whether we speak of the determination of the Sun's *distance*, or of the *Solar Parallax*, the same astronomical problem is involved.

In place of the surveying method, which is for the above reason inapplicable in the case of the Sun, notwithstanding that it is most useful and efficient when applied to terrestrial measurements, or even to determine the distance of the Moon from the Earth, other methods, less direct, but more interesting, are therefore employed. Of these, that which depends upon observations of transits of the planet Venus across the Sun's disc is the most generally known and the most full of popular and historical interest, although the estimation of its scientific value has, as we shall presently see, of late decidedly diminished.

To two young Englishmen, Jeremiah Horrox, Curate of Hoole, Lancashire, and William Crabtree, a clothier, or draper, of Broughton, near Manchester, and especially to the former, whose early death cut short a career of the highest promise, belongs the credit of observing the first recorded transit of this planet, on November 24th, 1639 (O.S.) Horrox calculated the date before its occurrence, and communicated it to his friend Crabtree. These two enthusiastic students had for some time been in the habit of corresponding with each other, and with Samuel Foster, who was afterwards Professor of Astronomy in GRESHAM COLLEGE, and they arranged to observe the transit by a method which may still be found useful by those who are not possessed of elaborate mechanical appliances, viz., by allowing the light of the Sun to pass through a very small hole in a shutter, so as to form a distinct image upon a screen in a darkened room. Horrox tells us that at a quarter past 3 p.m., an opening in the clouds rendered his observations successful. "Oh, most gratifying spectacle!" he says, "the object of so many earnest wishes. I perceived a new spot of unusual magnitude, and of a perfectly round form, that had just entered upon the left limb of the Sun, so that the margins of the Sun and spot coincided with each other, forming the angle of contact." * Crabtree was also fortunate in obtaining a brief view of the transit, by which he confirmed the observations of Horrox.

The orbits and periodic times of the Earth and Venus are such that two transits of the latter occur at an interval of only eight years, after which considerably more than a century elapses, in which there is none. Then there are two more, eight years apart; then another long interval, and so on; every alternate two taking place in December and June respectively.†

The transit of December 1639 having been preceded by one which had passed unobserved in December 1631, the next two did not take place until June 1761 and June 1769. In 1716 Dr. Halley had explained the importance of an idea originally suggested by James Gregory, in his *Optica Promota*, published in 1663, that these transits might be utilised for the determination of the Sun's distance. As the time approached, the English Government, as well as those of other countries, consequently made liberal arrangements for the necessary observations. In 1761 the observers were not very successful; but in 1769 they were much more fortunate, and a discussion of all the results which they obtained indicated that the Sun's

---

* Grant's "History of Astronomy," p. 421. See also Horrox, "Venus in Sole visa," translated in his Memoir by the Rev. A. B. Whatton.

† The plane of the orbit of Venus being very slightly inclined to that of the Earth, and thirteen of its revolutions round the Sun occupying only one day less than eight of those of the Earth, the positions of the two relatively to the Sun are almost exactly reproduced after the lapse of eight years. There is, however, a slight residual difference of place, but so small that it permits the occurrence of two successive transits separated by the above short interval. It is, however, sufficient to prevent the similar occurrence of a third. And the very slowness with which the discrepancy of relative position increases, is at the same time the cause of the rapid succession of two transits, and of the long interval before another pair can take place. If we were to suppose two men to start at the same moment to run round and round two concentric circular courses, the one running steadily at 7 miles per hour, and the other at 6 miles per hour ; then, if the first course were 18 miles round, and the second 25 miles, it is evident that the first would take $33\frac{3}{7}$ hours to go thirteen times round his course, and the second would take $33\frac{1}{3}$ hours to go eight times round his longer course

distance must lie between 92,000,000 and 96,000,000 miles. About fifty years afterwards, Encke, as the result of an elaborate investigation, came to the conclusion, that the most probable distance thus deducible was between 95 and $95\frac{1}{2}$ millions of miles; and this value was thenceforth generally adopted by astronomers.

Some subsequent investigations have, however, indicated that a right understanding of the records of 1769 might have given a considerably smaller value; and other methods of solving the problem having also favoured this supposition, the transit of December 8th, 1874, was almost impatiently awaited. We shall be better able to estimate the value of the conclusions obtained from it, if we pause for a few moments to explain the exact nature of the observations which are made at such times, and the especial difficulties which they involve.

It is found that different places upon the Earth may see the planet *begin to enter* upon the Sun's disc, *or leave it*, at moments which may differ by more than twenty minutes. It also follows that if such places be so situated that they can see both the beginning and the end of the transit, *the whole duration* of it as witnessed from them may by no means be the same. In fact, this difference of duration may exceed half an hour, as was the case in 1874. This effect is a *joint result of two causes*.

i.e., at the end of $33\frac{1}{4}$ hours they would have both very nearly completed a whole number of rounds, except that the first would need $\frac{2}{21}$ of an hour, or rather less than 6 minutes, more to finish his thirteenth round. If, therefore, when they started, they were in a straight line, passing through the common centre of the two courses, they would at the end of $33\frac{1}{4}$ hours be within 6 minutes of passing across the same straight line exactly together again. But the *smaller* the defect from an exact agreement of time in their so passing, the longer would be the interval and the greater the number of rounds before the agreement would be perfect. This illustration may perhaps help to explain the somewhat similar behaviour of the Earth and Venus, although the length of the circumferences of their orbits is immensely greater than that of the short courses suggested for the runners; and the actual speed of Venus would bring it slightly before, instead of slightly behind the Earth, at the end of its thirteenth revolution. Otherwise, the comparative speeds and the lengths of the courses are very nearly in the actual ratios of those of the two planets.

Of these one is the rotation of the Earth, by which any place upon its surface is moved with a velocity whose magnitude and direction at any time depend upon its latitude and longitude. This produces an apparent motion of Venus in the reverse direction across the Sun's disc, and may either tend to shorten, or to lengthen, the transit according as the place of observation is carried round upon the same side of the Earth's axis as Venus, or upon the opposite side. An observer at either pole of the earth, who would be unmoved by its rotation, would of course experience no such effect.

The other cause by which a difference of duration is produced is the difference of the directions in which, owing to their distance apart upon the Earth, any two places, at any given moment, look at the planet. The lines joining them to it will consequently neither meet the Sun's edge at the same point of its circumference, nor at the same instant of time. Two such places, indeed, look at the planet in directions which differ just as do those in which the surveyor, in the method previously indicated in Figs. I. and II., looks at a distant object from his stations. It is when, and where, those directions intersect the edge of the solar disc that the transit is seen to begin, or to end. One place may from this cause see the planet enter upon the disc sooner than another, and leave it later, or *vice versâ*. And its path across the Sun seen from the one station will appear to be longer, or shorter, than that which is seen from the other; and will, in the former case, at the middle of the transit, pass correspondingly nearer to the centre of the Sun's disc.

Nor is it hard to understand that the angle between the lines of sight in which Venus is seen from the two places of observation will be as much greater than the difference of the directions in which they look at the Sun, as Venus, when seen in transit, is nearer than the Sun is to the Earth, *i.e.*, about 3 times. But even such a difference of direction as this is still much too small for the application of the surveying method; *

* We are, moreover, unable to use the actual displacement of the position of Venus (caused by the difference of the directions in which it is looked at), as we are considering its displacement upon or *relatively to*

otherwise, upon the occurrence of a transit, we might by that means calculate our distance from Venus, and a simple sum in proportion would give our distance from the Sun.

It will, however, be a very different matter if, instead of observing the angles corresponding to two such directions of observation, we can use, as the foundation of our calculations, an interval of time which may amount to twenty or thirty minutes. Such an interval may either be the difference of the *whole duration* of a transit, or the difference of the moment when it *begins* or *ends*, as seen from two places upon the Earth.

If only its beginning *or* its end is observed at each station, the method is termed that of *Delisle*. This method, however, is found to involve the disadvantage, that, in order to deduce the Sun's distance, the longitudes of the stations must be determined with extreme accuracy. And this is often a matter of very considerable difficulty.

If, on the other hand, the difference of the whole duration of the transit, as seen from two selected localities, is used, the method is called that of *Halley*. In this case both places must see its beginning *and* its end. They will, in general, see it throughout its continuance, although it may happen in places not far from the poles, that the beginning may occur before sunset, and the end after the next sunrise, a short Arctic night having intervened.

The Halleyan method very much restricts the choice of localities, but it does not need the accurate determination of their longitudes. Its efficient application involves the use of stations so far apart in north and south latitude, that access to some of them is usually very inconvenient, if not impossible. The difference of latitude helps to separate the paths by which the planet is seen to cross the Sun, and consequently to increase the difference between the durations of its transit. It must not, however, be forgotten that this difference of duration will also be much affected (and sometimes to a still

the Sun, which is itself correspondingly displaced. The above $3\frac{1}{3}$ times is therefore reduced to $2\frac{1}{3}$ times.

greater extent than by the utmost possible change of the observer's latitude) by the corresponding effect of the Earth's rotation acting, in various latitudes and longitudes, as we have previously described. In selecting stations, both of these considerations must be, therefore, taken into account.

The methods of Delisle and of Halley have consequently their respective advantages and disadvantages. Either may give, as the result of careful calculations, a very accurate determination of the Sun's distance, *provided it be possible to observe satisfactorily the exact moment at which the disc of the planet enters, or leaves, that of the Sun.* But in attempting to do this, most annoying difficulties were experienced in 1761 and 1769. And again in 1874 similar troubles recurred. It is found, owing to effects which depend upon the diffraction of light, and perhaps, also, upon some other causes (not yet fully explained), connected with slight imperfections in our instruments, and with certain peculiarities of human vision, that it is impossible to say exactly when the edges of the two discs just touch externally (which is called *external contact*), or even to say exactly when the whole of the planet's disc is just inside that of the Sun (which is called *internal contact*). If we take, for example, the beginning of the transit, it may in the first case be hard to be sure that the contact exists, until all at once it suddenly seems that one disc has encroached some considerable distance upon the other. In the second case, when the decisive moment is just approaching, the planet may appear to be gradually elongated into a somewhat pear-like shape, its edge for a while clinging to that of the Sun ; then, perhaps, a dark filament is noticed joining the two, which more or less suddenly fades and vanishes away, leaving the planet's disc so far within that of the Sun, that a very appreciable space separates the edges of the two.

Very similar phenomena occur in a reversed order at the end of the transit. The bewildered observer cannot decide at which instant of these proceedings the actual contact has taken place. He is also often confused, in the case both of internal and external contacts, by a narrow line of light surrounding Venus, supposed to be produced by its atmosphere, which, at the time

DRAWINGS OF THE TRANSIT OF VENUS, DECEMBER 8, 1874.

Fig. III.—Ingress observed at Rodriguez, drawn by Lieut. Hoggan.

Fig. IV.—Ingress observed at Kerguelen's Land, drawn by Lieut. Corbet.

Fig. V.—Egress observed in Egypt, drawn by Miss Newton.

The black oblongs represent the sky bounding the Sun's edge, of which the portion shown is too small to exhibit any curvature.

of contact, unites in a very troublesome manner with the bright surface of the Sun.*

Upon the whole it follows that the value of transits of Venus for the determination of the Sun's distance is very much diminished. The final results of the most elaborate calculations vary, to a very important extent, according to the interpretation which we put upon the language used by the observers in describing what they have seen, and the deductions consequently drawn as to the exact instants at which they really beheld the occurrence of the contacts.

It may suffice to state that the first, and somewhat hurried, result of the British expeditions sent out to observe the transit of December 8th, 1874, which was presented to Parliament in 1877, gave 93,300,000 miles (afterwards altered to 93,375,000) as the most probable distance of the Earth from the Sun, with a possible error in excess or defect of about 140,000 miles. The above was a mean value of the results given by observations of the beginning of the transit, and of its end, four times as much weight (owing to certain circumstances) being attached to the former as to the latter. These results, however, differed by more than 1,150,000 miles.

But Mr. Stone (now Radcliffe Observer at Oxford), from a careful and elaborate discussion of these same observations, came to a very different conclusion, viz., that the solar distance could hardly fall short of 91,500,000, nor exceed 92,500,000 miles; and that somewhat less than 92,000,000 miles was the most probable value. A further discussion at Greenwich, by Colonel (at that time Captain) Tupman, to whose superintendence of the British Expeditions science is very deeply indebted, gave a value between 92,700,000 and 92,800,000 miles. Afterwards, however, from a still fuller investigation of the meaning of the observers' language, he deduced a mean result of very nearly 92,400,000 miles. But he was obliged to allow that

---

* Some of the effects which we have described may be seen in the diagrams upon the opposite page, which are copies of three of those made by observers in December 1874, and published in the Admiralty Report of the transit presented to Parliament. In the second of these views, the atmosphere of Venus is very noticeable.

the observations only determined the probable value to be somewhere between 92,000,000 and 92,700,000 miles.

There is no doubt that the discordances of these results chiefly depend upon the different interpretations that have been given to the actual wording of the reports of the observers. In saying this, we do not for a moment wish to disparage the value of their work, or of the observation of transits of Venus for the purpose in question; but it must be confessed that there is at present an amount of uncertainty in the figures obtained, which is disappointing, but which seems likely to continue, unless we can in some way improve our instruments, or the method of their use.

It must also be a subject of great regret that the numerous photographs of the transit which were taken by the British Expeditions in 1874, have not been found to depict the edges of the Sun and of the planet with sufficient sharpness to allow of any very accurate calculations being made from them.

It has, however, been recently suggested by Mr. Maunder, of the Royal Observatory, Greenwich, that in the transit of 1882 Venus may pass sufficiently near to some clearly defined sun-spot, to allow the measurements of its position in a photograph to be made with greater accuracy from the edge of the spot, instead of from the edge of the Sun's disc.

We may still obtain some further information from a comparison of the British results with those of other nations; of which, as yet, we know comparatively little, but it seems, so far as transit observations are concerned, that an uncertainty amounting to at least 500,000 miles will most probably remain. A preliminary discussion of the American photographs by Mr. D. P. Todd, in the *Journal of Science*, of June 1881, which, owing to the special method by which they were taken, are, we believe, capable of a more precise interpretation than those of the British Expeditions, seems to indicate a value of a little over 92,000,000 miles.*

Upon the whole it is very noticeable that all these values, deduced from the transit of 1874, are much nearer to the

---

* It is also gratifying to know that the American photographs of the transit of Mercury in 1878 have proved very successful.

92,000,000 miles which Mr. Stone obtained from that of 1769, than to the 95,000,000 which Encke published in 1824, and which was for more than thirty years generally accepted. It is also a remarkable fact that successive endeavours to solve the problem by other methods, subsequent to Encke's investigation, all indicated a much smaller value than his.

We must not stop to describe these various attempts at any length. Two of them depend upon certain small irregularities respectively occurring in the motions of the Moon, and in the apparent position of the Sun during each lunar month, which involve, in a special manner, the distance of the latter. Others are connected with minute perturbations in the orbits of Mars and Venus. From these Leverrier, in the year 1872, very confidently announced that the Sun's distance must in all probability be between 92,200,000 miles and 92,300,000 miles.

Another very interesting method is founded upon observations of Jupiter's satellites, combined with experimental determinations of the velocity of Light. If, by such appliances as those invented by Fizeau,* or Foucault,† we can measure the velocity of Light, we may estimate our distance from the Sun by noting how much sooner, or later, than would otherwise be the case, the eclipses of these satellites are seen to take place, according to the distance of the Earth at any given time from Jupiter.

It is true that if the Earth is so placed that the Sun is almost exactly between it and Jupiter, or, in other words, if the Earth is at its greatest possible distance from Jupiter, that planet and its satellites will be invisible, owing to their apparent proximity to the intense brightness of the Sun's lights. But if the Earth be not very far from such a position, as at E_1, the satellites and their eclipses may be detected, and (as is shown in Fig. VI.) the light coming from them must, in order to reach the Earth at E_1, traverse a distance not much less than the diameter of the Earth's orbit, in addition to that which it would traverse to reach the Earth, when, as at E_2, it is exactly

---

* See Arago's "Popular Astronomy," vol. ii.
† See Ganot's "Physics."

between Jupiter and the Sun. From the observations which have been made, it is found that about 997 seconds is the time which the light of the satellites occupies in travelling across the full diameter of the Earth's orbit. And if we take 997 times the number of miles which the experiments previously referred to give as the velocity of light per second (viz., about 186,380 miles), we obtain a value for that diameter of about

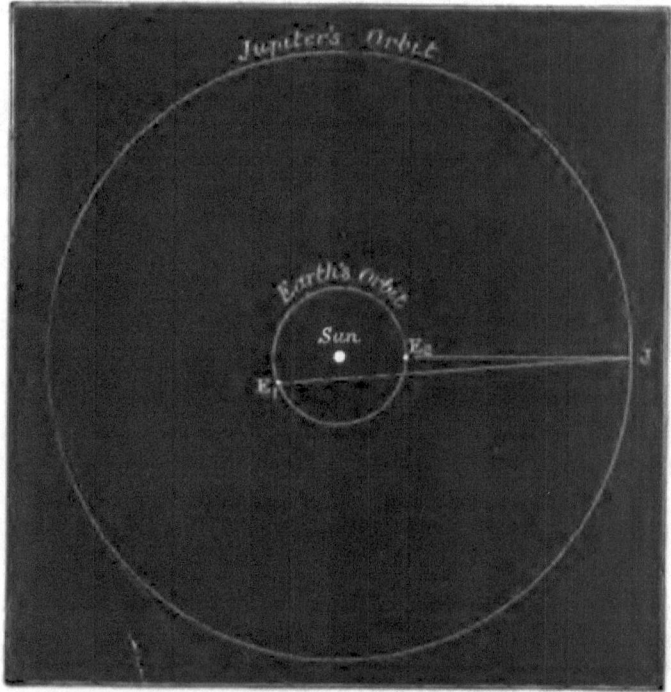

Fig. VI.—The different distances traversed by light in passing from Jupiter to the Earth at different times.

185,820,000 miles, one half of which, or about 92,910,000 miles, is the corresponding distance of the Earth from the Sun. The method, however, has not been found at present to be susceptible of very great accuracy, owing to certain difficulties connected with the calculation and observation of the eclipses, and results differing by as much as $1\frac{1}{2}$ millions of miles have been obtained from it; but its importance will doubtless increase as the theory of the motions of Jupiter's satellites is more carefully studied and investigated.

The experimental determination of the Velocity of Light has also been used in connection with a certain small apparent displacement of the positions of the fixed stars, which is called their *"Aberration."* This effect is analogous to that by which a shower of rain, falling vertically on a calm day, appears to slope towards a person moving rapidly through it. It depends upon the comparative value of the velocity of the light which comes from the stars to an observer, and of the velocity with which he is at the same time carried through space as he receives it, by the motion of the Earth in its orbit. The former velocity being known, the latter can be calculated from the amount of any star's aberration or apparent displacement. And the velocity of the Earth in its orbit being thus deduced, we can tell at once how far it will travel in a year; *i.e.*, we know the length of the circumference of its orbit. It is, then, a very simple matter to evaluate the radius of that orbit, or, in other words, the Earth's distance from the Sun. About 93,000,000 miles appears to be the most probable value which this method gives.

Observations of the planet Mars have also been made when it has been especially near to the Earth (as, for instance, in 1862 or 1877), with a view to the solution of this same great problem. The method at first adopted was to notice from *two different observatories* how much nearer Mars appeared to be to certain stars as seen from the one, than as seen from the other. This effect is due to the difference of the directions in which the two observatories look at the *planet*, the *stars* being so much farther away that no appreciable change is produced in their apparent positions. From such observations we may proceed to calculate the distance of Mars from the Earth by a modification of the surveying method already described (see Figs. I. and II.); and it is found that we are able to do so with very considerable accuracy.

The nearest distance of the planet Mars from the Earth being not much less than $1\frac{1}{2}$ times the nearest distance of Venus, this last statement may appear to be inconsistent with our previous remark, that the method referred to is not applicable to the case of a transit of Venus. But the inconsistency does not really exist. We can apply the method to what at first sight appears to be a more difficult case, because we are able to make a long series of observations of Mars night after night. We then take the mean of the results given by them, and in this way eliminate many sources of error.

The distance of Mars having been thus obtained in 1862, it at once followed by a well-known proportion which exists, as a consequence of the law of gravity, among the planetary distances, and which was first discovered by Kepler, that the corresponding distance of the Sun must be about 92,200,000 miles. But the method used had one great disadvantage. It necessarily involves the comparison of the work of *different* observers and of *different* instruments at the two observatories. In such delicate calculations, this is a very serious cause of error. It may, however, be avoided by a most valuable modification of the method of observation, the principle of which, we believe, was originally suggested by Cassini and Flamsteed, and to which the Astronomer Royal (Sir G. B. Airy) drew special attention in a paper read by him before the Royal Astronomical Society in 1857, as that which in his opinion was likely to give the most accurate result attainable. It may be briefly stated as follows:—

That a locality should be chosen for the observation of Mars when the planet is especially near to the Earth, such that it may be observed for a considerable number of nights soon after sunset, and its apparent distances from certain stars be noted. The corresponding distances should again be observed just before sunrise. In the meantime the place of observation, if near to the Equator, would have had its position in space altered, owing to the rotation of the Earth on its axis, by an amount which in 12 hours would be not much less than 8,000 miles, and in 9 or 10 hours would amount to between 6,000 and 7,000 miles. This would be a movement (unlike that of the Earth in its orbit) which would be quite *independent* of the distance either of the Earth, or of Mars, from the Sun. It would move the place of observation over a space as great, or almost as great, as that between any two such observatories as we have previously mentioned.

The stars, as we have stated, are so distant that no appreciable effect can be produced upon their apparent positions by any such change in the observer's place. But it is found that very appreciable differences are seen in the apparent distances between Mars and neighbouring stars, when they are measured soon after sunset and shortly before sunrise. From these differences the known amount of the observer's displacement may, by suitable calculations, be compared with his distance from Mars. That distance is consequently deduced with much accuracy, and, as explained in page 16, the distance of the Earth from the Sun is then also known; but with this very great advantage, that all the calculations made are founded on observations taken in *the same place, by the same observer, and with the same instrument.*

A form of telescope, originally invented to measure the Sun's apparent diameter, and called a Heliometer, is found to be the best with which to take the interval between the planet and a star. It was with a valuable instrument of this description very kindly lent by Lord Lindsay (now the Earl of Crawford and Balcarres) that Mr. Gill (since appointed Her Majesty's Astronomer at the Cape of Good Hope) made a special journey to the Isle of Ascension, where he resided for six months in the year 1877. In spite of many difficulties he obtained an excellent series of observations, involving about 350 sets of measures. The *locality* was chosen because it fulfilled the conditions which we have described as most important. The *year*, because Mars was nearer to the Earth than it will be again at any time during the present century. There is no doubt that very great weight is due to Mr. Gill's observations, and to his calculations founded upon them. He has employed all possible care in their computation, and has been ably assisted in the accurate determination of the places of the stars used, by the principal observatories of this and of other countries. He concludes that the distance of the Sun is very nearly 93,080,000 miles; the probable error in excess or defect not being more than about 127,000 miles.

It may be interesting to notice that a similar method of observation has also been applied to some of the minor planets. Even when nearest to the Earth the distances of those which have been thus observed are very much greater than that of Mars, but their minute star-like discs conduce to great accuracy in the measurements made, and the results have agreed very satisfactorily with those obtained from Mars. Two, named Victoria and Sappho, will approach sufficiently near to the Earth to be useful in this way during the summer of 1882. It seems likely that the repetition of such observations upon all suitable occasions may in time lead to a very accurate result.

Upon the whole, it appears that the most probable conclusion to which all our investigations at present tend, is, that *the Sun's distance from the Earth is not far from 93,000,000 miles*. This is the value which we shall adopt throughout these lectures. But we cannot, at present, feel any certainty of its exactness within a $\frac{1}{100}$th part of the whole value. Nor can we ever hope to be able to determine so large a measure with *absolute accuracy*. The problem, as we have previously stated, is not only a very fascinating one, but is also very difficult.

If by repeated observations of transits of Venus, and by the comparison of various other methods, we may be able, at some future time, to obtain a value true to within a $\frac{1}{1000}$th part, we shall be well content, although this will still leave an uncertainty amounting to nearly 100,000 miles.

It is to be regretted that the nearest distance of the planet Mercury from the Earth is so much greater than that of Venus, that its transits are of very little use for the purpose in question. On the other hand, those of Venus occur much less often than we could wish. There will not be another, in succession to that of December 6th, 1882, until the pair which will take place in June 2004 and June 2012.

When the vast importance of the Sun's distance as the great *unit of astronomical measurement* is considered; when we remember that its value is involved in almost every calculation connected with the solar system, and that it is in terms of it that we measure our distances from the fixed stars, and the magnitudes of the orbits of binary and other stellar systems; when we note that an error in our estimate of its value of only 90,000 miles (or about a $\frac{1}{1000}$th part) affects our calculation of the distance of the nearest star to the extent of about 20,000,000,000 miles; we surely ought to use every possible means to secure its accurate determination.

Englishmen will, we trust, grudge no expense that may be necessary for the observations of 1882, nor for the due investigation and application of any opportunities that may occur for the use of other methods.

The results of December 8th, 1874, although very important, may not have been quite so successful as it was hoped they might be. But the experience then gained, both with regard to eye and to photographic observations, may prove to be exceedingly beneficial in 1882. We may meet with a success surpassing our best expectations. In any case, the true spirit of science is to neglect no opportunity and to spare no pains in acquiring knowledge. It is to those who woo her in this spirit that she reveals her most precious secrets.

LECTURE II.

THE SUN—*continued.*

> "Welcum the lord of lycht, and lamp of day,
> Welcum fostyr of tendir herbys grene,
> Welcum quyknar of floryst flowris scheyn,
> Welcum support of euery rute and vayn,
> Welcum comfort of alkynd fruyt and grayn,
> Welcum weilfar of husbandis at the plewys,
> Welcum reparar of woddis, treis, and bewys,
> Welcum depayntar of the blomyt medis,
> Welcum the lyfe of euery thing that spredis,
> Welcum be thy brycht bemys, gladyng all."
>
> GAVIN DOUGLAS.

WE now proceed to describe some of the discoveries connected with the physical constitution of the Sun which have recently been made, in spite of the immensity of its distance from us.

Until about thirty years ago very little was known of the great orb of day, except that it was a huge body of some 860,000 miles in diameter; of a mean density rather less than $1\frac{1}{2}$ times that of water; attracting the Earth and the other planets, and ruling their orbits, by the force of gravity; intensely hot; surrounded by a certain halo of glory, and occasionally showing some red coloured protuberances when its disc was hidden behind the Moon during a total eclipse; at other times presenting to the view a surface of extreme brilliancy, which, however, was frequently defaced by the appearance of *dark spots*, easily detected in a telescope, and occasionally large enough, as both ancient and modern records prove, to be seen with the naked eye. It was also observed that the central portions of its disc appeared to be considerably brighter than the outer parts; and that the spots, to which we have just referred, indicated a revolution of the Sun, about an axis of its figure, in about 25⅓ days. The spots were often seen to be more or less surrounded, or preceded in their appearance, by irregular markings of special brightness, to which the name of *faculæ* was given. It was, moreover, evident that they were in many cases darkest in their central portions, and that those portions were depressed below the surrounding surface, which was proved by the effect of the gradual foreshortening produced by the Sun's rotation, as it carried them

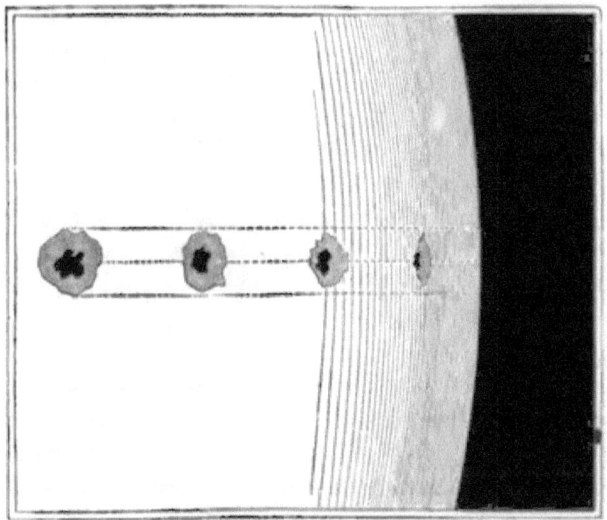

Fig. VII.—The change in the appearance of a spot caused by the Sun's rotation proves that its central part is depressed below the surrounding surface.

from the edge to the centre of its disc, and *vice versâ*. This effect is illustrated in Fig. VII.

One attempt, and only one of any importance, had been made up to the time to which we refer, to fathom the mystery of these spots. It proved, however, to be very erroneous, although it originated with so great an astronomer as Sir William Herschel. He put forward the hypothesis, that the very dark central portions of the Sun-spots might be parts of a solid nucleus existing within the fiery envelope which dazzles our sight, and that this nucleus appeared to be intensely black by comparison with the surrounding brightness. It was also suggested that, if the inner surface of the fiery envelope (which is generally termed the *Photosphere*, or sphere of light) were of a dense and cloudy nature, and very slightly pervious to heat, the supposed nucleus within it might be a convenient, or even a comfortable, abode for animated beings.

This hypothesis with regard to the Sun-spots was at the time by no means irrational. But it has since been entirely disproved. A vast advance has been made in photometry, or the art of measuring the intensity of any given light. Appliances have been invented by which the illumination of any special portion of a spot can be examined, and the whole power of a large telescope be brought to bear upon any given part of the solar surface, whether occupied by a spot or not, without danger to the observer. And it has been announced, by Professor Langley, as the result of very careful observations, that the central portions of a spot, which, as the result of contrast, seem to be of the blackest darkness possible, are really brighter than the arc of a powerful electric light, and radiate more than half as much heat as an equal area of the ordinary surface of the photosphere.

It is proved that in a Sun-spot we look down into depths of fiery vapour, or incandescent matter; that we see lower layers of the glowing envelope of the Sun in a peculiar state of disturbance, but that we see no solid nucleus, nothing that is really dark.

In endeavouring to explain the conclusions to which recent investigations have led us, we shall frequently have occasion to refer to the use of the Spectroscope, the invention of which has provided an instrument of extraordinary power for the physical study, not only of the Sun, but also of stars, planets, nebulæ, comets, and meteorites. Its construction is founded upon the great discovery of Sir Isaac Newton, that a beam of sunlight, shining through a small hole in a shutter into a darkened room, will be spread out into a lengthened band of coloured light, if it pass through a wedge or prism of glass. Such a band is called a *Spectrum*. The colours in it are gradually shaded into one another, beginning with red at one end, and passing through a succession of tints, which are generally described as orange, yellow, green, blue, indigo, and violet.

It is, in fact, found that the original beam of white light is compounded of different colours. In passing through the prism these are differently deviated, or bent, from their original direction; the red constituent being deviated least, and the violet most. But such a spectrum is termed an impure spectrum, because the light of each individual colour, having a special capacity of its own of being bent out of its course by its passage through the prism (which is technically denoted by the term *refrangibility*), forms its own image of the hole which determines the shape of the original beam; and all the different images so formed, side by side, successively, overlap one another, except at the extreme ends of the spectrum.

The spectroscope is an instrument by which this overlapping and consequent mixture, or impurity, of the light can be prevented. An arrangement is made in it which forms a beam of light bounded by an extremely narrow rectilineal slit, the image of which slit is a very fine straight line, and the light is made to pass through one or more prisms of glass, or of some other suitable substance, in such a course as secures the utmost sharpness and the best possible definition in the images thus formed.

Each constituent portion of the light consequently forms a very fine, bright, straight line, in a position corresponding to the extent to which that particular portion is bent out of its original direction. And all these fine straight lines thus formed side by side, and parallel to one another, constitute what is termed a *pure spectrum*, whose length is in a direction perpendicular to them all. They are in fact so fine that, in a *perfectly* pure spectrum, they would not in the least overlap one another. The light at any point of a pure spectrum is no longer more or less mixed, but is such as belongs to that particular position only. Other appliances connected with the spectroscope enable us to observe the spectrum thus formed through suitable eye-pieces, so as to magnify it, to a greater or less degree, as we may desire.

In a pure spectrum, thus obtained, it is easy to understand that there will be a dark gap, if the light that would form an image in any special portion of it be wanting in the original beam. This is found to be the case in so many parts of the solar spectrum that, in an instrument of sufficient power, thousands of these dark gaps may be seen. In other words, the coloured band of light is crossed in a direction at right angles to its length by thousands of fine dark lines, some of which are, however, observed to be much finer than others.

It is the discovery of the meaning of these lines which has vastly increased our knowledge of the Sun's constitution.

Experiments with the light of solid bodies, as well as with that of certain liquids, or compressed gases, so highly heated as to be luminous, show that a similarly coloured band, or spectrum, is obtained from them, but without any such interruptions. Such an one is termed a *continuous spectrum*. But it is once more found that if such light, instead of being allowed to fall directly upon the spectroscope, is first passed through the vapour of some other substance (such vapour being cooler than the previously mentioned source of light), *dark lines immediately appear in what would otherwise be a continuous spectrum*.

We therefore conclude that in the outer regions of the Sun there are various vapours which originate the dark lines which, as we have previously stated, are invariably seen athwart the spectrum of the solar light.

But we are able to infer much more than this. Further investigation has enabled us to identify certain dark lines with certain particular vapours. And this has been done, not simply by sending light that would otherwise give a continuous spectrum through various vapours, and then noting for comparison the positions of the dark lines produced, but in a much more interesting and easy manner, which we now proceed to explain.

Without entering into any elaborate theoretical discussion, it may suffice to state that experiments prove that, if the light of a gas not specially compressed, or of that produced by vaporizing a solid or liquid substance at such a temperature that its vapour is luminous, be looked at through a spectroscope, its spectrum is in general found to be utterly different from the continuous spectrum that is obtained from the light of an incandescent solid substance, or in certain cases from a very dense liquid or gas.

What we see in such experiments is only a certain number of *isolated bright lines* in place of the long continuous band of colours. These bright lines may be comparatively few (as, for instance, in the case of the luminous vapour of sodium), or may number some hundreds, as in the case of the vapour of iron. But in every case they occupy certain fixed positions, relatively to the whole space that a continuous spectrum would occupy.

Our next procedure is to record these positions with great accuracy. We then recur once more to our first experiment, and pass in succession such light as would otherwise give a continuous spectrum through certain of the gases which we have been using, the light in this case being taken from a source whose temperature is higher than that of the vapours through which it is now passed. As we have already stated, black lines will at once appear across the spectrum.

And the very remarkable discovery has been made that the black lines, produced by the passage of the light through each vapour which it traverses, *exactly correspond in position with the bright lines which the light of that vapour would by itself produce.* Moreover, a consecutive passage through two or more such vapours, or a passage through a mixture of them, is found to give dark lines corresponding to the bright lines which belong to all those vapours.

If therefore we first of all observe the *bright lines* in the spectra of various highly-heated and luminous gases, or in the luminous vapours of various substances, and then identify in the solar spectrum *dark lines* in positions corresponding exactly to those of these bright ones, we are at once enabled to conclude, that in the outer regions of the Sun the solar light has passed through these identical vapours, and that each of them has absorbed out of the continuous spectrum which otherwise would be seen, those portions which correspond to its own constitution, and has formed dark lines exactly in the positions where by itself its light would give bright ones.

Of late very careful experiments have been made, which tend to show that, under certain circumstances, some, or even all but one, of the bright lines belonging to a given substance may disappear from the spectrum of its vapour; and in like manner, that some, or all but one, of the dark lines produced by the passage of light through that vapour, may be wanting. It follows, that if we recognize only *one dark line* in a spectrum, which by its position corresponds to a given substance, we may be justified in saying that we have evidence of the presence of the vapour of that substance in the sun.*

The spectroscope has thus proved to us not only that the light of the Sun, which comes from that layer or surface of it by which it is ordinarily visible to us, is originally such as would give a continuous spectrum of colour varying from red to violet; but it has taught us that in so coming it passes through overlying vapours, which are less heated than the underlying layer, and which are not ordinarily visible to

* This statement ought perhaps to be still more qualified. It is possible that the one line observed in such a case may not indicate that the substance referred to exists in the Sun as a whole. It has been suggested by Mr. Lockyer, that the intensity of its heat, which far exceeds any with which we can experiment, may dissociate into simpler constituents what we are accustomed to call an elementary substance. If so, the one line, or the few out of more lines belonging to the ordinary spectrum of a substance, seen in the Sun, may indicate the existence of some simpler constituent of the substance. In an elementary lecture such as the present, we must pass by many such points as these, which are nevertheless of the highest importance and interest. Nor must we discuss the investigations which have of late been carried on by many distinguished men of science, both in England and elsewhere, as to the various orders and kinds of spectra given by the same substance under different molecular conditions, or under changes of temperature and pressure; the phenomena of long and short lines, of basic-lines, of columnar and banded spectra, etc. We must also refrain from describing the method of obtaining spectra by means of the diffraction of light incident upon surfaces ruled with exceedingly close parallel straight lines; such spectra (termed *diffraction spectra*) being in many cases superior to those produced by the use of prisms. For information on these points we refer our readers to the proceedings of the Royal Society, the British Association, and other learned societies.

us. It has enabled us to recognize that the dark lines produced by these vapours belong to various substances with which we are familiar upon the Earth; some of the most important of which are Hydrogen, Sodium, Calcium, Manganese, Magnesium, Iron, Nickel, Barium, Cobalt.

The presence of some of these is indicated by only a few dark lines, corresponding to a greater or less proportion of the bright lines which are seen in experiments with their vapours. On the other hand, more than 450 dark lines have been recorded in the solar spectrum which correspond to bright ones belonging to the vapour of Iron. And these, although so numerous, agree exactly, each to each, and not only in position, but in relative thickness and intensity.

In fact, we are able, by means of this wonderful instrument, to bridge over space, and to analyse the constituents of the great luminary of our system, in spite of its distance of nearly, or even more than, 93,000,000 miles. It gives us a power, even to hope for which might well have seemed an absurdity to the astronomers of any previous generation.

We have not, however, by any means exhausted our description of its marvellous revelations. It not only shows and explains to us the dark lines in the ordinary spectrum of the Sun, but on certain rare occasions it enables us to see these lines transformed into bright ones, with dark intervals between them.

It will easily be understood from the preceding explanation that we ought to be able to see them thus, if we could look at the light of the vapours superimposed upon the photosphere of the Sun, apart from the light of the photosphere itself, which in general passes through them. Although these vapours are at a lower temperature than the light of the brilliant layer beneath them, they must be intensely heated, and sufficiently luminous to give a bright-line spectrum if their light could be seen independently.

It is therefore very interesting to learn that in 1870 Dr. Young, Professor of Astronomy in the college of New Jersey, was watching the total eclipse which occurred in December of that year, and was looking at the edge of the Sun through a spectroscope just when the dark body of the Moon had hidden the last narrow crescent of the light of the photosphere, the region immediately above it being therefore visible for two or three seconds without that light shining through it. He found that when the ordinary light of the solar spectrum had just vanished away, its place was instantly taken by a multitude of fine bright lines. The dark lines by hundreds or thousands became instantaneously bright, as by a magical transformation, and remained so during the brief interval occupied by the Moon's edge in passing over the layer of vapours. This layer is in consequence sometimes termed Young's layer.

The same beautiful effect has been noticed in subsequent eclipses, and was distinctly seen in that which was recently observed by an unusually large number of skilful astronomers in America, in July 1878. The average depth of this region of the Sun has been supposed to be very small, measuring only 400 or 500, or at any rate less than 1000 miles in thickness; in agreement with the very few seconds in which the Moon's edge passes over it. If so, it lies so close to the photosphere, that it is only by the interposition of the Moon, during a total eclipse, that we are able to isolate its light, which, at other times, is overpowered by that of the photosphere.*

But it is not difficult to understand that, if *some* of the vapours which constitute it rise up, here and there, to a much greater

* It has, however, been recently suggested by Mr. Lockyer, that more accurate observations may show a considerable variation between the bright-line spectrum which briefly flashes into view during a total eclipse, and that which would accurately correspond to a reversal of the dark lines in the ordinary solar spectrum. It may be that the above mentioned bright-line spectrum partly has its origin in regions lying considerably above the so-called reversing layer, and that special peculiarities may occur in lines corresponding to those which are seen to exhibit the most remarkable changes in the ordinary observation of spots and prominences.

It appears, from the brief reports of the Eclipse of May 17th, 1882, which have arrived while the above is passing through the press, that Mr. Lockyer's observations in Egypt, conducted with a special reference to this question, have confirmed his above-mentioned suggestion, which was put forward upon certain theoretical grounds.

distance in sufficient quantity, and at a luminous temperature, it may be possible at other times to detect their light with a spectroscope by carefully inspecting the circumference of the Sun at a moderate distance from the visible boundary of its disc. There will be much brilliant glare of ordinary solar light arising from the illumination of the Earth's atmosphere immediately around the direction in which the Sun is seen, which will be greatly in our way. But the principle of the construction of the spectroscope is such, that, if we increase the number of prisms used in it, we may lengthen out the spectrum of that glare more and more, and make it increasingly faint.

At the same time, if we are looking at the light of any mass of vapour rising up from the edge of the Sun, the bright lines caused by that vapour will not get fainter as we increase our battery of prisms, but will only be further separated from one another. It is consequently found that when we direct our instrument to the neighbourhood of the Sun's apparent edge, we can often detect the bright lines belonging to masses of vapour rising up from it, or suspended over it. We are thus able to view, not only during a total eclipse, but *at any time*, much of what is going on in regions where it might have been supposed that the blinding brilliancy of the adjacent photospheric light would have for ever utterly baffled our observations. The credit of this discovery, which was made in the year 1868, is due to Messrs. Janssen and Lockyer.

It would, however, be a very laborious process to find out the shape and boundaries of any such masses of vapour by slowly moving the spectroscope about, and noticing exactly when the bright lines corresponding to the vapour in question might appear or disappear, as the field of view of the instrument might pass across the vapour at different levels in succession. By so doing we might, no doubt, obtain some such result as is indicated in Fig. VIII., in which a series of sweeps in directions parallel to the Sun's edge are supposed to be made, and the general shape and extent of the vapour to be approximately deduced by the length of the parallel bright lines successively seen. For the sake of contrast, the surrounding region in this figure is blackened, although it would actually, as seen in the instrument, be only somewhat less bright than the bright lines themselves.

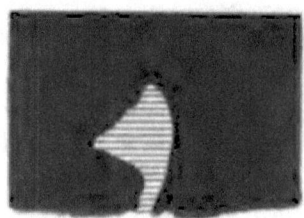

Fig. VIII.—Showing how a series of parallel bright lines seen at different levels may indicate the form of a mass of vapour rising above the general surface of the Sun.

But it was most fortunately found by Dr. Huggins (*see* Fig. IX.) that if the slit of the spectroscope be opened wide enough to embrace the whole image of such a mass of vapour, *a coloured figure of the whole of it* is formed in the neighbourhood of each of the bright lines which the spectrum of any portion of it would produce. And these coloured images are sufficiently bright to be seen *at all times* in an instrument of adequate power. It is only necessary, on any given day, to examine all round the Sun, just outside the edge of the photosphere, and the shape or movements of any such masses of gas or vapour as we have described will be visible. Sometimes there are many of them; at other times very few. The following is a copy of the original sketch in the Proceedings of the Royal Society, made by Dr. Huggins, of the first prominence, or protuberant mass of solar vapour, seen by him with a spectroscope having its slit widely-opened on February 13th, 1869.

Fig. IX.—The first prominence seen by Dr. Huggins with a spectroscope having a widely-opened slit.

The shapes of these prominences are often most fantastic their changes of form most rapid, and their movements of enormous magnitude. One of the most remarkable ever seen was observed in October, 1880. It rose, from a height of about 212,000 miles, to the extraordinary elevation of at least 350,000 miles in less than half an hour, the average speed with which the matter in it was carried upwards during that interval being therefore between 4,000 and 5,000 miles per minute, which would indicate a probable initial velocity of expulsion from the surface of the photosphere of 25,000, or more, miles per minute.

In some instances huge masses many thousands of miles in breadth, or height, have been seen to be utterly broken up and dispersed into fragments in the course of a few minutes. The most elevated are principally composed of hydrogen vapour, the small density of which is consistent with the great elevation to which they attain. In the accompanying plate are some drawings taken from those published in the *Memoirs* of the Italian Spectroscopical Society. The uppermost represents part of a view of the whole circumference of the Sun observed by Professor Tacchini at Palermo in 1871, the portion selected including the highest prominence then visible, which was of remarkable elevation and of extraordinary shape. Some other interesting formations of less elevation are also shown. The complete drawing is in vol. viii of the above *Memoirs*, plate cxi. The next lower view is that of a prominence observed by Tacchini on December 11th, 1871 (vol. i, plate v). In it a large portion of the prominence is seen floating at a considerable elevation above the photosphere, without any visible connection with it. The lowest figure is from vol. i, plate xi, and represents a drawing made by the late Father Secchi, at Rome, on July 11th, 1872, 6h. 20m. p.m. One prominence in it is particularly noticeable. It seems to have rushed up at first with enormous violence, until its upward velocity having diminished, it was wafted along by a horizontal current almost like the smoke from some tall factory chimney.

The spectroscope would at the same time show blue and yellow images, exactly-corresponding in shape to the red images, and situated in those parts of the spectrum in which the blue and yellow bright-lines of the prominence-matter are found. The red images, being much the most brilliant, are shown in our plate. Out of many hundreds of drawings we have endeavoured to select some of the most interesting forms; but in actual observation there seems to be no end to their ever-changing variety.

Such prominences as these are found to be identical with the beautiful rose-coloured appearances which, as we have already mentioned, are visible (even to the naked eye) during a total eclipse of the Sun. A few observations of them made at such times are recorded from the beginning of the seventeenth century onwards. But it was not until the year 1851 that Airy, then Astronomer Royal, first pointed out that the way in which the Moon's body passed over them, gradually uncovering them on one side of the Sun while it gradually concealed them on the opposite side, proved undoubtedly that they belonged to it, and not to the Moon; as to which, even at so recent a date, considerable doubt existed. In fact, until the occurrence of the eclipse of that year comparatively little attention had been paid to them. Since that date, however, they have been most attentively watched in every total eclipse that has taken place, and it has been found that in some eclipses those seen have been much larger and more abundant than in others.

Even when they were known to belong to the Sun, it was impossible, before the invention of the spectroscope and the discovery of the best method of its use, to determine their true composition and character, much less to see and watch them day by day and hour by hour. Nothing, perhaps, can better illustrate the remarks with which we commenced our first lecture with regard to the rapidity of recent progress in astronomy, than the contrast of our present knowledge of these remarkable formations with that of thirty years ago.

Since their daily observation has been found to be possible it has been noticed that they are met with in all latitudes of the solar globe. They are, however, most abundant over two zones, extending from about 10° to about 20° on either side of the Sun's equator, where Sun-spots are also most frequent. But instead of being hardly ever met with over the equator, as is the case with the spots, they are not much less abundant there than at some

PLATE II.   SOLAR PROMINENCES IN 1871 & 1872.

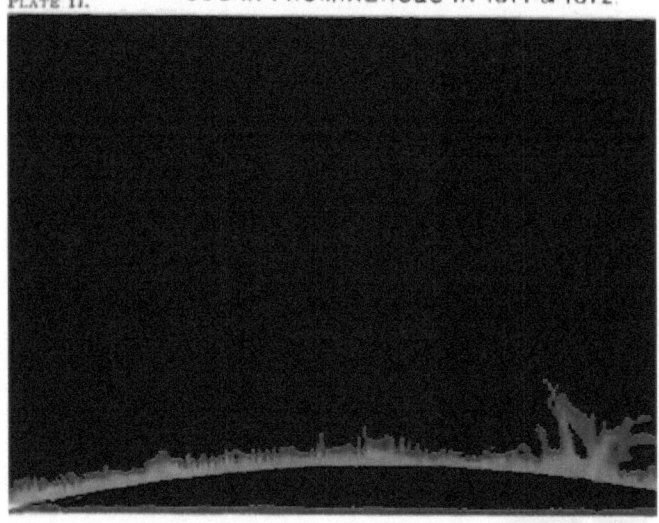

W. H. WESLEY, *Lith.*

From the "Memoirs of the Italian Spectroscopical Society."

distance from it. Again, spots are not found beyond a north or south latitude of 40°, but the prominences, after diminishing very considerably in frequency at a latitude of about 50°, appear to renew their frequency as we approach the poles of the Sun. These facts are excellently represented in a diagram on pp. 141 and 200 of Professor Young's treatise upon the Sun, published in 1881, in which are shown, on the one side, the comparative positions and frequency of the prominences, according to 2767 observations made by Secchi in the year 1871; while on the other side the results of Carrington's observations of 1386 Sun-spots between the years 1853 and 1861 are in like manner depicted.

When the rotation of the Sun carries a spot round to its edge, it is often noticed that there are prominences connected with it, or surrounding it; but a still more constant connection is found to exist between prominences and the locality of the bright corrugations of the solar surface called faculæ, which are not only generally seen to precede the appearance of a spot, or to follow its disappearance, but are also found in regions lying beyond the two spot-zones, and apparently unconnected with any spot. The spectroscope, moreover, from time to time gives evidence of the existence of a mass of vapour, such as may constitute a prominence, rising up in front of the dark background of a spot, even when the spot is far removed from the edge of the sun. This is shown as follows.

It is found that not only is the spectrum of the Solar Light generally darkened in the locality of a spot (a proof that dense masses of cooler vapour are absorbing and obscuring the light of the photosphere), but over special portions of a spot certain bright lines are frequently seen to shine out in the particular parts of the spectrum to which they belong. They are such as would arise from immense and intensely heated masses of those vapours which form the prominences. In this way prominences may be discovered over, or partly over, spots in any part of the spot-zones, although it was at first thought, when the spectroscope was applied to the uneclipsed Sun, that the utmost that it would accomplish would be to detect their existence when they might be seen upon the edge of its disc.

We ought, however, to mention an important distinction between the constitution of such prominences as are connected with spots and of those which occur upon other parts of the Sun's surface. The former are generally of an eruptive character, and contain, especially in their lower portions, various other metallic vapours in addition to that of hydrogen. The latter are of a more quiescent and cloud-like nature, and consist almost entirely of hydrogen and of one other element (named *helium*), whose nature is at present unknown, but which gives a bright line in its spectrum near to the yellow lines of sodium. They are probably formed either by the gentle (and not eruptive) emission of hydrogen from the regions beneath them, or by some action which causes masses of gas floating in the solar atmosphere to change from a dark to a luminous condition.

We have made special reference to the connection between Sun-spots and prominences, because we believe that an outburst of the latter is very frequently the origin of the former. Not that a prominence actually breaks out in the exact locality of a spot, but rather (as Professor Young has recently suggested) that the projection of prominence-matter around a certain locality causes a diminution of pressure beneath it. This is most likely followed by a falling-in of some of the clouds of the photosphere, which may be supposed to consist of more or less cooled and condensed matter previously carried up in a gaseous form from regions below. Into the cavity currents of rising gas and some of the smaller filaments of the photosphere would then be drawn, occasionally with a vertical or whirlpool-motion, as well as a good deal of the heavier vapours of the neighbouring prominence-eruption. There is also, in all probability, a considerable indraft of matter, so far cooled as to be in a fine state of division, which may be somewhat vaguely compared to smoke.

In this way, although there are many difficulties left unexplained, and no theory hitherto suggested is really satisfactory, we may perhaps obtain a tolerably true idea of the nature and origin of a Sun-spot. If so, its beginning takes place deep down in the gaseous body of the Sun, where some disturbance commences, which probably soon raises a

PLATE III.   SUN-SPOTS & FACULÆ, APRIL, 1870.

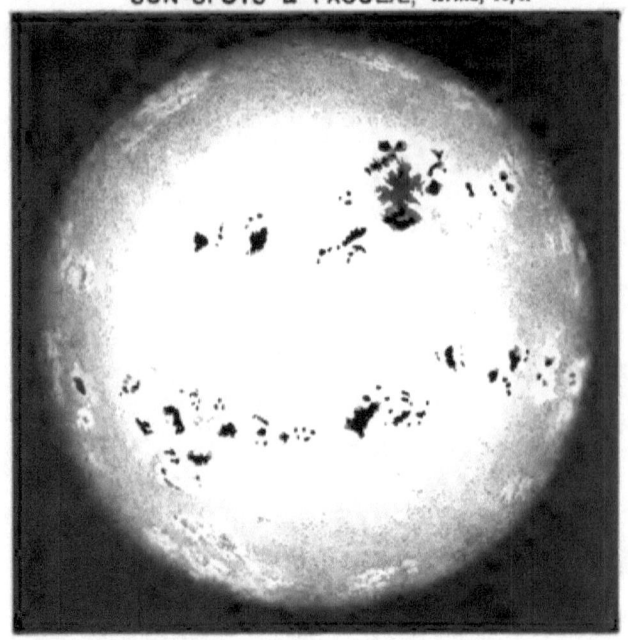

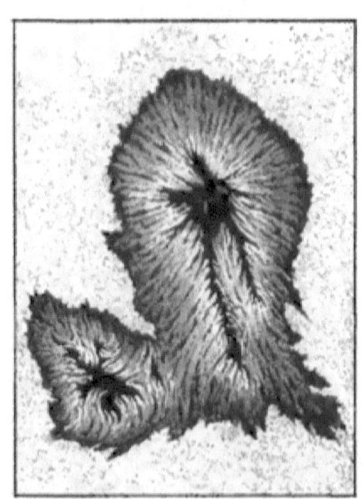

W. H. WESLEY, *Lith.*

SUN-SPOTS, MAY 27 & 28, 1880.

From the "Memoirs of the ITALIAN SPECTROSCOPICAL SOCIETY."

group of faculæ upon the photosphere. If it continue, a fierce eruption, or series of eruptions and explosions, next occurs, in fiery fury passing all our powers of imagination, and presently, as above described, a spot, or group of spots, appears.

The drawings in Plate III. may serve to illustrate the preceding remarks. The first is a general view of the Sun, taken from a sketch made by Tacchini at Palermo in April 1870, and published in vol. vi. of the *Memoirs* of the Italian Spectroscopical Society previously referred to. It shows the occurrence of faculæ in other parts of the solar surface as well as in the two parallel zones on opposite sides of the equator, in which a large number of spots of various sizes are seen.

The other two drawings are elaborated and enlarged from those made by Professor Riccò, at Palermo, of two spots observed on May 27th and May 28th, 1880, and published in vol. ix. of the above *Memoirs*.

They clearly show the dark umbra, or central portion of the spot, as well as the surrounding penumbra, or less dark portion. An attempt has also been made to indicate the way in which the filaments of the photosphere are apparently drawn inwards towards the umbra, so that by their convergence the lower part of the penumbra is brighter than its upper portion.

Indications may also be seen, in the left-hand and right-hand drawings respectively, of the way in which isolated portions of the photosphere occasionally float across, or form a bridge over, the depression, or (to use Professor Young's graphic term) the *sink*, of which a spot consists.

Both within the cavity of a Sun-spot and in the prominences, the spectroscope enables us, by certain distortions and slight changes in the position of the bright lines which it shows, to determine the velocity and direction of the motion of some of the huge masses of vapour involved. It is in this way, as well as by the general appearance of a spot, that we are assured of their occasional rotatory movements; or, to use more popular language, that we see in them veritable cyclones of fire.

We are aware that a theory has recently been suggested that some of all this vast turmoil and almost inconceivably rapid movement of huge and intensely heated vaporous masses is only apparent, and that the phenomena observed may be chiefly of an electrical character, and the light produced somewhat similar to that of discharges in vacuum tubes, or of the Aurora Borealis.

There is a certain amount of plausibility in this theory, and it is one which should not be lost sight of, especially when we remember the undoubted coincidences which have more than once been noticed between certain brilliant appearances on the disc of the Sun and disturbances of the Earth's electrical condition. It may also apply to the formation of some of the more quiescent prominences seen at a considerable distance from the solar equator, and especially to those which appear from time to time like rose-coloured clouds, without any apparent connection with the photosphere beneath. But so far as our present knowledge goes, we are decidedly inclined to accept the eruptive theory for most of the prominences, and to attribute a similar origin, as suggested by Secchi and modified by Professor Young, to the Sun-spots, although there can be little doubt that the phenomena connected both with their formation and continuance must also involve violent electric action.

The theory which we have thus attempted to elucidate is in one respect exceedingly important, inasmuch as it puts the spots before us as evidences of great *Solar Disturbance*. At times when they occur in abundance and of unusual magnitude, we are therefore not to consider that, a corresponding portion of the Sun being darkened, its light and heat will be upon the whole proportionally lessened. We are rather to understand that processes of such great disturbance are going on, as may, in all probability, intensify the total amount of light and heat evolved.

If this be the case (and there seems little reason to doubt it), we are at once led to enquire whether any law of periodicity can be detected in the number or size of the spots or prominences. If so, it may also be asked, is any corresponding effect produced upon the Earth? Our answer to the former of these two questions is, that there is from some cause or other an undoubted approach to regularity in the increase and decrease of the phenomena in question; the average time between two successive maxima being believed to be very nearly $11\frac{1}{9}$ years.* It is also noticed that the increase from a minimum to a maximum is more rapid than the subsequent decrease to another minimum; the former division of the period being about two years longer than the latter.

But at present we can only say that the above conclusion is that which is indicated by our observations. We cannot in any wise profess to explain why it is that this alternate increase and decrease of disturbance occurs. We also notice that, although it is upon the whole tolerably regular in its action, there is in connection with it a *frequent and very considerable deviation from absolute regularity*. To such an extent is this the case, that Professor Wolf states that the interval between two successive maxima may be occasionally as little as $7\frac{1}{2}$, or as much as 16 years. The last minimum, which occurred at the end of 1878, or at the beginning of 1879, took place about $1\frac{1}{2}$ years later than the date which an average interval from the previous maximum would have assigned to it. The next maximum may be expected to occur near to the beginning of the year 1883, unless it should be correspondingly delayed.

In answer to our second question we can state, that there seems to be an undoubted connection between one terrestrial phenomenon, and the number or size of the spots and prominences which we have been describing. We refer to the magnetic condition of the Earth. Very careful records of the daily variation of the declination of the magnetic needle, and of the horizontal component of the Earth's magnetic force, have been made at Greenwich and elsewhere, and if curves drawn to show these daily variations be compared with one

* A period equal to, or very closely agreeing with, the 11·86 years which Jupiter occupies in the circuit of its orbit, has recently been suggested by M. Duponchel (see *Comptes Rendus*, tome xciii, pp. 827 and 950); but the shorter period mentioned above is that which is generally accepted, as having been obtained after a very elaborate investigation of all accessible records by Professor Wolf of Zurich. He, however, considers that the result is doubtful to the extent of about $\frac{7}{9}$ths of a year.

which represents the total area of Sun-spots day by day observed, they are found to be most remarkably similar. Any sudden irregularity in one of them is answered by a corresponding irregularity in the others, although the oscillations on the Sun-spot curve may often slightly precede those in the magnetic curves.

In a very important memoir by Mr. W. Ellis, Superintendent of the magnetical and meteorological department of the Royal Observatory, Greenwich, published in part ii. of the Philosophical Transactions for 1880, the following conclusions are deduced from the observations of the preceding thirty-seven years:—

1. That the diurnal ranges of the magnetic elements of declination and horizontal force are subject to a periodical variation, the duration of which is equal to that of the known eleven year Sun-spot period.

2. That the epochs of minimum and maximum of magnetic and Sun-spot effect are nearly coincident, the magnetic epochs on the whole occurring somewhat later than the corresponding Sun-spot epochs. The variations of duration in different periods appear to be similar for both phenomena.

3. That the occasional more sudden outbursts of magnetic and Sun-spot energy, extending sometimes over periods of several months, appear to occur nearly simultaneously, and progress collaterally.

The close agreement of the three curves shown in a diagram for the whole period of thirty-seven years is most interesting. We append a reduced copy for the ten years from 1867 to 1876 inclusive.

It should be mentioned that the numbers in the scale on the right hand correspond to Dr. Wolf's determination of those which best represent the relative frequency of the spots, observed during the years in question. For the curves B and C, twice one of the upper graduations (each of which corresponds to ten of Sun-spot number), or one of the lower graduations of double width, is equivalent, respectively, to one minute of arc of declination, or 0·003 of the horizontal force.

As a rule, displays of the Aurora Borealis are found, as

might be expected from their electrical character, to be accompanied with disturbances of the Earth's magnetic condition. We are not therefore surprised to find that, so far as we can judge from our very imperfect records of their occurrence, there seems also to be some correspondence between the number of Sun-spots observed, and the number of Auroral

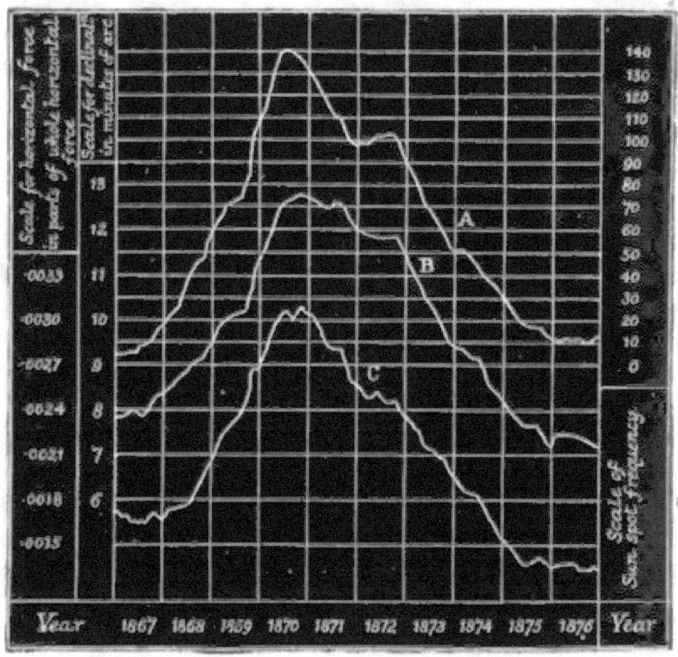

Fig. X.—The curve A represents the Sun-spot frequency; the curves B and C the variation in the diurnal range of the magnetic declination and horizontal force respectively at the Royal Observatory, Greenwich, during the years 1867 to 1876.

displays.* But we hardly like to speak positively with regard

* During April 1882 some unusually large spots appeared, three being visible to the naked eye, in the middle of the month, at the same time,—a most unusual occurrence. During the same period, however, the prominences seen were but small in size, nor were there any deserving special notice in the neighbourhood of the great spots. On the other hand, faculæ were unusually abundant. On the 17th and 20th days of the month the Earth was disturbed by violent magnetic storms, which, in this instance, may have had a special connection with the outburst of the great spots we have referred to, in exact coincidence with one of which a grand auroral display was also observed in America.

to any connection existing between Sun-spot frequency and any other terrestrial phenomena. Attempts of a somewhat elaborate character have been made to show that there is a correspondence between the number of Sun-spots in any given year and the amount of rainfall, the occurrence of cyclones and wrecks, good and bad vintages or harvests, the average yearly temperature, or those commercial panics which certainly have occurred in several consecutive instances after an interval of ten or eleven years, and which might in any case be reasonably expected to follow at a certain distance after a succession of unfavourable seasons.

Our great difficulty, however, at present, is to form any just estimate of the rainfall, or of the weather, of the Earth as a whole. Local variations are exceedingly perplexing; and it frequently happens that a particularly hot and dry season in one part of the Earth is coincident with one of just the opposite character in another quarter of the globe. There is, nevertheless, good reason to hope that we may in time make such progress in the study of meteorology, that comparisons of this kind may be more practicable than they now are; nor can there be any doubt that it is most important to maintain an unbroken and accurate record of Sun-spots (which may probably be best effected in some country where the Sun is more constantly visible than it is in England), and to compare such a record with every possible climatic or magnetic indication. We may, by such a comparison, at any moment discover some new law connected with the spots, or some new relation existing between them and other phenomena. We may, for aught we know, be even now on the very verge of such a discovery, and a little more perseverance be all that is needed to put it within our grasp. But, as yet, we have no *positive proof* that our seasons are directly affected by them, or that we are likely to be able to predict the weather, a few weeks or months in advance, by means of their observation.

There are some other points of interest connected with Sun-spots to which we must merely allude. We refer, for instance, to the supposed effects produced upon them by the relative positions with regard to the Sun, or to one another, of the planets Mercury, Venus, and Jupiter. Or to the possible connection between the period of Sun-spot frequency and the orbits of certain **streams of meteoric** bodies round the Sun. It has been suggested that such streams may, in those **parts** of their orbits **in** which **from** time to **time** they make their **nearest** approach **to it,** produce **disturbances** in the photosphere, either by their close proximity, **or by actual contact with it.**

We prefer, however, to pass by such questions as these, which are for the most part of a decidedly hypothetical nature, in order to speak of matters which are far more certain. And in so doing, we are **very** glad to be able to state that photography, which has failed **to give all** the results which had been **hoped for** in connection with the transit of Venus, has gained a triumph surpassing our best expectations in regard to **the** observation of the physical condition of the Sun. It not **only** secures an instantaneous and permanent record of the Sun-spots of any given day and **hour, such as** for a considerable number of years past has been regularly taken at Kew and Greenwich, by means **of** which we can at our leisure measure the area occupied by **them;** it not only enables us during the brief duration of a total eclipse to obtain with great rapidity a series of pictures of **the** solar prominences, and of the other surroundings of the **Sun,** far more beautiful and accurate than could otherwise be depicted; but it has been found possible by M. Janssen, at the physical observatory of Meudon, near Paris, to secure photographs of the Sun of a size and delicacy far surpassing anything previously attempted.

Some of his first great successes were gained by an exposure of the photographic plate, in connection with a telescope specially **adapted to the** purpose, for so short a time as the $\frac{1}{3000}$th **part of** a second. By this means images of the Sun, exceeding twelve inches in diameter, were obtained without any subsequent enlargement. **Of late** he has secured negatives nearly three feet in diameter, and has occasionally used much shorter exposures. The rapidity with which the picture is taken, and the absence of any subsequent enlargement of the negative, **as well as** the extreme care used in its development, effect a delicacy in the delineation of the solar surface which would be altogether confused and blurred by a longer exposure.

A certain structure of that surface, which had previously been noticed or suspected, with powerful telescopes under circumstances of very good definition, is most distinctly evident in these photographs. They show that comparatively small formations, which have been named *Solar Granules*,* are thickly congregated all over it.

In some parts these are very distinct, and tell us of a region where comparative calmness reigns. In other parts they are blurred and distorted, and indicate that a fiery storm is raging, or that a prominence is bursting forth from the photosphere. Their grouping frequently indicates the vortical action of some terrific solar cyclone, or tornado. Incessant changes take place in their forms and aggregation. What they actually are, we cannot, as yet, assert with any certainty. It may be that they are like the crests, or summits, of the waves of a fiery sea ; or, more probably, condensations of certain matters from uprising columnar-shaped currents of vapour ; or that they are in some way connected with small prominences or eruptions. But the photographs, two small specimens of which may be seen in Professor Young's work upon the Sun, and in certain French publications, while others upon a much larger scale are to be found in vol. ix of the *Memoirs* of the Italian Spectroscopical Society, undoubtedly show that from certain portions of the surface, occupying a comparatively moderate amount of the whole, we receive a much more intense light than from the remainder. It is very possible that we may find the study of such photographs, and the indications which they afford of the general condition of the Sun's surface, to be quite as important as the number and size of the spots, or of the larger prominences previously described.

Next to those regions of the Sun of which we have so far spoken, and with regard to which we have shown that the spectroscope has taught us so much, there come in order two

---

* Some slight indications of their appearance may be seen around the two large Sun-spots in Plate III.

PLATE IV.

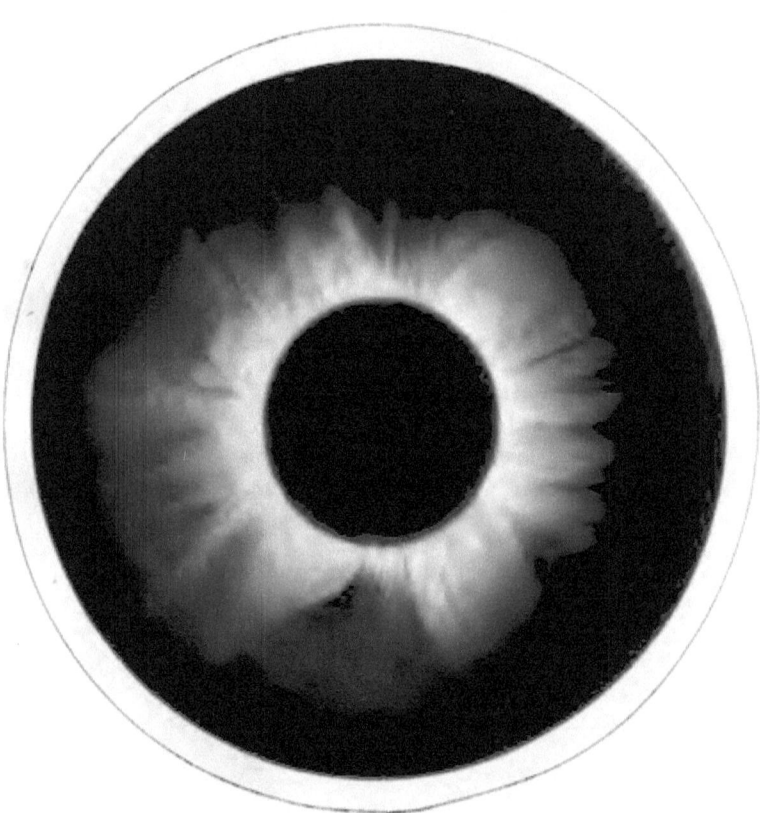

Woodbury process.

TOTAL ECLIPSE OF THE SUN,
Dec. 12th, 1871.

From a Drawing made from Negatives taken at Baikul by
Lord LINDSAY's Expedition.

successive envelopes or strata, further removed from its centre, which are called respectively the *inner* and the *outer corona*. These are only visible during total eclipses. At other times their light is not sufficient, either for spectroscopic observations, or for those made with a telescope. But during the totality of an eclipse, very useful and interesting information may be obtained with regard to them. It is then easily seen that the photosphere is surrounded by a brilliant region, the lower part of which, although more or less irregular in its boundary, is somewhat distinctly defined, and extends to a height which does not in general exceed 300,000 miles. The outer portion is much more irregular in its shape, and is frequently prolonged, especially in the neighbourhood of the Sun's equator, into the form of immense rays or streamers, which under favourable circumstances may be traced to a distance of at least 10,000,000 of miles.

We have much pleasure in putting before our readers a reproduction, by the Woodbury process, of a most beautiful steel engraving, published in vol. xli of the *Memoirs* of the Royal Astronomical Society (commonly called the Eclipse Volume), which was compiled and edited with extraordinary care by Mr. Ranyard, late Secretary of the Society. It represents the total eclipse of 1871, and is a copy by Mr. Wesley, the very able Assistant Secretary of the Royal Astronomical Society, of a drawing made from the original negatives photographed during the eclipse in connection with Lord Lindsay's expedition at Baikul. We know of no view in which a more beautiful and interesting amount of detail is depicted in the Sun's surroundings. Nothing, however, has been inserted which was not visible in at least three of the negatives taken. The general effect of the original will be best obtained by holding the Woodbury-print between twelve and eighteen inches from the eye.

During an eclipse the spectroscope indicates that the light of the corona is partly derived from solid matter, or from matter which gives the same kind of spectrum as that which is solid; but that it also consists to a considerable extent of luminous gas, a large portion of which appears to arise from some element which does not, so far as we are at present aware, exist upon the Earth.

It is probable that the Zodiacal Light, which is often faintly visible, especially within the tropics, shortly after sunset, or shortly before sunrise, and which appears to extend almost, if not quite, as far as the Earth's orbit, is another still more highly rarefied solar appendage, or outer atmosphere, of a lenticular shape, inasmuch as it has been traced so near to the Sun that it is certain that the long streamers of the corona must penetrate it to a considerable distance.

Interesting speculations have been started with regard to the nature of these long streamers or rays of the outer corona. Their appearance varies very much in different eclipses. One suggestion is, that they may be caused by the elongated orbits of swarms of meteoric bodies revolving in near proximity to the Sun. Such groups are known to travel in the orbits of certain comets, and to produce the phenomena of shooting stars when they become highly heated by friction as they pass through the Earth's atmosphere. In the neighbourhood of the Sun they are no doubt very abundant, owing to his overwhelming power of attraction and aggregation; and, as they pass through the corona, we can understand that portions of some of them may remain in an intensely heated state of incandescence, while others may be dissipated into luminous gas.

If the above suggestion be accepted, it is also probable that a considerable amount of meteorites of various sizes, many perhaps vastly larger than any with which terrestrial traditions are acquainted, are frequently drawn into the photosphere itself, causing thereby a considerable amount of disturbance, and possibly to some slight extent acting as fuel for the Sun. We believe, however, that some other explanation than the above must be sought for the origin of most, if not of all, of the long coronal rays and streamers. Possibly they may be connected with a repulsive force existing in the Sun, or be phenomena of an electrical character, or arise from the same cause (whatever it may be) which produces the tails of comets. It is very deeply to be regretted that the occasions when we can observe the corona are so few and of such brief duration.

As to what is the actual temperature of the solar photosphere we can say very little. It has been calculated that less than one two-thousand-millionth part of the heat emitted by the Sun reaches the Earth. Its effective emission of heat has been variously estimated to be equivalent to a surface temperature less than 2000° Centigrade, and rising up to 5,000,000° Centigrade,—so different have been the conclusions at which various investigators of this difficult subject have arrived.

The latest attempt to contrast the Sun's temperature with that of any terrestrial source of heat is, we believe, that of the American astronomer, Professor Langley. So recently as the year 1878, he compared the radiation of heat from the solar surface by suitable apparatus with that from the surface of a mass of molten liquid 30,000 to 40,000 pounds in weight, and supposed to have a temperature of 1800° to 2000° C. (exceeding that of melted platinum), which he watched as it issued from the mouth of a Bessemer steel converter. The result was that he concluded "the *minimum* value to be assigned to the solar radiation must be more than eighty-seven times that from an equal area of the molten metal. But he also states that the true value may be indefinitely greater."

This, however, still leaves us without any positive information as to the *actual temperature* of the general surface of the photosphere, or of different depths within it. We can only conclude that it must greatly surpass any degree of heat with which we can experiment in our laboratories; and that there is no *à priori* improbability in the supposition, previously referred to, that it may be sufficiently high to resolve into a more elementary atomic state, or into different molecular combinations, some bodies, which, at the temperatures attainable upon the Earth, have been hitherto considered to be elementary. At any rate, it cannot be denied that in regard to the effects of great pressure, or of intensity of temperature, a vast field of spectroscopic research, as yet but slightly invaded, lies open to the student. It has also, we believe, been recently suggested that by collecting the heat of the Sun in huge reflectors, and bringing it to bear in enclosed chambers upon so-called elementary substances, results may possibly be produced similar to those which occur in the enormously heated regions of the Sun itself.

Another very important and most interesting subject for investigation is the means by which the Sun's heat is prevented from diminishing at a much more rapid rate than we find to be the case. Of its very slow diminution we feel no doubt, but the solar radiation is so enormous as to indicate that the loss of much of the heat given off must in some way be compensated.* Our space only permits us to remark, that those who believe in the existence of very long past geological periods have found a difficulty in imagining the *original* temperature of the Sun to have been sufficiently high to provide for the cooling which must in the meantime have taken place. Even if we make allowance for the greatest imaginable supply of meteoric, or

---

* Since the above was written, Dr. Siemens has suggested, in a paper read before the Royal Society, a new theory of solar radiation and action, according to which aqueous vapour and carbon compounds are constantly being thrown off from the Sun's equatorial parts by the centrifugal effect of its rotation, and (if we understand him rightly) are supposed to be as constantly drawn in again by currents tending to its poles, after having travelled to vast distances, and while there, in a very attenuated state, having been dissociated by the solar heat radiated into space. He terms this outdraft and indraft a *fan-like action* of the Sun's rotation, and considers that on approaching it again the gases would become condensed and heated, and burst into flame on entering the photosphere; while after a time the products of their combustion would be once more ejected from its equator. In this way very much of the heat radiated would be continuously regenerated, and the loss of that which does work upon the planets, and would not be so regained, would be very small. We do not venture to criticize this theory, although it certainly seems to us to involve some important dynamical as well as other difficulties. It will certainly need much consideration before its true bearing upon the question of the maintenance of the Sun's heat can be adequately weighed. Our readers who wish to study it may consult the Proceedings of the Royal Society; also an article in *Nature* of March 9th, 1882, which is illustrated by a very suggestive diagram ; and a fuller article in the *Nineteenth Century* magazine of April, 1882.

other, fuel, it seems to some who have carefully considered this question to be impossible for the existing temperature to have remained as high as it is, unless at some past time our Sun has collided with another Sun. If so, the two bodies might not only have coalesced into one, but the united mass would have been raised by the collision to an almost inconceivable degree of heat. On the other hand, it has been lately suggested that if the Nebular theory of the Solar System be received, the gradual shrinking of the Sun from the dimensions of the original nebula may have generated a far greater amount of heat than has been considered possible, because this shrinkage may be far greater than is generally supposed. There may be, far within the fiery photosphere which bounds our view, a comparatively small and denser nucleus—a nucleus not seen, as was once supposed, through the apertures of Sun-spots, but far beneath the lowest vaporous depth that we can fathom. On the other hand, some of the greatest authorities on physics maintain that the Sun must be vaporous and of comparatively light density throughout.

Such questions are of the deepest interest. They are connected with some of the most important problems that the astronomer or the geologist can attack. They bear upon the future of our system, its stability, its endurance, or its rapid decay. They may be as yet little removed from the domain of pure hypothesis, but they must not be lightly put on one side. It is by further developments in the use of the spectroscope, especially in connection with the study of Molecular Physics, that we hope we may be able to solve the mysteries of the present temperature of the Sun, and of the other so-called fixed stars. If we can do this, we may also be better able to judge of what their past has been, of what their present is, of what their future may be.

Our discussion of the Sun has necessarily been very superficial. We have, as it were, merely hovered around it, to take a brief glance into a few of its marvels. What we have seen may, however, have sufficed to show that there is in the Sun subject-matter for the study of many a lifetime,—mysteries and wonders of never-ceasing interest—which, while they bid us look beyond themselves to the Power that made them, may well allow us to exclaim with Southey :—

> "I marvel not, O Sun ! that unto thee
> In adoration man should bow the knee,
> And pour the prayer of mingled awe and love ;
> For, like a god thou art, and on thy way
> Of glory sheddest, with benignant ray,
> Beauty, and life, and joyance from above."

PLATE V.

Woodbury process.

PHOTOGRAPH OF THE MOON,
From RUTHERFURD'S original negative,
of March 6, 1865.

LECTURE III.

THE MOON.

> "How like a queen comes forth the lovely moon
> From the slow-opening curtains of the clouds;
> Walking in beauty to her midnight throne!"
>
> CROLY.

THE Moon, although by no means important in size, in some respects surpasses in interest every other heavenly body. In the soft, mellow light of a moonlit summer's evening there is a fascination peculiar and almost indescribable. In the principal shadings upon its disc visible to the naked eye we see the origin of the myth of the old Man and his bundle of sticks, which is diffused through nearly every savage as well as every civilized nation; in the wondrous scene of valley and mountain, of landslip and crag, of cleft and crater, exhibited in our telescopes, we behold details which would be altogether invisible at the vastly greater distance of the nearest of the planets; in the accurate determination of its exceedingly complicated orbit, and of the many irregularities of its motion, the most accomplished mathematician may tax his utmost skill; while in its ever-recurring phases, in its tidal action, in its imagined power over the weather, as well as in other respects too numerous to mention, the Moon is closely connected with the events and business of everyday life. Surely we may assume that our readers will join us in feeling a special interest in its study.

That study may be most conveniently divided into two parts; the one connected with the theory of the lunar orbit, the other involving the physical observation of the lunar surface. Of these, the former, although of less popular interest, is in other respects of at least equal importance.

The determination of the Moon's orbit, and the investigations connected with the estimation of its size and weight, and of the tidal effects which it produces, are necessarily of a mathematical character; and consequently (apart from their own intrinsic value) are of great benefit as a mental exercise to those who study them. On the other hand, the physical observation of the Moon's surface suggests many considerations as to its past history, and the relation of that history to the past and the future of the whole Solar System, which are of the highest interest. We will therefore endeavour duly to apportion our remarks between these two branches of lunar study. At the same time we must ask the kind indulgence of those of our readers who may find the first and more abstruse portion of our subject less interesting than the second.

As in the case of the Sun, so in that of the Moon, the first thing to be determined is its distance from the Earth. And, fortunately, that distance is so much less than the Sun's that we are able to employ the surveying method of observation, explained in Lecture I., pages 2 and 3. In actual practice, the application of the method involves many refinements which are necessary in order that its use may be satisfactory. But the *principle* of the surveying method is independent of all such refinements. It simply depends upon the determination of the difference of the directions in which two places of observation upon the Earth look at the Moon at any given instant. Their distance from one another being known, the calculation of the Moon's distance from either of them is reduced to an ordinary problem in trigonometry. And the distance of the Moon from the Earth's centre may then be found with equal facility. By the employment of observations made at such localities as Greenwich and the Cape of Good Hope, the mean distance between the centres of the Earth and of the Moon is proved to be about 239,000 miles. This is a fact with which, in these days of widespread education, we may perhaps assume that most persons are well acquainted. But it is much less generally known how *greatly this distance varies from its mean value*, or, in other words, how much the Moon's orbit, relatively to the Earth, differs from a circle.

In fact, the greatest and least possible distances of the Moon from the Earth are about 253,000 and 222,000 miles; the difference amounting to about 31,000 miles, or to nearly $\frac{1}{7}$th part of the least distance.

The Moon's actual orbit is so intricate that we will not attempt to explain any of the various perturbations by which it is affected. They chiefly arise from the difference between the magnitude and direction of the Sun's attractions upon the Moon and upon the Earth; which difference again depends, not only upon the form which the Moon's path would have, if unperturbed, and the Moon's ever-changing position relatively to the Earth, but also upon the deviation of the Earth's own orbit from a circular form, and the inclination of the Moon's orbit to it. Subsidiary perturbations moreover result from those which may be termed primary, while some arise from the attraction of the planets and other causes. If the Sun and planets did not exist, the Moon's path relatively to the Earth would be an ellipse, in which the greatest distance would be about $\frac{1}{8}$th part greater than the least distance. When, however, the perturbations are allowed for, the fraction approaches the value $\frac{1}{7}$th, as above stated.

One interesting result which of necessity follows from any change in the Moon's distance from the Earth is a corresponding change in its apparent size. The angle subtended by its diameter at an observer's eye (or at the Earth's centre), will very nearly vary inversely as its distance. The apparent width of the disc may therefore sometimes be a $\frac{1}{7}$th part greater than at other times.

But its apparent area may vary to a much more important extent. At one time it may be represented by the square of the number 7, *i.e.* by 49; at another time it may correspond nearly to the square of 8, which is 64. These numbers, 49 and 64, being almost in the ratio of 3 to 4, it follows that on some occasions the apparent area of the Moon's disc may be nearly $\frac{1}{3}$rd part larger than on others. Its full phase never exactly occurs when it is at its greatest and least possible distances from us, but the increase of its light at such times may approximate within a moderate amount of the above ratio. An especial proximity of the Moon to the Earth may occasionally have helped to produce an increase in its light, which some of our readers may have altogether attributed to the clearness of the night, or to the Moon's elevation above the horizon.

The Moon's average distance from the Earth is about $\frac{1}{400}$th part of that of the Sun (more accurately, about $\frac{1}{355}$th). Its *apparent* size is, however, always nearly the same as that of the Sun, although it may occasionally appear to be rather smaller, instead of, as in general, slightly larger than the great globe whose light it reflects. It therefore follows that its diameter cannot differ much from $\frac{1}{400}$th part of the Sun's diameter of about 865,000 miles, in order that the difference in their real sizes may be so nearly compensated by the difference in their distances. The most accurate calculations give about $\frac{2}{3}$ths of a mile less than 2160 miles as the value of the Moon's diameter.

The Earth's diameter being rather more than 7,900 miles, the Moon's is therefore rather less than $\frac{2}{7}$ths of that of the Earth. And, the surface and volume of a globe respectively varying as the square and cube of the diameter, a little arithmetical calculation shows (by squaring and cubing the above fractions) that the Moon's surface is somewhat less than $\frac{4}{15}$ths, and its volume somewhat less than $\frac{8}{343}$rds of the surface and volume of the Earth respectively. More accurate values are found to be between $\frac{1}{15}$th and $\frac{1}{14}$th for the surface, and about $\frac{1}{49}$th for the volume. The fractions $\frac{2}{7}, \frac{2}{27}, \frac{2}{99}$ * are easy to remember, and sufficiently accurate for all ordinary purposes.

The Moon's mean density is generally stated to be about $\frac{3}{5}$ths of that of the Earth, or about $3\frac{2}{3}$ times that of water. There is, however, a considerable amount of vagueness involved in making any such statements as to the densities of the heavenly bodies. If, for instance, we speak of the Earth's

---

* These are suggested by Mr. Proctor, in his valuable work upon the Moon, in which the various methods of obtaining its weight, which we shall presently mention in this Lecture, will be found more fully explained.

mean density as approaching six times that of water, we mean that a globe of the size of the Earth, *if solid, and of uniform composition throughout*, would have to be made of some substance nearly six times as heavy as water, in order to be of the same weight as the Earth. But we cannot be at all certain that the earth is solid *throughout*, there being many reasons in favour of a contrary supposition. We are, moreover, sure that it is not composed of materials of uniform density.

It is consequently more advisable, in an elementary discussion such as the present, to speak of the *weight* of the Moon as a whole, independently of any consideration as to its solidity or uniformity of constitution, rather than to attempt to compare its mean density with that of the earth or of water.

Let us therefore now proceed to notice some of the principal ways in which the Moon may be weighed, all of which we shall find agree in proving that its weight is to that of the Earth in the proportion of about 1 to 81.

One method of investigation which has been successfully used depends upon the fact, to which we have already briefly alluded, that, if two bodies suspended in space, as are the Earth and the Moon, mutually attract one another, the smaller will not simply revolve round the larger, but *both* will revolve round the common centre of gravity of the two. If one be much larger than the other, this centre of gravity will be much nearer to that one, and may even be within its surface. Indeed, in the actual case of the Earth and the Moon, the point in question is situated at a distance of only about 3,000 miles from the Earth's centre, *i.e.* about 1,000 miles within its surface.

Moreover, two such bodies will revolve as if attracted by a mass placed at the above point equal to the sum of their masses. These statements depend upon an elementary but very important theorem in dynamics. Keeping it in mind, we watch the Moon's movements. We neglect that part of its motion which it has in common with the Earth around the Sun; and, as accurately as we can, we allow for any comparatively slight difference between the Sun's attractions upon it and upon the Earth. The rest of the Moon's motion may be described as that which it has in its apparent path round the Earth. We therefore observe how long it takes to circuit once round our globe, in the course of what is termed a sidereal month. Then, as we are already supposed to know its distance from the Earth, we can, from the above duration of its circuit, not only calculate the corresponding velocity of its movement, but also estimate with what power it must, at each moment of its course round the Earth, be drawn by some force or other out of a rectilineal path into one having the curvature necessary to carry it round at the distance in question.

The theorem, however, which we have mentioned teaches us that this curvature will be such as the attraction of a mass (or weight) equal to the *sum* of those of the Earth and Moon would effect; and we know the Earth's weight, and how much of this effect it would by itself produce. The *additional effect* which is observed consequently enables us to discover what proportion the Moon's weight bears to that of the Earth. A good many precautions are necessary in the practical application of this method; the final result is, however, very satisfactory.

Another investigation depends upon a *monthly motion of the Earth* about the common centre of gravity of itself and the Moon, which is also involved in the theorem to which we have already referred. This movement of the Earth causes it to be alternately slightly more or less advanced in its annual path round the Sun at intervals of half a lunation, than otherwise would be the case. From careful observations of the consequent alteration in the direction in which we look at the Sun at such times we are able to judge how far the Earth is in this way moved. Or instead of observing the effect produced upon the direction in which we look at the Sun, we may note the much greater effect which (owing to the greater nearness of the planet) the movement in question will have upon the apparent direction in which we see Venus, at times when it is especially near to the Earth.

In either case we find an effect produced which shows that the Earth is displaced by a distance, roughly speaking, equal to about $\frac{1}{82}$nd part of its distance from the Moon. It follows that the centre of gravity of the Earth and the Moon divides the line joining them into two portions, which are approximately as the numbers 1 and 81, or that the Earth's weight is about 81 times that of the Moon.

A third method, the explanation of which is too complicated for our present discussion, is related to certain effects of the Moon's attraction upon the protuberant matter of the Earth's equatoreal regions, which depend upon the inclination of her orbit to the equator. These effects produce periodic nutations, or variations, in what is termed the precessional motion of the Earth's axis, from the amount of which nutations the value of the Moon's weight may be obtained.

A fourth method, more simple, although by no means so accurate, depends upon the determination, as the result of long-continued observations, of the mean heights of spring and neap tides in the open ocean, where they are unaffected by local causes. It is found that the average heights of such tides are nearly in the ratio of the numbers 41 and 15. From this it results that the Sun's power to produce a tide must be to the Moon's as 28 to 13; their effect, when they act jointly, and cause what is termed a spring tide, being represented by the sum of these numbers, or 41; and their effect in a neap tide, when they are opposed, being represented by their difference, which is 15. But a simple deduction from the law of gravity requires that these effects must also be in the ratio of the weights of the Sun and Moon divided by the cubes of their respective distances. These distances being assumed to be known, a by no means difficult calculation*

---

* The calculation involved may be very roughly indicated as follows:— The Moon's tidal power is to the Sun's.

$$\text{as} \ \frac{\text{Moon's weight}}{\text{Moon's distance cubed}} \ \text{is to} \ \frac{\text{Sun's weight}}{\text{Sun's distance cubed}}.$$

But the Sun's distance from the Earth is about 390 times the Moon's, which number cubed equals very nearly 60,000,000. The two tidal powers being found by observation to be as 28 to 13, it follows that the Sun's weight must be 60,000,000 times $\frac{13}{28}$ths of the Moon's weight, i.e., rather less than 28,000,000 times. But the Sun's weight is known to be about 330,000 times the Earth's. Dividing 28,000,000 by 330,000 the Earth's weight comes out about 85 times the Moon's, a result not very different from the true value.

gives about $\frac{1}{82}$ as the fraction which the Moon's weight is of that of the Earth.

By such various processes, the results of which agree so very satisfactorily, is this important and interesting value determined, so that we are able to state with the greatest possible certainty that the Earth's weight is almost exactly $81\frac{2}{5}$ times that of the Moon. We may, on another occasion, attempt to show that we are even able to achieve the wondrous feat of weighing some of the fixed stars, notwithstanding their almost inconceivable distance from us. At present let it suffice to have explained that there are *several* independent methods by which we can weigh the nearest of the heavenly bodies, albeit even that one, comparatively near as it is, is separated from us by an interval which never falls short of 220,000 miles.

Thus far we have spoken of the Moon's distance from the Earth and of its weight, and have only incidentally assumed that it is the Earth's satellite, accompanying it in its annual journey round the Sun. Our next attempt shall be to form a somewhat more accurate conception of the Moon's actual path in space, while month by month it appears to us to circle round the Earth in the opposite direction to that in which the hands of a watch go round when we look at its face, or with a rotation which is termed from *west* to *east*, and which corresponds in its direction with that of the Earth on its axis, and of the Earth and all the other planets in their orbits round the Sun.

That this is the case may easily be seen, notwithstanding that the axial rotation of the Earth from west to east introduces some little difficulty into the observation by making the Sun, Moon, and stars all apparently go round the sky from *east* to *west* in rather less than twenty-four hours. For if the Moon, while so appearing by this latter effect to move from east to west, be carefully watched, it will be noticed that it is at the same time actually travelling from west to east amongst the stars with a speed which is of no small magnitude. If it be seen to be at a certain distance east of some bright star at a given hour, one hour afterwards it will be further to the east of the star by rather more than its own diameter; in two hours by more than twice its own diameter, and so on; and this rate of motion takes it once round the heavens from west to east in about $27\frac{1}{3}$ days.

The Moon's mean distance from the Earth being rather less than 240,000 miles, and the circumference of a circle of that radius being about 1,500,000 miles, it follows that its average speed in journeying round the Earth must be at the rate of rather under 55,000 miles per day, or of about 2,280 miles per hour. But the Earth itself is all the while revolving round the Sun, with a speed of about 66,000 miles per hour. In order to keep in close attendance upon it, the Moon must therefore, in addition to its above-mentioned velocity round the Earth, also possess a velocity equal to that of the Earth round the Sun.

The Moon's motion is consequently compounded of two velocities; one of about 66,000 miles per hour round the Sun; the other of about 2,280 miles per hour round the Earth.

We are at once struck with the vastly greater magnitude of the former. It shows that, so far as regards its position in space at any given time, the Moon's motion round the Sun is much more important than its motion round the Earth. We are consequently led to enquire whether the Sun or the Earth exerts the greater influence on the Moon. And a little calculation will show, in opposition to what might at first sight have seemed probable, that the Sun has more than twice as much effect as the Earth in attracting the Moon.

The weight of the Sun is about 330,000 times that of the Earth; but its distance from the Earth being about 390 times as great as that of the Moon, its attractive effect is thereby proportionally diminished, the diminution depending upon the square of the distance. If we square the number 390 the result is 152,100. It consequently follows that instead of the Sun's power over the Moon being 330,000 times that of the Earth, we must, in order to get the true value, divide this number by 152,100. The quotient is 2 and a fraction, which proves that the Sun's power over the Moon is after all more than twice as great as that of the Earth.

In either case the power exerted bends the Moon's path

from the straight line, or tangent, along which, if left alone, it would at any moment proceed, into the curve in which it actually moves. The power of the Sun to produce this effect enables it to pull the Moon away from such a straight line towards itself in each successive second of time, through a distance equal to about $\frac{117}{1000}$ths (or nearly $\frac{3}{25}$ths) of an inch. In one minute the effect would not be 60 times as great, but, according to the law which holds in such cases, 60 times 60, or 3,600 times ; so that the pull of the Sun upon the Moon would in one minute deflect it from a rectilineal path through a distance of about 35 feet. The Sun is of course constantly exerting a similar power upon the Earth, and curving its orbit almost exactly to the same extent. But as far as the *Earth's* attraction upon the Moon is concerned, apart from that of the Sun, the orbit of the Moon round the Earth is only curved so much as to draw it towards the Earth out of motion in a straight line to a distance equal to about $\frac{54}{1000}$ths (or about $\frac{3}{27}$ths) of an inch per second, or about $16\frac{1}{3}$ feet in a minute.

Remembering this, let us first of all consider the circumstances of the Moon's motion when it is in the position of New Moon. Let the Moon be situated at M (Fig. XI.), which

Fig. XI.—Showing how the attractions of the Sun and of the Earth respectively curve the orbit of the Moon.

we may, for simplicity, imagine to be exactly in the straight line, ES, joining the Earth (which is supposed to be in the direction ME) and the Sun (which is supposed to be in the direction MS). The Moon has, as we have explained, a velocity of about 66,000 miles per hour from west to east round the Sun. In one minute it would consequently move from M to F, in a direction perpendicular to MS, through a distance of about 1,100 miles, if no attraction of the Sun or Earth continued to act upon it. But we have stated that the Moon has also a velocity of about 2,280 miles per hour from west to east round the Earth. As M is on the opposite side of the Earth to that which it occupies with regard to the Sun, it is easy to see that its rotation from west to east round the Earth will not carry it from M towards F, but towards Q, in exactly the opposite direction, through a distance of about 38 miles in a minute. The combined result is a motion equal to the difference of these two, or of about 1062 miles in a minute towards F. Let this distance be MP. During this minute, however, the Earth's attraction would, as we have mentioned, bend the Moon's path towards itself (*i.e.* in a direction approximately parallel to ME), through a distance of about $16\frac{1}{3}$ feet. Let this be PR. Then R will represent the position which the Moon would have at the end of a minute after leaving M, on the supposition that the Sun's attraction were not acting upon it.

But the Sun would curve the path in the opposite direction, and with more than double power, so as to draw the Moon towards itself, through about 35 feet in the minute of which we are speaking. We therefore measure RN equal to 35 feet, and N will be very approximately the real place to which the Moon will attain.

The Earth will in the meantime have moved onwards, so as to be still within its proper distance of about 240,000 miles from the Moon; but this is not important as regards our present explanation. The essential point which is involved is this, that although in the diagram the lengths of RP and PN are necessarily immensely exaggerated, and EM and MS are drawn very much shorter than they really are, the concavity of the Moon's path from M to N will clearly be turned towards the Sun.

And if this is the case when the Moon is New, the truth of the statement, which we may now make, that the Moon's path is *always* concave to the Sun, may be easily realized. For under no other circumstances will the Earth tend to draw it so directly *from* the Sun as in the case we have just considered. At the time of Full Moon it will, in fact, draw it directly *towards* the Sun, and make the concavity of its path towards that luminary so much the more pronounced. In other positions the effect of the Earth's attraction will be of an intermediate character. The position of New Moon is the critical one, and if the path is then (as we have proved to be the case) concave towards the Sun, it must always be so.

The orbit of the Moon round the Earth is indeed so small compared with that of the Earth round the Sun, that we may even speak of them as both going round together very nearly in the same orbit,—the Earth at a mean distance of about 93,000,000 miles and with a speed of about 66,000 miles per hour,—the Moon with a speed which is, periodically, about 2,300 miles greater or less than the Earth's, and varies from about 64,000 to 68,000 miles per hour; while its distance alternately exceeds or falls short of that of the Earth by about a quarter of a million of miles. This variation of distance is, however, so small in comparison with the radius of the Earth's orbit, that it is almost impossible to represent it upon any scale with which a diagram can conveniently be drawn. It is a comparatively trifling quantity in contrast with the changes to which that radius is itself subject in the course of a year, as the result of the ovalness of the Earth's orbit, by which in midwinter of our northern hemisphere we find ourselves about 3,000,000 miles nearer to the Sun than we are in midsummer.

To those who may find some difficulty in realizing how the Moon can really go round and round the Earth month by month while it is all the time moving in an orbit round the Sun, which, like that of the Earth, is always concave to it, an additional diagram may perhaps be of use. Let M_1, M_2, M_3, M_4, M_5 in Fig. XII. represent the Moon's orbit; E_1, E_2, E_3, E_4, E_5 that of the Earth. When the Earth is at E_1, let us suppose the Moon to be at M_1; when the Earth is at E_2, E_3, E_4, E_5 respectively, let the corresponding places of the Moon be M_2, M_3, M_4, M_5. At M_3 the Moon will be New; at M_1 and M_5 it will be Full; the portion of its path which we have drawn corresponding to an interval of a month.* The curves are intended to represent the adjacent orbits of the Earth and Moon, both being concave

* The month in question is termed a synodic month, and exceeds a sidereal month (or the interval in which the Moon performs one circuit round the Earth) by rather more than 2¼ days. It may be seen from

to the Sun, and in comparison with its immense distance, close to one another. Then it may be understood that at M_1 the Moon will be moving onwards faster than the Earth, and will keep

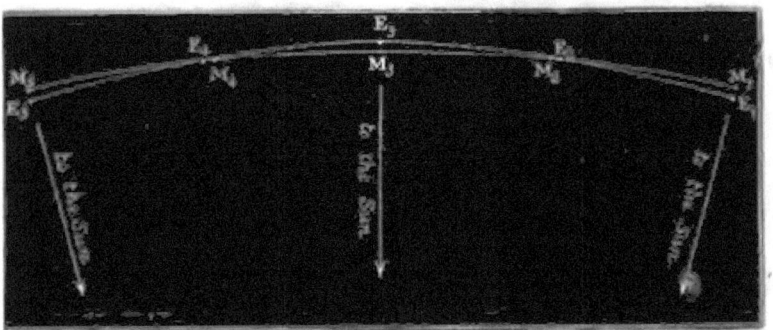

Fig. XII.—Showing how the Moon's orbit is **always** concave to the Sun.

in advance of it until, half-way between the time of Full and New Moon, they occupy the relative positions M_2 and E_2. After this, the Moon's speed will gradually fall more and more short of that of the Earth, and the Earth will gain upon it until, at the occurrence of New Moon, their places are M_3 and E_3. The Earth next keeps in advance, although the Moon's speed is gradually augmenting, until, after the lapse of another quarter of a lunation, they are to be found respectively at M_4 and E_4. Once more the Moon will gain upon the Earth until at the next Full Moon their positions are M_5 and E_5. During the middle half of the month the Moon's path is seen to be farther away than the Earth's from the Sun; during the other two quarters it lies within it.

A subsidiary diagram, Fig. XIII., drawn upon a larger scale, shows that an observer upon the Earth would at first see the Moon in the direction EM_1, presently in the direction EM_2 (corresponding to the places E_2 and M_2 in Fig. XII.), next in the direction EM_3, then in the

Fig. XIII.—The apparent path of the Moon seen from the Earth.

Figures XII. and XIII. that the Moon has to describe this additional portion of another revolution between being twice in succession Full or New, in consequence of the onward motion of the Earth around the Sun.

direction EM₄, and so on. Although, therefore, its orbit round the Sun is, as we have stated, nearly the same as that of the Earth, it is abundantly evident that the Moon will appear to an observer on the Earth to travel round and round the latter month by month.

We have endeavoured to elucidate this point somewhat fully, not only because the mental effort needed to comprehend it may, we hope, be useful to some of our readers, but because it may assist them in realizing the vast scale upon which the Solar System is constructed, by showing how insignificant a matter the Moon's orbit round the Earth really is. The point is, moreover, one which is insufficiently explained in some astronomical text-books, in which diagrams resembling Fig. XIV. have been used to represent the paths of the Earth and of the Moon. We would fain hope that those who have perused the

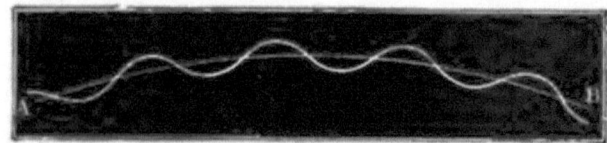

Fig. XIV.—False representation of the Moon's orbit.

preceding explanation will henceforth understand how much the real orbit of the Moon differs from any such wavy curve.

We may sum up the conclusions to which we have so far attained by stating that the Moon, in regard to the Earth, may be termed its *Satellite* or attendant; but in regard to the Sun, it should rather be termed a *Planet*, its orbit round the Sun being so little changed by the Earth's attraction that it always remains concave to that great central luminary, like that of the Earth itself, although the degree of concavity is slightly less than in the Earth's orbit when the Moon is New, and greater when it is Full. The very slight difference in the two orbits also involves the fact that the Sun attracts them both with nearly the same power, and that consequently the Moon does not desert the Earth, although the Sun's power over it is more than twice as great as that of the Earth. It could, indeed, only do so, if at any time the *difference* between the Sun's power over it and over the Earth were greater than the Earth's attraction upon it. But this difference being always very small, the two keep in near proximity to each other while they pursue their respective paths around the Sun. Even if the Earth ceased to exist, the Moon's orbit round the Sun would continue to be very little different from what it now is; nor would the Moon, in that respect, much feel the absence of its larger companion.

We must now pass on from the theory of the Moon's movements to that of its *Phases* and *Librations*. We are all accustomed to the regular course of the former as, month by month, we watch the alternate waxing and waning of the full circular disc from the narrow crescent, seen two or three days after the Moon is New, whose delicate beauty almost tempts us to exclaim :—

> " I would that aspect never might be changed.
> Nor that fine form, so spirit-like, be spoiled
> With fuller light.
> Keep the delicious honour of thy youth,
> Sweet sister of the Sun, more beauteous thou
> Than he sublime ! "

But the explanation of these Phases is not a very simple matter, unless certain mathematical propositions can be assumed or certain experimental demonstrations be given. It is easy enough to understand that one-half of the Moon's surface is always illuminated by the Sun, and that the whole of that illuminated half is seen at the time of Full Moon, while the dark half is turned towards us at the time of New Moon. The next point to be realized involves, however, slightly more difficulty, viz., that when we look at the illuminated hemisphere of the Full Moon, the appearance presented, owing to its distance from us, is not that of a hemisphere, but that of a flat, circular disc, which is, in mathematical language, the projection of the hemisphere upon a plane perpendicular to the line of sight.

But to come to something still less easy, let us suppose that just one-half of the hemisphere, which we see, is illuminated. The boundary between the light and dark parts seen will in this case be a circle passing half-way round the spherical surface of the Moon. But, as the whole hemisphere previously appeared like a flat circular disc, we shall in like manner now see the half of it, as though it were bounded by a straight line passing through the centre of that disc. In other words, the bounding edge of the bright portion when the Phase is half-full, which is really a semi-circle passing half-way round the Moon, will look like a straight line, because we

Fig. XV.—Moon full.   Moon half-full.

are unable to realize the protuberance of the central portions of its globe towards us. What we see consequently appears, as in Fig. XV., to be one-half of a circle, although it is really one-half of a hemisphere.

Perhaps the following perspective drawing may help to illustrate this still more clearly. Let PMNQ in Fig. XVI.

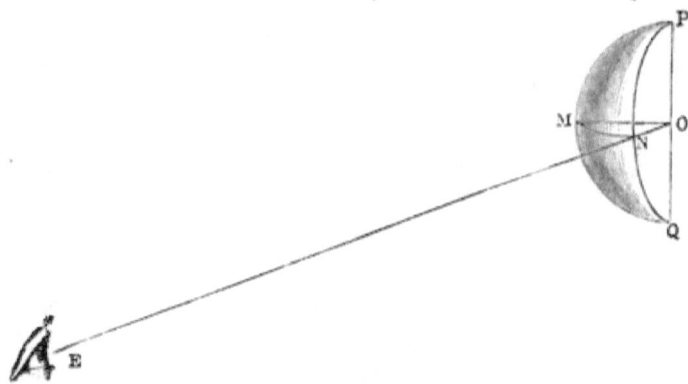

Fig. XVI.—One-fourth of the Moon's spherical surface is seen from the Earth as a flat semi-circle.

represent in perspective that half of the Moon's surface which is supposed to be seen illuminated. An eye at E, looking along E N towards O, the Moon's centre, is unable to judge of the small difference between the distances EP, EN, EQ, and, the semi-circle PNQ consequently appearing as if it were projected into the straight line POQ, the illuminated half-hemisphere PMNQ is seen as the semicircle POQM.

THE MOON.

If we next suppose the Moon, after having been reduced from Full, half-way towards being New, to continue to approach nearer and nearer to the latter position, one boundary, PMQ, of the enlightened part will remain unaltered, while the other, PNQ, will always be a semi-circle on the Moon's surface passing through P and Q. But PNQ will be gradually turned more and more away from our line of sight, and will draw nearer and nearer to PMQ. As it does so it will necessarily appear to form a curve, on the disc of which we have previously spoken, which will always pass through P to Q, but will gradually change from the straight line POQ, first having the form of a very much flattened ellipse, then of one more curved, and so on; until at last, just before New Moon, it will nearly coincide with the circle PMQ.

The appearances will, in succession, be such as are shown in

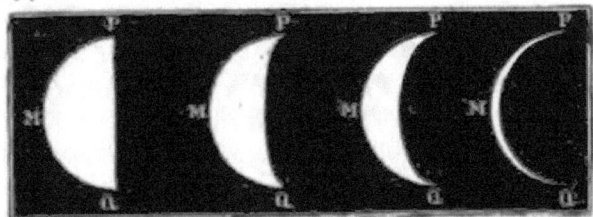

Fig. XVII.—Phases of the Moon.

the drawing in Fig. XVII., taken in order from left to right. The same drawings, taken from right to left, may illustrate the increase of the phase of a New Moon, during the first quarter of a month. A similar explanation will also account for the appearances which the Moon presents when it is more than half full, during which time it is technically said to be gibbous.

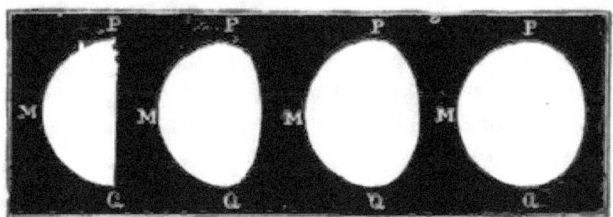

Fig. XVIII.—Phases of the Moon.

In this case the elliptic boundary is simply turned the other way, as in the drawings in Fig. XVIII., which, according as they are taken in order from left to right, or from right to left, may serve to indicate the changes of phase during the second and third quarters of the month respectively.

To those who are familiar with the theory of projections, or even with the ordinary rules of perspective, the above explanation involves no difficulty. To others, we fear, it may be somewhat troublesome to realize that a boundary, which is really circular, may at one time appear to be a straight line, and at other times an ellipse of varying degrees of ovalness.

The next point which calls for special attention is, that the Moon always turns nearly the same half of its surface towards our view; the reason being that it turns once round upon an axis which is nearly perpendicular to the plane of its orbit, while it travels once round the Earth. We are aware there are some who still deny this truth, even as there are those still to be met with whose ideas of parallax encourage them to deny the spherical shape of the Earth. With such we do not wish to argue. But for the sake of others we remark, that to prove the Moon's rotation upon its axis, it is sufficient to ask any one to walk round a circular table, while all the time he keeps *his face turned towards its centre.* If when he begins to walk round he is looking east, he will be found, when half-way round, to be looking west, and at two intermediate positions north and south respectively. When half-way round he has turned his face from east to west, or in other words, he has turned himself exactly half-way round about a vertical axis; and by the time that he has gone once round the table, he has also turned himself through one entire revolution, and has passed through the four positions, in which his face has respectively looked east, north, west, and south; or east, south, west, and north; according to the direction in which he has walked round the table. In exactly the same way the Moon must turn once round, or very nearly once round, upon its own axis, while it goes once round the Earth, in order to keep very nearly the same face turned towards the Earth's centre.

It is simply because this motion round its own axis is uniform, while there is some considerable amount of variation in its speed in its orbit round the Earth, that it does not always show us *exactly* the same face. In fact, according as the Moon travels, during any given time, rather more quickly or more slowly than would accord with its mean rate of motion, we see a little more, which we should not otherwise see, upon one side, or the other, of the disc usually visible to us. This effect is technically termed a *Libration*.

But in addition to this it is also found that the Moon's axis of rotation is not quite perpendicular to the plane in which she travels round the Earth. The result of which is, that, in different parts of her path, we see somewhat beyond each pole alternately, just as the Sun shines, or an observer upon it would see (although to a much greater extent), beyond the two poles of the Earth, alternately, in the course of a year. In this way we once more obtain a view, at different times, of somewhat more than exactly one-half of the Lunar Disc.

There is also a *third* kind of Libration, which deserves mention, called the Diurnal Libration. As any place of observation is carried round by the rotation of the Earth upon its axis, it is thereby changed in position to a considerable extent in the course of twelve hours; in fact, the change for a place upon the Equator (as we have explained in discussing the determination of the Sun's distance by observations of Mars; Lect. I., page 16) is nearly equal to a distance of 8,000 miles, or to the length of the Earth's diameter. And from these two different positions there may be seen two slightly different views of the Moon, which if put together would include rather more than one-half of its surface.

A similar effect may be obtained by two observers situated at a considerable distance apart upon the Earth. They will see, at the same instant, the one a little beyond one part of the edge of the disc which would bound the view of an observer at the Earth's centre, and the other a little beyond another part of it. By the combined effect of all these causes, opportunities occur, which enable us in a long succession of observations, made upon suitable occasions, to see nearly three-fifths of the whole surface of the Moon. It must, however, be allowed that a good deal of the outlying portions thus seen is so near to the edge of the disc, that it is much foreshortened, and very unfavourably situated for observation.

As a summary of these remarks, we may therefore state that it is not strictly true to say that the Moon always turns the same face to the Earth. But it is true that it revolves once upon its axis (with almost absolute uniformity), during each one of its revolutions round the Earth; and that, if its orbit round the Earth were exactly perpendicular to that axis, and its motion in it exactly uniform, an observer at the centre of the Earth would always see the very same half of its surface. If we are only speaking in general terms, we may be permitted to use the phrase, that it always turns the same face to the Earth. In a more accurate statement we must allow for the libratory effects produced by the two causes we have named, and for the small additional effect produced by the movement of an observer in space, owing to the Earth's daily rotation, or to his own journeying to different parts of its surface. It is by allowing for these effects that we find that an additional $\frac{1}{7}$th of the Moon's surface may be seen, although in general very obliquely; which $\frac{1}{7}$th added to $\frac{1}{2}$ makes up altogether nearly $\frac{3}{5}$ths of the whole.

LECTURE IV.

THE MOON (*continued*).

> "Meanwhile the Moon,
> Full orb'd, and breaking through the scatter'd clouds,
> Shows her broad visage in the crimson east,
> Turn'd to the Sun directs her spotted disk,
> Where mountains rise, umbrageous dales descend
> And caverns deep, as oblique tubes descry;
> A smaller earth, gives all his blaze again,
> Void of its flame, and sheds a softer day."
>
> THOMSON.

To many lovers of astronomy there is a charm in the study of the physical condition of the Moon's surface, far surpassing any pleasure they may feel in the complicated theory of its movements. Even to the naked eye there is evidence of the existence of various configurations of form and feature, abundantly sufficient to excite a wish for a closer view; while in a good opera-glass many irregularities may be noticed in the inner edge of the lunar crescent,—mountain peaks which tower up to catch the Sun's beams, while the intervening valleys lie invisible in the shade. But a small telescope discloses a scene of marvellous beauty; and one of greater power reveals innumerable details full of the deepest interest.

It is from three to eight days before, or after, it is new, that the most striking views of the Moon are obtained; because the light of the Sun, at such times, shines so obliquely upon the formations near to the edge of its bright crescent, that all inequalities of the surface are thrown into prominent relief by the intensely black and sharply-defined shadows which fall behind them. On the contrary, when the Moon is nearly full, although certain delicate markings are then best seen, irregularities of level are almost obliterated, since the sunlight no longer meets them obliquely, but shines so nearly vertically upon them that their shadows almost disappear. The marvellous relief of mountain and valley is gone, and the scene is comparatively tame and uninteresting, except to an advanced and experienced student.

It may be well if we attempt to put before our readers some faint idea of what may be seen in a powerful telescope when one of the larger lunar formations is in a favourable position for observation. In very many instances a mighty mountain range rivets our gaze, of vast dimensions, and of an approximately circular contour. It rises to a height, in places, of 10,000 to 12,000 feet, or more, above the general level around it. We very likely notice that in many parts this circular chain of mountains is formed of successive ridges or steps, which tell us that at intervals in its past history huge landslips have rushed in headlong rout and ruin down its sides. We may also see the shadows of many sharp and jagged peaks into which the uppermost edge is riven.

Within the mountain circle there lies in general a comparatively level space, usually depressed far below the regions outside it; so that, while the ring of mountains may rise some 10,000 or 12,000 feet above the exterior level, this interior plain may be 20,000 to 25,000 feet below the summits that tower above it. Hundreds of such approximately circular formations may be detected, ranging in size from a very few miles to 120, or even 150, miles in diameter. In many cases we also see that there are one or more hills, or elevations, of an apparently conical form, in the central parts of the interior plain.

In order to become fully acquainted with any individual ring of mountains, repeated opportunities must be utilized. The labour of the necessary observations is, however, compensated by the unceasing variety of the beauty and general aspect of the scene. We watch the sunlight as it gradually rises upon the ring, and throws the ever-changing shadows of its eastern peaks across it. We follow it as it steals down the interior of the opposite or western side, until presently, as it shines in its meridian splendour vertically down upon the interior plain, almost every shadow disappears. Then the lengthening shades of the jagged western summits begin to creep up the eastern slopes. Soon the last lingering rays of light vanish from the scene, and we wait for another lunation in which to watch afresh.

The most important points which may thus be observed in these formations, are their vast diameters, the great height and terraced character of the ring by which they are bounded, the frequent occurrence of a deep depression of the interior plain below the surrounding level, and of conical elevations in its midst. Most of these characteristics, as well as many other interesting features and formations upon a smaller scale, may be seen in the view of the lunar mountain Copernicus and its surroundings (see Plate VI.), which by Mr. Nasmyth's kind permission we present to our readers. It is reduced from that published in his and Mr. Carpenter's well-known and costly treatise upon the Moon, the original photograph having been taken from a plaster model most carefully formed, the appearance and the shadows cast by the different parts of which, when it was placed in a strong light, exactly agreed with the view of the corresponding portion of the Moon seen with a powerful telescope.

In addition to the large ring-mountains, to which our description has, so far, chiefly referred, there are some which extend in a longitudinal direction, and consequently more nearly resemble the mountain chains with which we are familiar upon the Earth. Two of the largest of these are respectively named the Lunar Alps and Apennines. If the Moon, when about half-full, be looked at through a telescope, which, by the reversing action of its lenses, will in our latitude make its northern parts appear to be the lowest, the Alps will be seen about $\frac{1}{5}$th, and the Apennines about $\frac{1}{3}$rd, of the way up its inner edge. The Apennines are the more extensive and remarkable of the two ranges, their length being about 430 miles, and their loftiest peaks some 18,000 feet in height; while the highest of the Alps (called, after its sister peak upon the Earth, Mont Blanc) scarcely exceeds 12,000 feet. Many other ranges of a similar character, but of less prominence and interest, may also be easily found. In the beautiful photograph by Mr. Rutherfurd, of New York, inserted at the beginning of Lecture III., and placed so as to correspond with the reversed image shown by a telescope, the Alps may be noticed a little above, and to the left of, the large ring-mountain (named Plato) which is seen close to the edge of the Moon's terminator (*i.e.*, the boundary of its enlightened portion), about $\frac{1}{8}$th of the way up it. The long range of the Apennines, sloping downwards to the left, is seen rather below the middle point of the same illuminated edge. Beneath them is a large circular mountain, Archimedes, to the left of which there are two smaller but similar formations, in the lower one of which (Aristillus) a central cone is very evident. Tycho, 54 miles in diameter, and nearly 3 miles deep (a mountain to which we shall presently make special reference), is not quite $\frac{1}{8}$th of the whole way down the same edge, its ridges on the left-hand casting a deep shadow, which reaches nearly to the base of its central cone. The shadow of the cone may also be distinctly seen in the photograph. We may be allowed to state that we have the greatest possible pleasure in exhibiting this enlarged copy of Mr. Rutherfurd's negative.* It will bear the closest examination, and may well be inspected by means of a magnifying glass. It is purely a photograph, an image untouched by human hand, or by the engraver's art. Every feature is depicted by the Moon itself, and we are confident in our opinion that no more perfect result has as yet been obtained in lunar photography.

In some parts of the Moon large and comparatively level regions are met with, of a darkish hue, and in many cases somewhat depressed below the general level. Several of these (*e.g.*, the Sea of Serenity, the Sea of Vapours, the Sea of Storms) are still called by names which were long ago given to them under the supposition that they were portions of water, or that they bore a certain resemblance to the seas of the Earth. This idea is now discountenanced, although it

---

* We owe the privilege of so doing to Messrs. Kegan Paul, Trench, & Co., who have most kindly provided the copies.

PLATE VI.

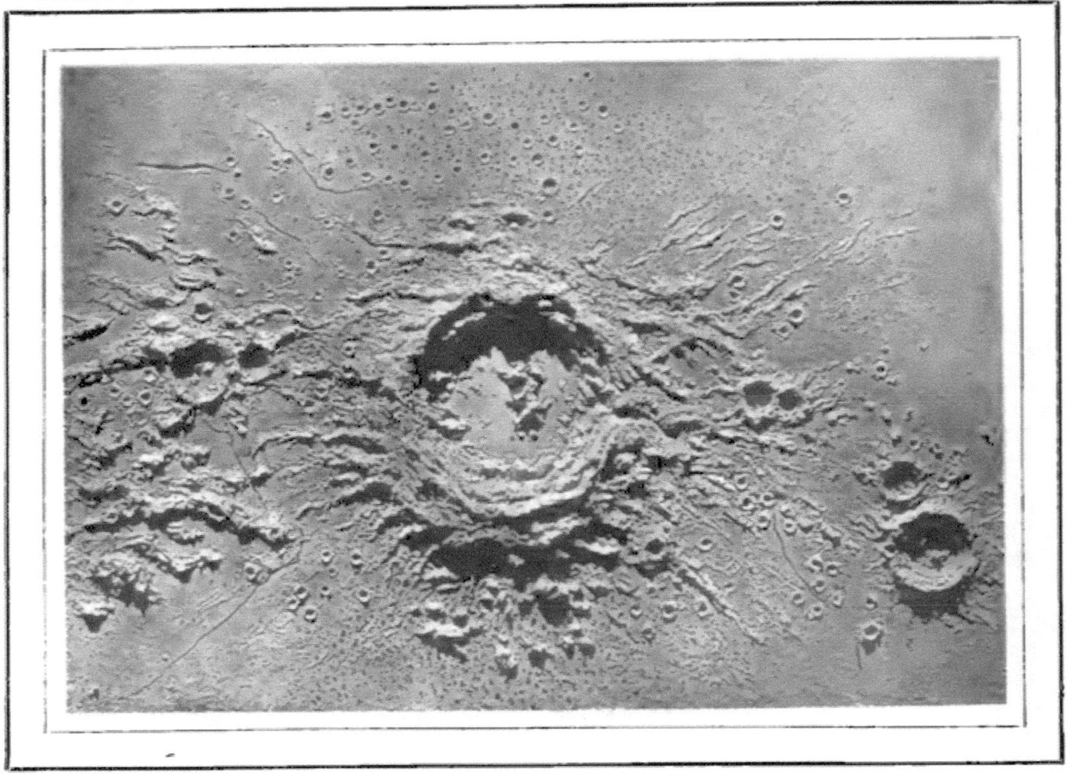

THE LUNAR MOUNTAIN "COPERNICUS."
From a Model by Mr. NASMYTH.

Woodbury process.

cannot be denied that some of them may very possibly be the beds of defunct oceans, by which, in ages long past, they may have been covered.

In addition to the large ring-mountains of which we were recently speaking, there are many circular formations, of which some beautiful examples are to be seen in the plate which contains the view of Copernicus, so small as to be more justly called craters; while to others, smaller still, the name of craterlets, or crater-pits, has been given. In certain regions these minute craterlets, varying from less than a mile to five or six miles in diameter, are exceedingly numerous.

It is very interesting, even with a small telescope, to compare the Moon with a good outline lunar map;* to identify the various mountains by name; to note those which are distinguished by central cones, or by other peculiarities; to observe the long pointed shadows of the summits of the Apennines; or to search for occasional instances of isolated peaks of a more or less conical shape. One of the most remarkable of these is named Pico. It is about 7,000, or, perhaps, 8,000 feet high, and may be found near to the lunar Alps.

There are other peculiarities of the lunar surface, more difficult to detect than those which we have hitherto mentioned; the most important being termed clefts, or rills.† Some of these present an appearance very similar to that which would be afforded by the bed of an extinct torrent or river, and may possibly resemble the cañons of North-West America. Others are so straight and regular that they might almost make us imagine that they must have been artificially constructed, were it not that many reasons discountenance such a supposition. They are traceable for great distances, and

---

* Such maps may be found in Webb's "Celestial Objects," and, upon a larger scale, in "The Moon," by Mr. Neison; the most elaborate of all being that of Dr. Julius Schmidt, 6 Paris feet in diameter, published in 1878, at the expense of the Prussian Government.

† This word, adopted from the German "*rillen*," is not intended in any wise to imply the existence of water, as would be the case with the more usual meaning of the English word.

frequently continue their course with an apparent disregard of all changes of level. Their origin may perhaps be more plausibly attributed to the cracking of a shrinking surface, than to any other known cause. About 1,000 of them have been observed, some being as much as 200 or 300 miles in length.

When the Moon is full, or nearly so, and its aspect is in many other respects comparatively uninteresting, several systems of long rays, or light streaks, are seen to originate in certain districts, or mountain formations. Some of them are 600 or 700 miles long, and one appears to extend to a distance of at least 2,000 miles, which is considerably more than ¼th of the Moon's equatoreal circumference. The most noticeable are those which radiate from the mountain Tycho; the position of which we have pointed out in our description of Mr. Rutherfurd's photograph. There are, however, others which are very remarkable. They seem to be of such a nature that, when the sunlight is shining fully upon them, they reflect more of it than the surrounding soil or rock, and consequently appear exceptionally bright. They pass onwards, like many of the clefts, or rills, up hill and down valley, over crater, ridge, and mountain; evidently showing that their origin has, in some cases, been subsequent to that of the other formations which they thus traverse.

The hypothesis has been suggested that they arise from the filling up, by the exudation of volcanic matter from depths below, of cracks radiating from the site of some vast explosion. When, however, we remember the very great distances to which they extend, it is hard to imagine how any catastrophe sufficiently terrible to have produced them can have occurred. Nor do we see other effects remaining which such a supposition would lead us to expect. No really satisfactory theory of their origin has yet been put forward. We can only say that the way in which they radiate from various centres is strongly suggestive of an explosive cause.

One of the most puzzling problems of the lunar surface is to explain the origin of such vast *circular*, or approximately circular, formations as we have previously described. Nothing at all comparable to them is found upon the Earth, although there is a considerable resemblance between certain regions (such as the neighbourhood of Vesuvius), in which we know that volcanic agency has been at work, and some of the smaller groups of craters upon the Moon. This has well been shown in a diagram by Messrs. Nasmyth and Carpenter, in which the district surrounding Vesuvius is contrasted with a similar portion of the Moon. We append from Admiral Smyth's *Speculum Hartwellianum* (now somewhat rare) a copy of a drawing of

Fig. XIX.—The great crater of Teneriffe.

the great crater of Teneriffe,—"eight miles in diameter, with its parasitic cones and peak, the latter 12,198 feet high,"— which is undoubtedly a volcanic formation. It may certainly bear interesting comparison with many of the lunar mountains, or craters.

We have upon the whole strong reasons for believing that the lunar formations are in like manner most probably of volcanic origin. Nevertheless, so great is the diameter of many of the larger circles of mountains, that it is hard to conceive that they can have been produced by central eruptions. And yet the existence of *central cones* in many of them, and even in some of the largest, favours this hypothesis. It is well, therefore, to remember, that the attraction of the force of gravity upon the surface of the Moon is not quite $\frac{1}{6}$th of its value upon the Earth, and that an eruption of matter from a central cone, with such a velocity and at such an angle that it would descend upon a level plain on the Earth at a distance of five miles, would, if similarly projected on the Moon, be carried rather more than 30 miles before it would fall. It is therefore a not altogether unreasonable supposition that such a centre of eruption may have deposited matter in a ring of 60 to 70 miles, or even more, in diameter. But it may nevertheless be urged in reply to this argument that it is by no means likely that such violent eruptive forces would be generated in the comparatively small mass of the Moon as in the much larger mass of the Earth, and that the diameter of the largest crater upon the Earth, which is, we believe, in the Sandwich Islands, does not exceed 10 miles.*

Other theories have consequently been elaborated as to the origin of these huge ring-shaped lunar formations. It has, for instance, been imagined that they may have been produced by the exudation of molten matter at different times during the cooling and solidifying of the Moon's crust, a process which would involve frequent changes of internal level and pressure, and vast subsidences and derangements of the surface.

On the other hand, if we maintain the previously mentioned theory, viz., that the circular ridge has originated in the discharge of matter from a central cone, we need not be surprised if such cones nevertheless seem in many cases to be wanting; for a subsequent eruption of less energy might discharge matter sufficient to fill up the surrounding circle to the level of the cone, or nearly so, which would cause it to be no longer visible.

We cannot, however, afford further space to the discussion of these theories. We will sum up our remarks by saying that there is every reason to believe, from the present appearance of the Moon, that in ages long past it has been the scene of

* See Miss Bird's account of her recent tour in these islands.

tremendous volcanic action. Such action appears to have been in itself much more violent than any which the Earth has experienced, while the effects produced upon the Moon by any given explosion would be much greater than upon the Earth, owing to the smaller attraction of the force of gravity by which they would be restrained. It is certainly the case that such disturbances seem to have been extraordinarily frequent upon the lunar surface. There is also much evidence of a *succession* of eruptions, for in many places craters, or ring-mountains, are seen overlapping and interfering with one another in a way which proves that the origin of some must have been anterior to that of those by which they have been disturbed.

If this be so, the question naturally suggests itself:—Is there any evidence of the existence at the present time of volcanic energy still in action upon the Moon? Our answer is:—There is either none, or, if there be any, it is very slight. We have no incontestable observation of any such occurrence. Certain spots upon the Moon, which appeared unusually brilliant to Schröter, Mädler, and others, may have lost some of their brilliancy. But it is most probable that they are spots which have a peculiar power of reflecting the Sun's light, or (when they have been noticed in the dark part of the Moon) the faint light which the Earth reflects to it, and which, near to the time of New Moon, causes the appearance which is popularly known as the New Moon in the Old Moon's arms. This is much more likely than that they are volcanoes in active eruption. The absence, or almost entire absence, of any appreciable amount of atmosphere or water upon the Moon (to the proofs of which we shall presently allude) also makes it very difficult to conceive that volcanic action can continue.

In connection with this question, we ought, however, to mention that some fifteen years since a considerable amount of discussion took place as to a change which was supposed to have occurred in a crater named Linné.* There seemed to

* This crater is situated in the locality of a bright spot in Mr. Rutherfurd's photograph, which may be seen in the midst of a darker region a little way to the left of the lower part of the Apennines, and slightly

be some grounds for believing that it had become much smaller and less easy to distinguish than it was in earlier observations. But it offers such different appearances under varying circumstances of illumination, that the evidence is not conclusive. The same statement might perhaps be made with regard to a more recently announced discovery, viz., that a new small crater has appeared near to a well-known one named Hyginus.* Some valuable observations and drawings of it, recently made by Mr. Neison and Mr. Green, have however given additional probability to the supposition that it is really a new formation.

In any case we must allow, that landslips and a gradual deterioration and crumbling of the lunar rocks must almost undoubtedly still continue as a result of the intense alternation of heat and cold which the Moon undergoes in the successive halves of each month; and it is possible that processes of this description, and of unusual magnitude, may have very much altered the appearance of Linné. We have good reason to hope that the careful attention now given to the Moon by many painstaking observers, combined with the great telescopic power at their disposal, and the accuracy of the lunar maps recently published, may before long enable us to speak much more positively as to the reality of such changes. But at present we can only say that the balance of evidence is decidedly against, rather than in favour of, the existence of active volcanic action upon the Moon, and affords little proof of the progress even of such changes upon its surface as are otherwise antecedently probable.

The Moon, in fact, gives to us the impression that it is a body which has long since passed through a condition similar, or somewhat similar, to that in which the Earth is at present. Since that time it has, we believe, grown colder

above a horizontal line passing by the upper edge of the circle of Archimedes (the position of which we have pointed out in our previous description of the photograph), and at a distance from its centre equal to rather less than five times its diameter.

* This may also be detected in the photograph by those who know where to look for it.

and colder. The water which in ages past doubtless existed upon it (or at any rate nearly all of it) has long since vanished, having perhaps been absorbed into the rocks which form the lunar crust. In like manner the atmosphere, formerly belonging to it, which, owing to the Moon's comparatively small mass, was probably of moderate amount, has disappeared, or so nearly disappeared as hardly to leave any certain traces of its existence.

Among the proofs which may be given for this last statement, the following may be mentioned. We every now and then see the Moon pass between us and some of the myriad stars that stud the firmament. When this occurs, the disappearance of the star behind the lunar disc and its re-appearance are sharp and sudden, indicating that there is no appreciable effect produced by any lunar atmosphere. Again, we meet with no definite evidence of its existence when we watch the progress of a solar eclipse.* If, as some think, we are occasionally able to detect signs of twilight upon the Moon, they are so exceedingly slight as to indicate that any atmosphere upon it cannot possibly possess a density greater than the 200th part, or it might perhaps be said the 500th part, of the density of that upon the Earth's surface.

The contrast between sunlight and shade upon our satellite must consequently be intense. The blackness and sharpness of the mountain shadows may well form one of the most remarkable features in a telescopic view of the Moon. If, however, we could transport ourselves thither, we may be sure that the reality would surpass the utmost bounds of our imagination; one step would take us from the deepest blackness of night into the fullest blaze of the unclouded Sun. There would be no diffused atmospheric radiance, no softening-down and toning-off of shade. In some positions an opposite mountain might reflect a little light to an observer standing in the midst of a dark shadow, but such cases would be exceptional.

The sky would have none of that beautiful blueness which

---

* Since the above was in print, news has however reached us, that some traces of its presence have been detected with the spectroscope by the French observers of the total solar eclipse of May 17th, 1882, in Egypt.

the watery vapour or other constituents of our atmosphere produce. It would be of the darkest black even in full daylight. The glare of the Sun would be bounded by the exact limits of its disc, so that its coronal and other appendages would be constantly visible. The stars, like minute but brilliant diamonds, in number far greater than we behold, would, by day and by night alike, shine without twinkling in the black vault of heaven. Interesting as all this might be, it would not upon the whole be very pleasant; and other consequences of the absence of a lunar atmosphere would be still worse. The heat of the Sun would be unbearable during the long lunar day of about 354 hours' duration, while the unmitigated cold of the succeeding night would be perhaps even more terrible.

In fact, Lord Rosse has estimated that the change in the temperature of the Moon's surface during each month exceeds 500° Fahrenheit; and that under the vertical rays of the Sun it is probably considerably higher than that of boiling water. Those who have ascended to great altitudes, and have felt the intensity of the Sun's heat in a rarefied atmosphere, and then have been so unfortunate as to spend the night unsheltered on some rocky elevation in the midst of Alpine glaciers, may have possibly gained thereby some very faint idea of the alternation of heat and cold upon the Moon.

The Earth, as seen from the Moon, would of course, month by month, go through phases complementary to those which the Moon exhibits to us. Full Earth would take place when we see New Moon, and *vice versâ*; and the sight would indeed be glorious. Moreover, as the Moon always turns very nearly the same face to the Earth, the latter would seem to a lunar observer to be nearly fixed in the heavens; while the outlines of continents and oceans would give a pleasing variety to its disc, which would appear nearly $3\tfrac{2}{3}$ times as broad, and more than 13 times as large in area, as that which the Moon shows to us. We must not, however, pursue such hypothetical observations any further. It will be more profitable to turn to a few remaining points of interest connected with those which can be made from our own standpoint on the Earth, after which we shall discuss the Moon's influence and usefulness in connection with matters terrestrial.

One important fact, which deserves special attention, is the great success which has been achieved in recent years in lunar photography. De la Rue, Rutherfurd, Brothers, Draper, Ellery, and others, have obtained most perfect photographs of the Moon in all its various phases; none however, in our opinion, surpassing such an one as that by Rutherfurd, which is reproduced in Plate V. Some, taken in different stages of the Moon's libration, and therefore affording pictures such as would be seen from different points of view, give an interesting appearance of solidity, and raise all the different formations in picturesque relief when viewed in a stereoscope.* It may, however, be mentioned, that, when so directed, they also produce an exaggerated elongation in the direction of that axis of the Moon which is pointed towards the observer.

It is to be hoped that lunar photography may be of considerable use, in enabling us, as time goes on, to test the occurrence of any changes of surface, which, as we have previously suggested, may arise from large slips or falls of land. Numerous photographs have also of late been most carefully measured by Professor Pritchard of Oxford, and by Dr. Hartwig of Strasburg, in order to determine the existence, or amount, of a slight physical libration or motion of the Moon, about an axis of its figure, which theory tells us must exist.

The Moon, owing to the large amount of light received from it, and its nearness to the Earth, is, with the exception of the Sun, the easiest of all the heavenly bodies to photograph, and in some respects even more easy than the Sun, in consequence of the overpowering amount of light which the latter affords. At the same time great care, a very well regulated instrument, and very refined processes are necessary to ensure success. Those, therefore, whose success has been so great as to be almost perfect deserve all possible praise.

* Transparencies from the very beautiful negatives of Mr. De la Rue thus arranged may, we believe, be obtained from Messrs. Beck and Co., Opticians, Cornhill, who have more than once very kindly lent some of them for the illustration of these lectures at Gresham College.

We must not, however, forget the many important respects in which the Moon is an object of interest to those who are not only unable to photograph her image, but who have not even a telescope to aid their view. Such observers may carefully watch the very considerable changes in the apparent size of her disc, by which, as explained in our previous lecture, it is increased from time to time by as much as one-third part of its least value;—times of which Othello says :—

> "It is the very error of the Moon;
> She comes more nearer earth than she was wont."

They may also notice that at such times the tides are correspondingly intensified, and that, if spring tides also occur when the Moon's distance is unusually small, and certain conditions of wind and weather act conjointly, the inundations of the Thames and of other tidal rivers may prove to be very serious.

It is also interesting for the unscientific observer to note how conveniently the Moon's path in the heavens is arranged, so that we enjoy a far greater amount of moonlight in winter than in summer. We do not mean, that, the nights being longer in winter, we notice the Moon's light more, in the absence of that of the Sun; but that the Moon is in mid-winter above the horizon for a longer time on those nights during each month when its phase affords a greater amount of light, and for the longest time of all when it is full, or nearly so; while in midsummer it is a shorter time above the horizon when full, or nearly so, than in any other part of the lunar month.

In fact, the increase in the hours of moonlight thus obtained in *winter* bears a close relation to that of the hours of sunlight in *summer*. This arises from the fact that the Moon's path, as seen amongst the stars, differs, month by month, very little from the great circle of the heavens called the ecliptic, which is the Sun's apparent annual path. And the Moon, when full, being very nearly in a straight line with the Sun and the Earth, but upon the opposite side of the latter, it follows that a full Moon will *at midnight* be always seen nearly in the same position in the sky as that in which the Sun was seen *at noon six months previously*. The path of the Full Moon across the sky during the *night* will, in fact, be very nearly that of the Sun during the *day*, six months before or after. And therefore, as the inclination of the Earth's axis to the plane of the ecliptic causes the Sun's path (in such a latitude as our own) to be much longer, and its noontide elevation much greater in *summer* than in *winter*, so will the path of the Moon when it is in its full phase be in like manner longer and its elevation greater in *winter* than in *summer*. The duration of daylight in the latitude of London varies from about $7\frac{3}{4}$ hours upon the shortest day, to about $16\frac{1}{2}$ upon the longest. That of the light of the full Moon, owing to the deviation of the Moon's path from the ecliptic, may, we believe, amount in midwinter to about 18 hours and in midsummer be barely 6 hours. It is certainly a very interesting fact that we thus receive most moonlight at the very season of the year when the long nights make it most useful.

An eclipse of the Moon is another important phenomenon which is very easy of observation, and one which in some respects may be very well seen without any, or with very little, telescopic aid. At such times the Moon first enters the penumbra, or outer portion of the Earth's shadow, from which the Sun's light is only partially excluded. Little notice, however, is taken of this; but, after a while, the dark umbra, into which (except by the refraction of the Sun's rays by the Earth's atmosphere) no solar light can penetrate, gradually encroaches upon the disc.

At the distance of the Moon's orbit, the diameter of the umbra is about three times that of the Moon, whose speed of motion round the Earth is (as we have mentioned in Lecture III.) such as to take it over a space equal to its own diameter in about one hour. It will therefore, if it pass centrally through the umbra, first of all occupy about one hour in becoming wholly immersed in the shadow. It will then remain for nearly two hours totally eclipsed before its opposite edge begins to emerge. Towards the end of the fourth hour it will again be entirely clear.

As an eclipse progresses, it is generally noticed that the edge of the shadow is hazy and more or less tinged with red.

And when the Moon is totally immersed a considerable amount of such reddish light may often be still detected, so that its whole disc may remain faintly visible. On other occasions, but more rarely, the totality is nearly absolute, and the Moon is for the time almost invisible. This variable result depends upon the amount and nature of the clouds and vapours overlying the bounding circumference of that hemisphere of the Earth upon which the Sun is shining, which produce important effects upon the amount and the tint of the solar rays which are refracted into its shadow.

Lunar eclipses are, however, sometimes only partial; at other times, although total, of comparatively brief duration; more frequently still, the Moon is full without suffering any eclipse. In fact, as we shall presently show, it never traverses the umbra of the Earth's shadow more than three times in a year, while the occasions are comparatively rare upon which its passage is so nearly central that the duration of an eclipse approaches its maximum possible value.*

It is sometimes said that the Moon's place, when full, is opposite to that of the Sun as seen from the Earth. But this cannot in general be strictly true, inasmuch as it would involve the occurrence of an eclipse at the time of every Full Moon, and would require the plane of the Moon's orbit to be exactly coincident with the plane of the ecliptic, in which the Earth moves round the Sun. These planes are actually inclined to one another at an angle of about 5 degrees. The Moon, when full, will therefore only be exactly in the prolongation of the straight line joining the centres of the Earth and of the Sun, if it happens at the same moment to be also passing through one of the two points in which the plane of its orbit intersects that of the ecliptic. Otherwise it will be at a perpendicular distance above or below the straight line in question, which will vary between certain small limits, defined by the above-mentioned inclination.

* The Moon may at times pass through the penumbra of the Earth's shadow, when it escapes the umbra. Such penumbral eclipses are occasionally interesting, although their occurrence is not noticed in the Almanacs.

The two points of which we have been speaking, in which its monthly orbit intersects the plane of the ecliptic, lie upon a straight line very approximately passing through the centre of the Earth, which is technically termed the line of Nodes.

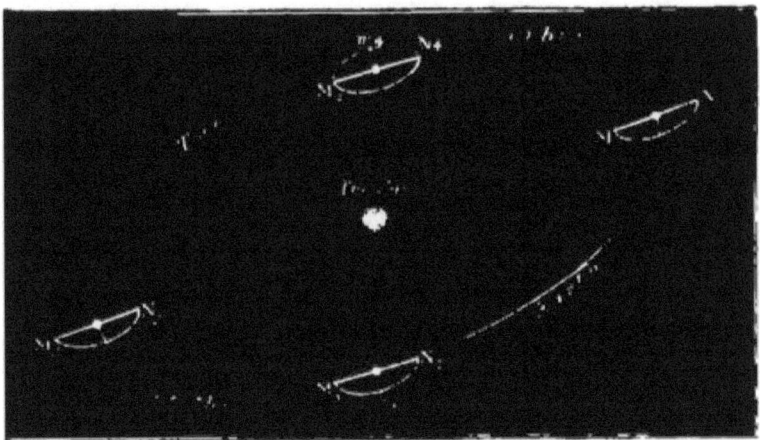

Fig. XX.—Showing that the direction of the line of Nodes of the Moon's orbit passes through the Sun at two opposite periods of the year.

The most important question involved in determining the frequency of lunar eclipses consequently is:—How often will this latter line also pass through the centre of the Sun? It can only do so at intervals of about six months, as may be

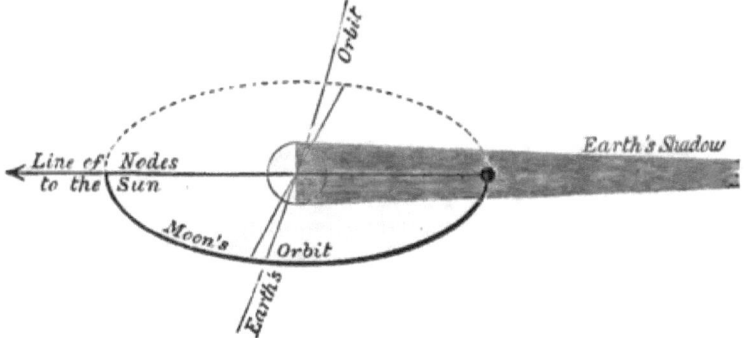

Fig. XXI.—Showing how the Moon may suffer a Total Eclipse of the longest possible duration.

seen in Fig. XX., in which the Earth, as it travels round the Sun, is supposed to carry the lunar orbit with it, the line of Nodes remaining parallel to itself, and only passing through the Sun in the two opposite positions M_1N_1, and M_2N_2. In the diagram the portion of the Moon's orbit which is dotted is supposed to be slightly below the plane of the Earth's path (or the ecliptic), and the other portion slightly above it.

If, then, upon one of the days upon which such a passage occurs, the Moon is full, as in the positions N_1 and M_2 in the above figure, a total eclipse of the longest possible duration will be seen, the Moon passing through the very centre of the Earth's shadow, as in Fig. XXI.

It is also found that if the Moon on such a day is within about eleven days before or after being Full, an eclipse of some kind, either partial or total, but of shorter duration, will take place

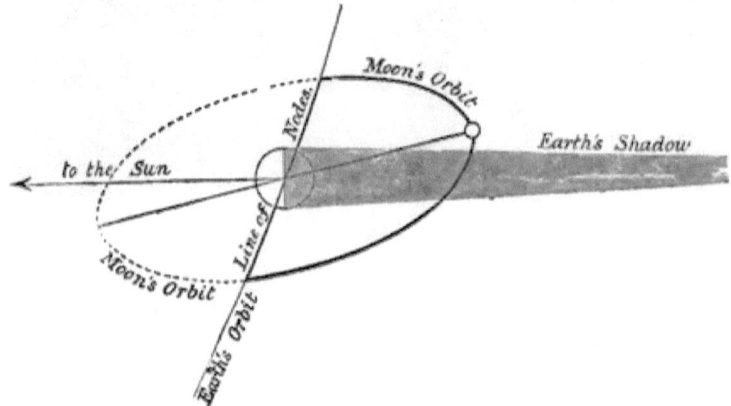

Fig. XXII.—Showing how the Moon, when full, may escape being eclipsed.

at the next following, or next preceding, Full Moon, when the Moon will still pass within so small a distance, above or below the straight line joining the centres of the Earth and of the Sun, that it will in some degree traverse the umbra of the Earth's shadow. At all other times the Full Moon will, in consequence of the tilt, or inclination, of its orbit to the plane of the ecliptic, in which the Earth circles round the Sun, pass altogether above or below the Earth's umbra, and escape eclipse, and it will pass clear of the shadow by the greatest possible distance when it occupies such positions as m_3 and m_4 in Fig. XX., which correspond to places of the Earth just half-way between those which coincide with the occurrence of the longest possible eclipses. We have endeavoured to illustrate such an escape from eclipse in Fig. XXII.

According to the above statement, it therefore appears that there may be one lunar eclipse, or there may be none near to the time of each passage of the line of Nodes through the Sun, according to the number of days by which such an occurrence is separated from the date of a Full Moon; and that if a lunar eclipse occur it will be within about eleven days of the Nodal passage.

The conclusion would consequently seem to follow that there may be two lunar eclipses, or none, in any given year. But owing to certain causes of a somewhat complicated nature connected with the lunar orbit, which involve a slow turning-round of the line of Nodes, it is found that the above-named interval of six months is reduced to 173 days. From this it results that there may be as many as three lunar eclipses in the course of a year, one at the very beginning of the year, one in the middle of it, and one just before its close.

Although we have done our best, by means of some diagrams different from those generally drawn, to render this explanation as lucid as possible, it is difficult so to do, owing to the perspective necessarily involved. A suitable model (such as may be seen at GRESHAM COLLEGE) may, however, be easily constructed, and will be found much more efficient and of great assistance both to teachers and students.

We must only refer very briefly to the Moon's connection with solar eclipses, especially as the beautiful phenomena then witnessed have been already described in Lecture II. (see Plate IV.), in connection with the Sun's coronal and other surroundings. The most important difference between a solar and a lunar eclipse is that the latter is visible from a whole hemisphere of the Earth at once, *i.e.*, from every part of the Earth upon which the Moon would otherwise be shining; while a solar eclipse is only seen from a small portion of the Earth at any given time. The shadow of the Moon, by which a solar eclipse is caused, is, in fact, never much more than long enough to reach the Earth. Sometimes, when the Moon is at its farthest from the Earth, and the Sun at its nearest, it even falls short of doing so. In such cases the shadow comes to a point before it meets the Earth, and the Moon appears to be rather smaller than the Sun. A ring of light therefore remains visible, even to observers who are exactly in the direction of the straight line joining the centres of the Earth, the Moon, and the Sun; and the eclipse is termed annular.

At no time can the maximum breadth of the Moon's shadow, where it falls upon the Earth, exceed 173 miles, nor will it often approach very near to this value. Whatever the width of the shadow may be, it rapidly sweeps across a zone of the Earth of corresponding breadth, with a relative speed, which (when the rotation of the Earth is taken into consideration) may only amount to about 20 miles per minute. While it is passing over any place a total eclipse of the Sun, the most impressive of all celestial phenomena, continues. The above figures (by dividing 173 by 20) would allow the greatest duration to be fully $8\frac{1}{2}$ minutes, but more accurate calculations give 7 minutes 58 seconds as the extreme limit, under the most favourable conjunction of circumstances. As an instance of one of long duration we may mention the very notable eclipse of August 17th, 1868, in which the totality lasted in the Gulf of Siam for nearly 7 minutes.

Upon either side of the zone of totality a wider zone exists, in which a partial eclipse is seen, of greater or less magnitude, according to the distance of the observer from the central zone. But the phenomena thus produced are of comparatively little interest or importance.

Solar eclipses occur upon the whole with greater frequency than those of the Moon. But the width of the zone of the Earth, in which the totality of any one can be seen, being, as we have explained, very narrow, a total solar eclipse is exceedingly rare *in any given locality*. The last visible in England was in 1724, in London in 1715. With the exception of a possible totality of a very few seconds, which may occur over a line drawn from the Isle of Anglesey across Northumberland on June 29th, 1927, and which, in any case, must be so slight as hardly to deserve the name, none will, we believe, be seen in these islands until August 11th, 1999, when a short totality will occur in the south-western counties. None will probably be visible in London for more than 500 years from the present date.* It will therefore doubtless be thought well to despatch both governmental and private expeditions to observe any eclipses of considerable duration which may be visible in other reasonably accessible localities during this period, and to secure every possible observation of the corona, or other solar appendages, which the obscuration of the photosphere by the interposition of the Moon may render it possible to make during the few brief minutes of totality.

The greater frequency of the occurrence of a solar than of a lunar eclipse, visible from some part or other of the Earth, independently of the question of its visibility in any given locality, arises as follows :—It is found that in order to permit the phenomenon to take place, the Moon need not be so near to one of the nodes of its orbit in the former as in the latter case ; in which it may be remembered that we stated it to be necessary that the Moon should be full within eleven days on either side of the date of a passage of the line of Nodes through the Sun, which gave a limiting period of twenty-two days, or less than a month. But for a solar eclipse the limiting period during which the Moon may be new somewhat exceeds thirty days. Consequently the Moon may be sufficiently near to one of the positions in question on the occurrence of *two successive* New Moons to cause a solar eclipse to be seen upon some part of the Earth ; that is to say, it will then pass at a sufficiently small distance above or below the line joining the centres of the Earth and of the Sun, to allow the extremity of its shadow to reach the Earth.

And as the limit of each period in which it may do this exceeds a month, so that it may, as we have just said, embrace two successive New Moons, it *must* include *one* New Moon. There *must* consequently be *one* solar eclipse, and there *may be* two, at each such epoch. But, as we have previously explained, the interval between the above periods is only 173 days, or about nine days less than six months. The sum of two such intervals therefore falls short of a year by about nineteen days. If then a New Moon occurs within three or four days of the beginning of a year, and causes a solar eclipse to take place, the

* See " Eclipses, Past and Future," by Rev. S. J. Johnson.

thirteenth New Moon of that year, just before its close, may also occur sufficiently near to a passage of the line of Nodes through the Sun to produce an additional eclipse for the year in question. It results that in any given year there *may* be five, and there *must* be two, New Moons, which will produce total solar eclipses. If there be five, they may be the first and second, the seventh and eighth, and the thirteenth; or the first, the sixth and seventh, and the twelfth and thirteenth.

One curious fact, little noticed in its connection with eclipses in ordinary text-books, is, we think, deserving of special atten-

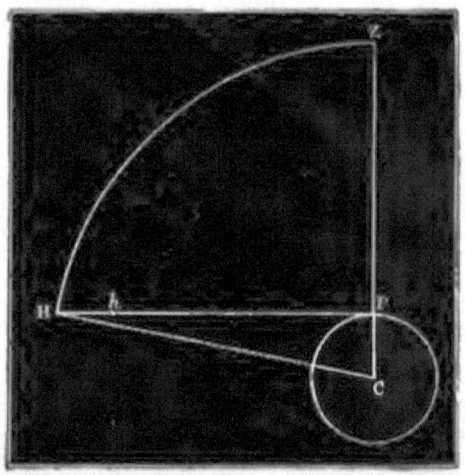

Fig. XXIII.—The Moon is nearer when seen in the zenith than when seen in the horizon.

tion. It is that the Moon really appears larger when overhead, or near to the zenith, than when seen in the horizon. On its rising there is no doubt that the first impression afforded is just the reverse. But this is only an optical delusion, possibly caused by the mind to some extent fancying its distance to be more comparable with that of terrestrial objects which are also seen near to the horizon. But by telescopic measurement, or even by looking through a tube devoid of glasses, any one may easily arrive at the truth. And a very elementary diagram may also suffice to show, that, if the Moon is at any time seen in the zenith of an observer, it is nearer to him than when seen in the horizon by a distance not far short of the length of the Earth's radius; such a passage through the zenith of a place of observation being possible (owing to the inclination of the Moon's orbit to the ecliptic) in localities situated within about $28\frac{1}{2}$ degrees of latitude on either side of the Earth's equator.

Let c in the above figure—in which, for the sake of simplicity, we may suppose the place of observation to be upon the equator of the Earth, so that the diagram may be all in one plane—be the centre of the Earth. Let H and z be the positions of the Moon on rising and when seen in the zenith; P the observer's position. Then cz equals cH, and PH does not fall much short of the same value. Consequently Pz is less than PH by a distance which is nearly equal to cP, the radius of the Earth. If the Moon were as near to P when in the horizon as when in the zenith, it would be at h instead of at H, Ph being taken equal to Pz. The diagram is drawn upon an exaggerated scale, the real value of cP (and approximately of Hh) being about 4,000 miles, which is about $\frac{1}{60}$th part of cz (or PH), the mean length of which is nearly 239,000 miles.

Under such circumstances the Moon will have an apparent diameter, when in the zenith, which will be about $\frac{1}{60}$th part wider than when it is in the horizon. And if at such a time a total solar eclipse occur, the totality will be prolonged, since the great distance of the Sun precludes any corresponding effect upon its apparent size.

The greatest excess of the Moon's apparent diameter when seen in the horizon over the Sun's is about $\frac{1}{15}$th part of either. In that case totality would last while that excess was travelling between the observer and the Sun. By suitable calculations it is found that it might continue for rather more than six minutes. But if the Moon be overhead, and its apparent diameter be increased by $\frac{1}{60}$th part, its excess over the Sun's diameter is made about $\frac{1}{4}$th greater than before ($\frac{1}{60}$th being one-fourth of $\frac{1}{15}$th), and the totality may consequently last $\frac{1}{4}$th longer than before, and approach a duration of eight minutes instead of being limited to six.

In such a phenomenon so considerable a prolongation may be of the utmost importance. And it is worthy of special notice that the unusual duration of the eclipse of August 17th, 1868, in the neighbourhood of the Gulf of Siam (in which locality, as we have previously stated, it lasted for nearly seven minutes), was to a very appreciable extent thus produced, owing to the Moon and Sun being nearly in the zenith, or, popularly speaking, overhead, at the time of its occurrence.

Our space forbids that we should refer, otherwise than very briefly, to the benefits which the Earth derives from the influence of the Moon upon its waters. For the same reason we must refrain from discussing the somewhat complicated dynamical investigations involved in a proper explanation of the origin of the Tides.* But we may very positively affirm that if the tidal disturbance of our seas and rivers were to cease; if they were to become stagnant, except so far as ocean currents and rainfall might affect them; most disastrous consequences both to health and commerce would ensue.

We can hardly over-estimate the vast benefit which the unceasing power of the tides confers, in sweeping away deleterious matter, in raising vessels into locks and harbours, in bringing merchandize up our great rivers, and in many other ways. It may also be that, some day, much of the immense energy provided by them, which is still unused, will be applied to drive machinery, or to compress air, or to generate electricity, which being economically carried to distant points of application, may be utilized to an extent at present little imagined.

Were it only for the beneficial influence of the tides, we might well be very thankful that the Earth possesses so influential a satellite. And this feeling is increased, when we bear in mind, that the knowledge of the Greenwich time at which the Moon is seen at a certain apparent distance from

* For a most lucid and interesting description of Mr. G. H. Darwin's theory of tidal evolution, affording reasons for believing that millions of years ago tides of much greater dimensions than those of the present time may have swept across the Earth; as well as for an explanation of the connection between the Moon's distance, the tides, and the gradual lengthening of the Earth's day, we refer our readers to Dr. Ball's learned and eloquent lecture recently published by Messrs. Macmillan and Co., and entitled "A Glimpse through the Corridors of Time."

some suitably selected star, recorded in tables, suffices, by comparison with the local time at which it is seen to be at the same distance from the star, to enable the mariner, who, far out of sight of land, is cleaving his way across the pathless deep, to determine his longitude with an accuracy which is of much importance, notwithstanding that modern chronometers have been so greatly improved that, independently of lunar observations, they afford a still closer approximation to its true value.

A due appreciation of the Moon should therefore in no wise be wanting in a country whose maritime and manufacturing interests are so important as those of England, or amongst the merchant princes of the city of London. It was, we think, very consistent with the distinguished position which SIR THOMAS GRESHAM held in our great metropolis, that in the foundation which he appointed by his will he directed that the Reader in Astronomy should pay special attention to the nautical branch of the science. Since his day, however, the confines of astronomical study have immensely widened, and nautical astronomy has been so much reduced to a series of routine calculations, by means of formulæ and tables, that it is hardly possible, in a lecture such as the present, duly to explain what is more fitted for a class of students in navigation. Still it must not be forgotten that if any improvement is to be effected in the accuracy of our knowledge of the Moon's position, or in the methods by which that knowledge may be practically utilized, it must be by a due encouragement of suitable observations, and of the progress of astronomical science in general.

There was formerly a small observatory erected by the Gresham Committee (having the star γ Draconis nearly in its zenith) in old GRESHAM COLLEGE, which in its day was very useful, and in which Dr. Hooke appears, without knowing the meaning of his discovery, to have effected the first detection of the aberration of light. We now rejoice in the existence, not only of the great national observatory at Greenwich, and of those connected with the universities of Oxford and Cambridge, but of several of comparatively small size, founded by private munificence, the work accomplished in which is of a high order in skill and accuracy. Nor can we help feeling some slight sentiment of pride that so much has hitherto been accomplished by Englishmen in the cause of astronomical progress. But we fear that England is now being somewhat left behind in the race. Although some splendid examples of individual enterprise have recently been seen, in the mighty Ealing telescope, or in the instruments and expeditions of the noble owner of Dun Echt observatory, in which a devotion to science shines forth which worthily rivals that of the late and of the present Lord Rosse,—we find that the Austrian, the French, the Australian, and the American Governments can afford to purchase instruments far surpassing in size those in any English national institution; while the recent munificent foundation of the Lick Observatory in California, under the will of a late wealthy citizen of that name in San Francisco, is a precedent so worthy of imitation, that we would fain hope it may ere long be followed by some in our own country who have money to spare and the desire to spend it well.

As to the connection of the Moon with the Calendar, it is worthy of mention that even if the date of Easter depended (which it does not) upon the true date of the occurrence of a Full Moon upon, or after, the 21st of March, the festival might fall five weeks earlier, or later, in different parts of the United Kingdom, if local time were taken into account. And in this way,—that, in one town, the Full Moon might occur a few minutes *before* midnight on March 20th by local time; in another, a few minutes *after* that same midnight, or, in other words, very early on March 21st. The result would be, if the former date were a Friday, to affect the date of Easter for the places in question to the extent we have indicated; so that for the latter Easter Sunday would be five weeks earlier than for the former.* A similar result would be still

---

* The rule at present is, that Easter Day is the Sunday next after the *ecclesiastical* Full Moon which happens on, or next after, March 21st. If, therefore, the date of the supposed full phase be 11.55 P.M., on Friday, March 20th, for any given place, the next following Full Moon would be on Sunday, April 19th, and Easter Day would be on April 26th. But if,

more likely to occur in different countries, whose clock times would vary according to the longitudes of their capitals, or of their principal observatories. The only way to secure the same date would be to agree to fix it according to the occurrence of Full Moon before, or after, the midnight of a given meridian.

But apart from this, it is also to be observed that the true place of the Moon is not used at all, but a hypothetical Moon is substituted, whose supposed position at any time may differ by a day or two from that of the real Moon, and change an early into a late Easter, or *vice versâ*. The movements of this imaginary moon are regulated in connection with certain ordinances of Pope Hilarius, of the date A.D. 463,* and are arranged in relation to what is termed the Metonic cycle, and to the present regulation of the calendar by the periodical introduction and omission of leap years, so as, by a very complicated system of Golden Numbers, Sunday Letters, Epacts, etc., to keep it from wandering too far away from the place of the real Moon, and in a long course of time to maintain a sort of average agreement with it.

But it would surely be much better, and more convenient to all, and especially to those whose holidays, or whose religious and other duties are connected with the date of Easter, if it could for the future be associated with the Moon only as a matter of *tradition* and *history*, and, together with the other festivals dependent upon it, be made a *fixed* † instead of a *movable* feast. The occurrence of Easter Day upon April 25th, in 1886, the latest possible date, may perhaps awaken

by local time, at another place some distance to the east of the former, the full phase should occur 10 minutes later, or at $0^h\ 5^m$ A.M. on the morning of March 21st, and that day should be a Saturday, Easter Day would be on March 22nd, or thirty-five days sooner. Local time, however, not being thus used, Easter Day, according to present arrangements, never falls on April 26th, although it may be as late as April 25th.

* See Sir E. Beckett's "Astronomy without Mathematics," p. 147 *et seq.* Also De Morgan in the "Companion to the Almanac" for 1845.

† That is of course with the limitation, according to the Nicæan rule and the general practice of Christendom, that Easter Day be observed upon a Sunday. We mean that it might be the nearest Sunday to, or (if it should be so preferred) the Sunday next before, or after, a certain fixed date.

the public mind to give more attention to this suggestion, which we thus venture once more to press upon their notice, although it has certainly been already put forward sufficiently often.

As regards other influences attributed to the Moon, we may remark that we do not believe that any proof has yet been given that it rules the weather, or that it has any important power over health or disease. Moon-blindness, and other similar effects, are in general caused by the cold produced by radiation during exposure, not to the Moon's rays, but to a clear and unclouded nocturnal sky. Any apparent synchronism of the Moon's movements with the periodicity of the paroxysms of lunacy, or of other diseases, is probably, in most cases, either fortuitous, or is the result in the nervous system of a belief in such an effect, the belief causing its own apparent confirmation. There may, however, also be some cases in which the comparative absence of darkness during the nights near to the time of Full Moon may help to increase sleeplessness, and consequently aggravate the symptoms of mental disorder.

But as regards the weather, it really seems absurd to speak of any connection between its changes and those of the Moon's phase. For the weather (at any rate in England) is almost always changing; and the Moon's phase does not change by jumps, four times in a month, from new to half-full, from half-full to full, and so on, but is also always and continuously changing.

Those who believe that the phases of New, Full, and Half Moon in general affect the weather, may also be so good as to notice that such phases recur at intervals of about every seven days. If therefore (as we believe we have seen stated) the change of weather be considered to follow, or precede, any of these special phases within two days, the probability is in favour of such apparent coincidences being very frequent, even if they really have nothing to do with the Moon. Or to put it somewhat differently, it is very likely indeed, that, on one of a certain group of four days out of every seven, there will be some decided change in English weather.

It may also be asked, in the case of occasional periods of

continued fine weather in our own land, and of still longer periods of such weather in other countries, during which the Moon passes through a whole series of phases, why is no change produced, if the Moon's changes have any real effect?

It must not, however, be forgotten that the Moon, when near to being full, reflects an appreciable amount of solar heat, as we have already mentioned that the careful experiments of Lord Rosse have proved; and it certainly seems likely that this heat, especially before it has traversed the lower regions of the Earth's atmosphere, may have a tendency to disperse and dry up terrestrial clouds. This supposition has been confirmed by the result of long-continued observations, which indicate that near to the time of Full Moon the nights are less cloudy than at other times. If so, as the result of the more unimpeded radiation of the Earth's heat, they will also be decidedly colder. Such is the only effect upon the weather which we consider has at present been shown, with any probability, to be connected with the Moon. Other effects, some of them perhaps depending upon atmospheric tides, corresponding to the lunar and solar tides of our waters, may exist, but are as yet unproved. It is possible that the delicate indications afforded by the large scale of the glycerine barometer (a comparatively recent invention), if carefully recorded for a sufficient length of time, may afford some information in this direction.

Here we must regretfully leave the Moon, with the hope that our readers may have been so far interested by the remarks which our space has permitted, that they may be encouraged to pursue the subject by the study of some of the various able treatises in which they will find it more fully discussed. It is one of which we think none should ever be weary, one to which each fair hour, when

> "The Moon
> Riding in clouded majesty, at length,
> Apparent queen, unveils her peerless light,
> And o'er the dark her silver mantle throws,"

should tempt us to recur with unflagging energy and unabated zeal.

LECTURE V.

PTOLEMY *versus* COPERNICUS.

> "When they come to model heav'n,
> And calculate the stars, how they will wield
> The mighty frame; how build, unbuild, contrive,
> To save appearances; how gird the sphere,
> With centric and excentric, scribbl'd o'er,
> Cycle and epicycle, orb in orb."
>
> MILTON, *P. L.*, viii., 78.

WE have given the first place in our discussion of the Solar System to the Sun, its all-important centre. We have described the intensity of its heat, the hugeness of its size, the enormous attractive power by which it rules the orbits of planet and satellite, of comet and meteorite, as they run their courses round it. We have awarded the second place to the Moon, owing to the special interest that belongs to its close proximity to the Earth, and the many details that we are able to observe upon it. The various planets next claim our attention. We shall find, however, that even those which approach the Earth most closely are so far distant that very few details can be detected upon them comparable with those which we see upon the Moon; while others are so remote that our acquaintance with them is in every respect most limited. Not even the largest of them can emulate the Sun, either in size or heat, or in the wondrous phenomena of photosphere and spot, corona and rose-coloured prominence. And yet they are each and all most interesting.

We can well imagine how the astronomer of ages long past, as he gazed at them from beneath the pure skies of the Chaldæan plains, and ever and anon saw Mercury glittering like a tinted diamond in the eastern or western twilight; or watched Venus wax and wane in brightness, until, at times, it was visible, day after day, even at full noon; as he noted the blood-red fire of Mars; the calm, but potent, radiance of Jupiter; or the dull pale light of Saturn,—thought that each held rule over its own house, or sphere in the heavens; that each exerted its own influence for bane or blessing on men's lives, on monarch and prince, on peoples and kingdoms.

We shall presently discuss the various planets in their order of distance from the Sun, and give a brief statement of some of the most important discoveries made with regard to them, both before and since the invention of the telescope. We will, however, by way of introduction, first refer to a few points of interest connected with the nature of the planetary orbits in general, and explain the difference between the Copernican system of astronomy and that which prevailed before the beginning of the sixteenth century.

We may assume that our readers believe with Copernicus that the Earth and all the other planets revolve around the Sun; a theory the simplicity of which, when rightly understood, is so beautiful, and in such striking contrast with the complexity of any other, as almost of itself to convince an unprejudiced mind of its truth. In order, however, rightly to understand its beauty, a considerable amount of care and study are necessary, and the student must understand how complicated the *apparent* movements of the planets are, before he can duly appreciate the way in which they are explained by the extremely simple *real* movements, attributed to them by the Copernican theory.

If we could watch them all from the Sun, or from some distant point of sight look down upon the planes of their orbits, and take, as it were, a bird's-eye view of the whole Solar System, it would be comparatively easy to discover the laws that rule their ever-changing places. But, instead of this, their movements appear to us to be most remarkably complicated, because we are forced to view them from the ever-moving standpoint of the Earth, as it speeds upon its rapid path through space; from which it results, as we shall presently more fully explain, that they sometimes appear to be travelling amongst the stars from east to west; at other times they seem to stand still; then once more they change their course, and apparently move from west to east. Their apparent movements are occasionally very slow, at other times much more rapid. We are in no wise surprised that they long ago received the

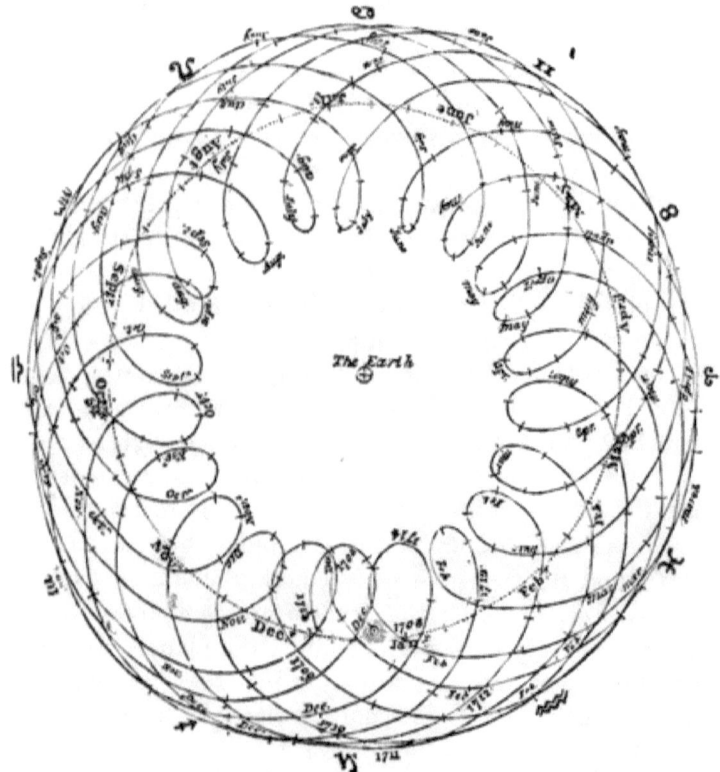

Fig. XXIV.—Path of Mercury relatively to the Earth from 1708 to 1715, after Cassini.

title of Planets,* or *wanderers* in the heavens; they certainly might well have been termed very erratic travellers indeed. So peculiar are the paths which they appear to traverse when watched by an observer upon the Earth, that we think it may be well, before we proceed any further in our discussion, to ask the attention of our readers to some diagrams,

---

* From the Greek πλανήτης, a wanderer.

in which they are illustrated. In Fig. XXIV., for instance, we have a copy, upon a reduced scale, of a diagram of the apparent path of Mercury, relatively to (or, in other words, as seen from) the Earth, between the years 1708 and 1715, which was presented to the French Academy of Sciences by J. D. Cassini, on August 7th, 1709; the dotted line, about half-way between the greatest and least distances of the planet, being intended to represent the apparent annual path of the Sun. It is of course to be understood that by the apparent path of any planet we refer to its movements *relatively to the fixed stars*, amongst which it appears to journey, as the result of its onward progress in its orbit, quite apart from, and in addition to, the apparent *daily motion* which it has around the Earth from east to west, in common with the Sun, the Moon, and all the stars, as the result of the Earth's rotation upon its axis from west to east.

In Fig. XXV., which is a copy of a portion of another by Cassini, and which, so far as it extends, is drawn to the same size as the original, we see a similar representation of the apparent paths of the two great planets, Jupiter and Saturn. The Earth, which, in order to save space, is not represented in the diagram, must be supposed to be situated some way beneath it, so that it would be in the centre of the series of loops, in the complete figure, which Jupiter would appear to describe around it in about 12 years, and Saturn in about $29\frac{1}{2}$ years. So far as we have copied Cassini's drawing, the apparent path of Jupiter is shown for about three years, beginning with the middle of the year 1716; and that of Saturn for about four years, from the middle of the year 1708; the monthly places of the planets being indicated by small dots.

Whether, then, we observe a planet whose orbit lies between the Earth and the Sun, such as Mercury (technically called an inferior planet), or one such as Jupiter or Saturn, whose orbit is outside that of the Earth (technically called a superior planet), it is found, as these diagrams show, that, in either case, the planet appears to pursue a lengthened onward course; then it pauses, in order, as it were, to double back for a shorter period upon its previous path; after which, having made another brief pause, it once more goes forward with increased speed, as though to make up for its lost time. It is well to remember that its backward motion is generally termed its *retrogression*, its forward motion its *progression*, and the pauses between the two, its *stations*, or stationary points. It is,

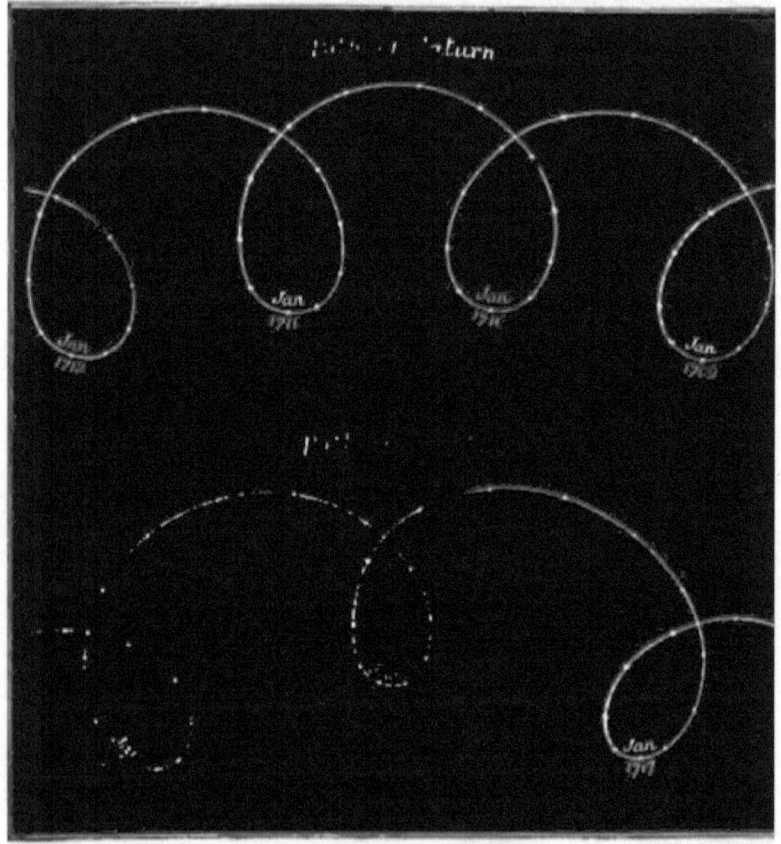

Fig. XXV.—Portions of the apparent paths of Jupiter and Saturn, as seen from the Earth, according to Cassini.

moreover, important to observe, that the diagrams also indicate that the reversed motion occurs when the planet is particularly near to the Earth.

In addition to the reduced view of the complete drawing of Cassini of the apparent path of Mercury during certain years, we will also give in our next three figures, *portions* (similarly reduced in scale) of his diagrams of the apparent paths of Mercury, Venus, and Mars, as they are seen from the Earth, inasmuch as we think that the complication of the whole of such a drawing as that in Fig. XXIV., makes it difficult to realize, so well as in a simpler figure, the actual character of the apparent curve described. We have also adapted the paths shown to more recent dates.

In Fig. XXVI. the apparent path of Mercury as seen from the Earth is drawn for a period of about one year; and a somewhat

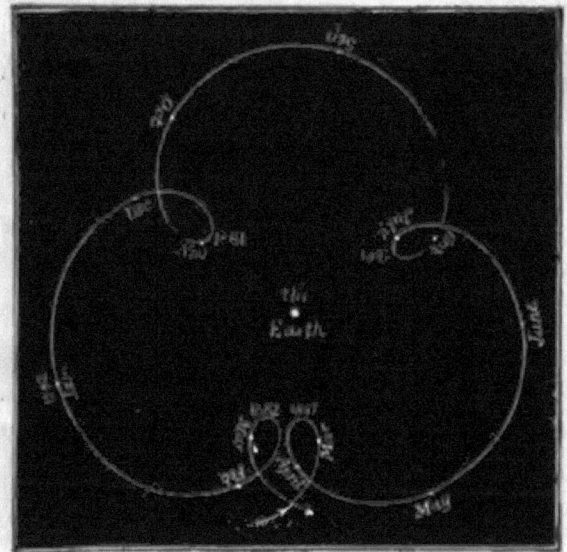

Fig. XXVI.—Apparent path of Mercury as seen from the Earth during part of the years 1881 and 1882.

careful consideration of the diagram may show that an interval of about 116 days occurs between the beginning or the ending of two successive epochs of the planet's retrogression, or between its being twice in the position of its nearest approach to the Earth.

In Fig. XXVII. the apparent path of Venus is represented for about two years, its monthly places being marked by small dots. This diagram also indicates that the interval between two successive nearest approaches to the Earth is very nearly $1\frac{3}{5}$ years (or about 584 days); one such approach taking place as the dots show, about the beginning of May 1881, and the next early in December 1882.

Fig. XXVIII. in like manner shows the apparent path of Mars, in 1877, 1878, and 1879; the average interval between its nearest approaches to the Earth, or (very nearly) between its being twice seen in succession in opposition to the Sun, being about 780 days.

We may remark in passing that the interval between two near approaches of Jupiter, or Saturn, to the Earth, differs but little from a year, as may be seen by a reference to Fig. XXV.,

Fig. XXVII.—Apparent path of Venus, as seen from the Earth, from February 1881 to February 1883.

it being for the former 399, and for the latter 378 days; while for Uranus or Neptune the excess over 365 days is still less.

Now it cannot be denied that these apparent paths are so complicated in their forms and periods of description, that the ancient astronomers and geometricians may have often been almost tempted to give them up in despair as inexplicable. And yet there is one additional peculiarity which they must have especially felt they were bound to explain;—one which is involved in all the above diagrams, although for simplicity's sake we have only distinctly indicated it in

Fig. XXV., in which the apparent annual path of the Sun is also shown. We mean that Mercury and Venus are always observed in close apparent proximity to the Sun, while all the other planets may be seen at any distance from it, and even in the very opposite part of the Zodiac, or belt of the heavens in which their paths lie, to that in which the Sun is found at any given time.

Let us therefore consider a little more particularly the various observations of the planets which it was possible to make before the invention of the telescope, so that we may the better

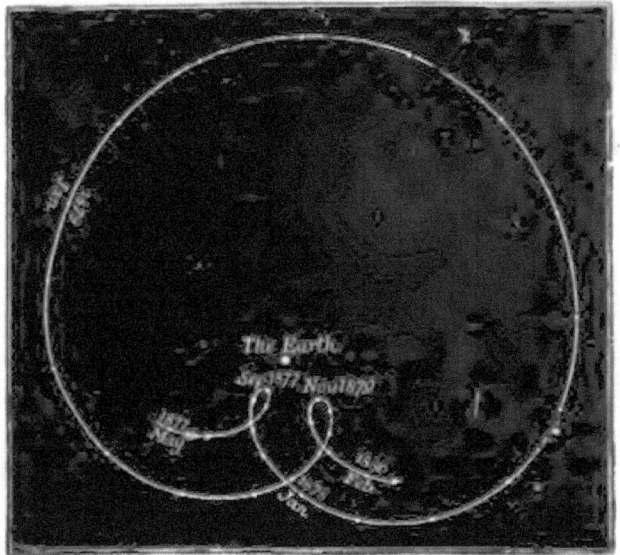

Fig. XXVIII.—Apparent path of Mars, as seen from the Earth, in 1877, 1878, and 1879.

understand what the ancients knew, and appreciate the skill with which the theory of the movements of the Solar System, which was invented and elaborated by Hipparchus, Ptolemy, and their successors, was contrived to explain what seemed to be so complicated.

To begin with the planet Mercury, it must be remembered that it was never very easy to see it, even when it was most favourably situated, owing to the strong glare of the morning or evening twilight. The first important step towards the determination of the true nature of its orbit was the discovery that the two bodies, which, as the result of the earliest observations, were believed to exist, and periodically to appear upon opposite sides of the Sun, never showed themselves together, and were really one and the same. To this earlier supposition we may attribute the traditions which have come down to us of the double appellations Set and Horus amongst the Egyptians, Rouhineya and Boudha amongst the Hindoos, Hermes and Apollo amongst the Greeks, which are said to have been used for Mercury; the name of Apollo (as the God of the Day) having been probably given to it when seen as a morning star before the same name was in later times appropriated to the Sun. So, in like manner, Venus, as a morning star, was called ἑωσφόρος, *i.e.*, the bringer of the dawn, or φωσφόρος, the bringer of light (in Latin "Lucifer"); and ἕσπερος (in Latin "Hesperus"), as an evening star.*

It was, however, much more easy in the case of Venus, than in that of Mercury, to recognize the oscillation from one side of the Sun to the other, because it was possible to watch the planet through so much longer a proportion of its path. Indeed, we cannot but think that, if the ancients had not been hampered by an intense belief in the immobility of the Earth, as the centre about which the Sun and all the other heavenly bodies revolved (a doctrine, the denial of which, even in the time of Galileo,† was thought to be the part of a heretic, owing to a

---

* See Homer, "Iliad," xxiii. 226 :—

$$\text{Ἦμος δ' Ἑωσφόρος εἶσι, φόως ἐρέων ἐπὶ γαῖαν,}$$
$$\text{Ὅν τε μέτα κροκόπεπλος 'πεὶρ ἅλα κίδναται ἠώς.}$$

("When the Dawn-bringer arrives, proclaiming light o'er the Earth, after whom the saffron-vested morn is diffused o'er the sea.")

And "Iliad," xxii. 318 :—

$$\text{Ἕσπερος ὃς κάλλιστος ἐν οὐρανῷ ἵσταται ἀστήρ.}$$

("Hesperus, which is placed the brightest star in heaven.")

Compare also Milton, "Paradise Lost," Book V. :—

"Fairest of stars, last in the train of night,
If better thou belong not to the dawn."

† In his recantation, when seventy years of age, Galileo was forced to say that he *abjured, cursed, and detested, the absurdity, error, and heresy of the motion of the Earth.*

misinterpretation of such texts as are quoted below),* they must have discovered, first in the case of Venus, and afterwards in that of Mercury, that these two planets were revolving round the Sun in orbits such as we now know that they possess; whatever they might have thought about Mars, Jupiter, and Saturn. But while they held to the above belief, it was necessary, in order to account for the apparent planetary movements, to invent a curiously complicated system, which

Fig. XXIX.—Cycles and Epicycles of Venus and Mercury.

is generally known as the Ptolemaic System of Cycles and Epicycles. This we proceed to explain.

So far as it applied to such planets as we now know to have orbits within that of the Earth, its chief features are shown in Fig. XXIX.; in which E is supposed to be the Earth, while the circle passing through s is the apparent orbit of the Sun around it. Between that orbit and the Earth, Mercury was

* Psalm civ. 5, P. B. Version: "He laid the foundations of the Earth that it never should move at any time." Psalm xciii. 2: "He hath made the round world so sure that it cannot be moved."

supposed to move in the following manner. First of all it was arranged that a point, P_1 (termed a *deferent*, or carrier), should be conceived to travel uniformly round E in a circle, or cycle, once in every $365\frac{1}{4}$ days. This large circle was called a deferent cycle. While this point, P_1, thus moved, it was imagined that Mercury was in some way whirled round it, once in every 116 days, in an epicycle, or superimposed and much smaller circle, which was carried along by P_1.

A similar arrangement, at a greater distance from the Earth, was assumed for the planet Venus, as is shown in the diagram. It was also supposed that the centres of the two epicycles of Mercury and Venus (viz., P_1 and P_2) moved so as always to lie in a straight line joining the Earth and the Sun.

It is not hard to see that, if suitable velocities be assigned to Venus and Mercury in their epicycles, such an arrangement would account for the principal peculiarities in their movements which we have hitherto described. They would evidently appear to oscillate within certain limits on either side of the Sun, which limits, as observation easily proved to be the case, would be greater for Venus than for Mercury, because of the larger epicycle belonging to the former. They would also alternately be brought nearer to the Earth, and removed farther from it. And, supposing that the velocity of either of them in its epicycle, when nearest to the Earth, was sufficiently great, it is evident that the planet might, in spite of the onward motion of the centre of its epicycle, appear to move for a time in the reverse direction, as the arrows in the diagram indicate. In this way their periodical retrogressions might be explained.

Any requisite variations of speed in describing the epicycle, as well as certain other irregularities of movement, were accounted for by placing additional epicycles upon those already mentioned; that is to say, instead of the planet Mercury moving in the epicycle shown in the diagram, another point was supposed to do so, which carried with it another smaller circle, or additional epicycle, in which the planet might move around it. In this way several epicycles were successively superimposed, as they were required. It was also found necessary by some of the astronomers of old to imagine that the Earth was not actually in the centre of the principal or deferent cycles, but somewhat eccentrically situated.

We need not, however, enter more minutely into these refinements of that ancient Ptolemaic system which has long been altogether extinct. Nevertheless we must allow that it displayed remarkable skill and ingenuity, and that it afforded a geometrical explanation of most of the peculiarities of the movements to which it was applied. At the same time, it was certainly altogether inconsistent with any true theory of the attraction of matter upon matter; an inconsistency which we believe the astronomers of those days would not have failed to perceive, if only they had been acquainted with the great law of gravity. As it was, they did not realize that the imaginary movement of points, such as those which were supposed to travel in the larger cycles, apart from the existence of attractive matter in them, was really an impossibility.

The Ptolemaic system would also doubtless have been given up long before it was, and probably before it had been so greatly elaborated by the heaping of epicycle upon epicycle, if its votaries had possessed any telescopic aid to their vision; for it would, in that case, have been possible for them to have measured the discs of Mercury and Venus from time to time, and it would have been found that no such arrangement as we have described could accurately correspond to their apparent changes in size, and their consequent changes in distance from the Earth; although, as far as rough observations with the naked eye were concerned, the agreement appeared to be satisfactory. It is also evident from our diagram that Venus and Mercury, according to the Ptolemaic system, would turn the unilluminated sides of their discs towards the Earth, when at their *furthest* from it, as well as when at their *nearest* to it; whereas a telescope at once shows that, in the former case, the discs are fully illuminated; which single fact is, in itself, sufficient to prove that they have passed to the opposite side of the Sun, although the true nature of their orbits is still more clearly demonstrated by telescopic observations of the whole course of their intermediate phases.

We do not forget that in the ancient system of the Egyptians, notwithstanding that the Sun and the other planets were supposed to circle round the Earth, an exception was made in the case of Mercury and Venus, to which a special rotation round the *Sun* as a centre was assigned. We have also read that Aristarchus of Samos, and Cleanthes of Assos, about 270 B.C., were (as it was then doubtless considered) so impious as to suggest that the Sun, rather than the Earth, was the centre of the celestial movements. But these opinions gained

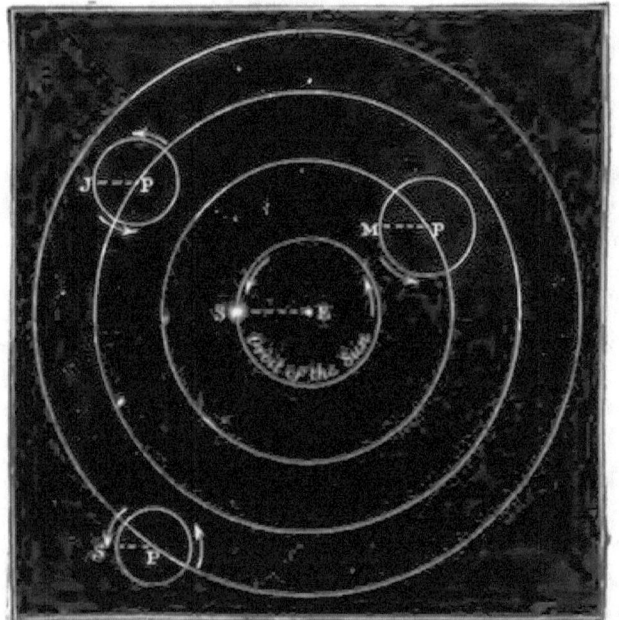

Fig. XXX.—Cycles and Epicycles of Mars (M), Jupiter (J), and Saturn (S)

little or no credit, and were no doubt looked upon as wild vagaries. The system of cycles and epicycles was therefore extended so as to embrace Mars, Jupiter, and Saturn. Its application to those planets is illustrated in our next diagram, Fig. XXX.

In order to represent their apparent motions it was found necessary to suppose that the centres of the epicycles respectively described their cycles in rather less than 2 years for Mars, in rather less than 12 for Jupiter, in rather less than

$29\frac{1}{2}$ for Saturn; while in each case the epicycle was described by the planet in *one* year, and in such a manner that the radius from its centre to the planet, viz., PM, or PJ, or PS, would point directly *towards* the Earth, whenever the planet was seen in a direction exactly opposite to that of the Sun (or, as it is termed, in Opposition), and directly *from* the Earth, whenever the planet was seen in the same direction as the Sun (or, as it is termed, in Conjunction).

It also followed, as a geometrical consequence of the above-mentioned conditions (the movements being supposed uniform), that the radius of each epicycle must *at all times* be parallel to a line joining the Earth and the Sun, as is shown in Fig. XXX.; but we are not quite certain how far this last fact was clearly understood; nor are we sure that the exact yearly period of Mars in its epicycle was fully realized. For, if all this had been duly appreciated, we should suppose that a simple application of such geometrical principles as the mathematicians of those days were perfectly acquainted with, would have led them to argue as follows :—The movement of each planet in its epicycle being the *same*, such movement is probably only *apparent*, and caused by a real movement of the Earth, according to the law, that, if an observer be situated upon a moving body, the apparent motion produced in anything which he looks at, will be just the same as if he were brought to rest, and a velocity, the reverse of his own, were communicated to the other body.

We have stated that each of the exterior planets was necessarily supposed to describe its epicycle in one year; nevertheless, it should be noticed that Jupiter and Saturn would occupy somewhat longer than a year in describing complete loops of their respective apparent orbits (such as we showed in Fig. XXV.); because the onward march of the centre of the epicycle would prevent the radius joining it to the planet, from pointing again to the Earth at the end of one epicyclic revolution. Before it could do so the radius would need to turn round through an additional angle equal to that which the centre had in the meantime described round the Earth. This may be easily seen in the subjoined view of one of the loops of Jupiter's apparent orbit, taken from Cassini's drawing, and somewhat reduced in size.

In the above diagram A, B, C, etc., represent successive positions of the point P, or centre of the epicycle (see Fig. XXX.); *a, b, c*, etc., corresponding positions of the planet, which are of course always at the same distance (viz., the radius of the epicycle) from the moving point P. By actually marking off the points *a, b*, etc., and drawing a curve through them, it may easily be seen how the peculiar looped form of the apparent

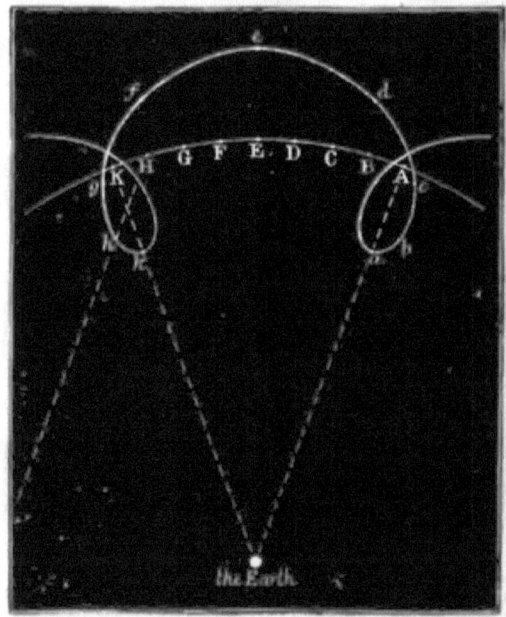

Fig. XXXI. Showing that Jupiter occupies somewhat more than a year in describing one complete loop of its apparent orbit.

orbit arises. The fact, however, to which we at present wish to draw attention is—that, when the centre of the epicycle has reached H, so that its radius H*h* is parallel to A*a*, the planet would have been once completely round it, and the time thus occupied, according to our previous statement, would be equal to one year; but it is clear that it would be necessary for the centre of the epicycle to move on to K, and the planet to *k*, before the radius would once more point to the Earth, and an entire loop of the apparent orbit be completed. The additional time involved, proves, in the case of Jupiter, to be 33 days, and in the very similar case of Saturn, 13 days.

We have thus far refrained from any special reference to the case of Mars, because it happens that the circumstances involved, and the figure necessary to explain them, are much

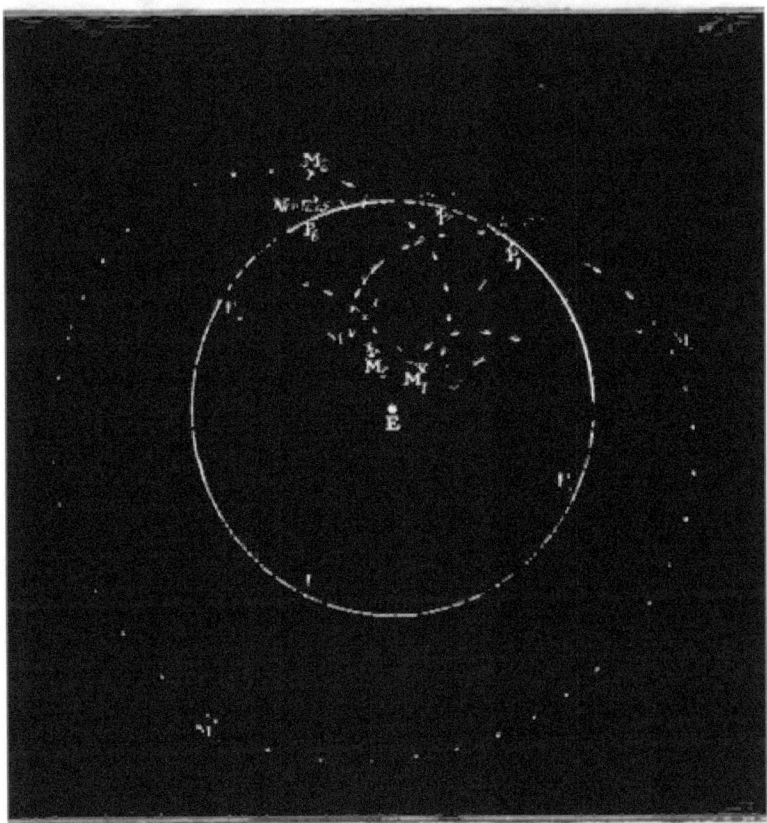

Fig. XXXII.—Showing, by means of a Cycle and Epicycle, that Mars takes more than two years to describe one complete loop of its apparent orbit.

more complicated. We will now, however, put before our readers a drawing, which may perhaps suffice to show, that, if Mars be supposed to revolve once round its epicycle in a year, while the centre of the epicycle rotates round the Earth in 687 days, the planet must perform more than two complete revolutions in the epicycle, and occupy *more than two years* in describing one complete loop of its apparent orbit; or in journeying between two successive positions in which the radius of the epicycle passing through it will also point exactly towards the Earth.

In the above figure, P_1, P_2, P_3, P_4, P_5 represent successive positions of the centre of the epicycle half a year apart. In each case the epicycle is shown drawn round its centre. M_1, M_2, M_3, M_4, M_5, are the corresponding positions of Mars. Now, if the position of P_1 be supposed to correspond with such a date as September 5th, 1877,* the line $P_1 M_1$ must point exactly towards the Earth, as Mars was then in opposition to the Sun. If P_2 be the place of the point P half a year afterwards, $P_2 M_2$, will point in an exactly parallel, but opposite, direction, Mars having gone half-way round the epicycle. Similarly $P_3 M_3$, and $P_4 M_4$, will, at the end of successive half-years, be parallel to the same direction. This will also be the case with $P_5 M_5$, at the end of two years, in which time P will have gone about $1\frac{1}{15}$th times round its cycle. By November 12th, 1879,* P will have reached the position P_6, the line $P_6 M_6$ pointing once more exactly towards the Earth, and Mars being again in opposition, after having occupied 798 days (or rather more than the average interval of 780 days) in describing one complete loop of its apparent orbit. By taking intervening positions of P and of Mars, it is easy, even for those who are unaccustomed to mathematical diagrams, to see that a looped curve, indicated by dots in the figure, and exactly corresponding in form to that of Cassini in Fig. XXVIII. (although reversed in position), will pass through the various places of the planet, and represent its apparent orbit as seen from the Earth.

We have taken some trouble to explain the way in which this curve may be drawn, because it is important to observe,

* Strictly speaking, owing to the elliptic form of the orbit of Mars, and also (although in a much less degree) to that of the orbit of the Earth, the actual nearest approach of the two planets does not take place exactly when Mars is seen in opposition to the Sun, as it would if the orbits were circular. In 1877 it occurred three days, and in 1879 as much as eight days, before the opposition. But the difference of distance involved is so slight, that we have thought it better not to attempt to indicate it in the curve in our diagram.

that the comparatively slow rate at which the Earth gains upon the planet Mars while it describes its real orbit around the Sun, causes the latter to be thus long in performing one complete loop of its apparent path, and makes the investigation of that path so much more complicated than in the case of the more distant exterior planets, Jupiter and Saturn, notwithstanding that the principle upon which the paths are drawn is in all the three cases exactly the same. We also wish our readers especially to notice that it is this complication in the case of Mars which has led us to doubt (as we have already stated) whether the sages of old properly grasped the fact that the revolution of these three planets in their epicycles, which the Ptolemaic system required, was, *in every case, of a year's duration;* or that the radius joining them to the epicycle's centre must *all the year through* be parallel to a line joining the Earth to the apparent place of the Sun, although we feel no doubt that they perceived that it must point directly to or from the Earth at those special times when the planet alternately appeared to be in conjunction with, or in opposition to, that great source of light from whose just claims they derogated, by unfortunately refusing to allow it to be the centre of every planet's motion.*

For Mars, and Jupiter, and Saturn, other subsidiary and smaller epicycles were supposed to be superimposed upon the larger ones, as in the case of Venus and Mercury; and by a further refinement upon the system as originally invented, the Earth was also considered to be somewhat excentrically situated in the midst of the whole. But into these points we will not enter; nor need we discuss at any length the way in which the alternations of the apparent brightness of the planets were, in some degree, supposed to be caused by their being carried round in connection with crystal spheres, which were believed

* For further information on this subject we refer our readers to Newcomb's "Popular Astronomy," in which it is very fully and excellently explained. Those who do not possess this somewhat expensive work may see it in the valuable library of the Corporation of London, which, we rejoice to know, is open freely to the public until 9 p.m. throughout nearly the whole of the year; and for which we believe that copies of every important work on Astronomy are invariably provided.

to be of considerable thickness, and in which, in some rather unaccountable way, each planet was made to travel, sometimes nearer to the outer surface and sometimes nearer to the inner surface of its corresponding sphere.

It may, however, be interesting to remark, as regards this last supposition, that one of the most extraordinary ideas of the ancient astronomers and one which proves how deeply fixed was their belief in the immobility of the Earth was, that, in place of attributing the apparent daily motion of the Sun, Moon, Planets, and Stars, to the rotation of the Earth upon its axis in a contrary direction, they believed this part of their movements to be caused by their connection with a series of crystal spheres (or rather, spherical surfaces), situated at successive distances from the earth, which turned round once in a day. These were supposed to be made of some crystalline substance, in order that it might be possible to see through them just as if they did not exist. This notion, as to the material of which they were made, may well be compared with the reason which was at the same time assigned for the inaudibility of the celestial harmony caused by the beautiful accord of their imaginary movements ; viz., that all mankind had been so continuously accustomed to it from their birth as to take no notice of it.

In, or upon, their corresponding spheres, and controlled by their cycles and epicycles, a certain liberty of movement was allowed to the Sun, Moon, and Planets ; while the fixed stars were supposed to be so fastened to their own appropriate sphere beyond that of the planets, that they kept their relative places unchanged. Outside all was located what was termed the *Primum Mobile,* or a sort of general grinding-machine, whose province was to keep the whole series in perpetual rotation. We, who now know so much better, can understand that it could only have been from the want of any true idea even of the planetary distances, much less of those of the stars, that the belief in the possibility of their describing day by day, round the Earth, the huge orbits which those distances would involve could have existed so long as it did.

Such, then, were the main features of the old Ptolemaic

system, which, dating (in its earliest form) at least from the second century before Christ, was so carefully elaborated, so persistently maintained. By way of contrast to it, let us describe the true theory of the Solar System, first published by Copernicus, upon his dying bed, in his great work "*De Revolutionibus Orbium Cœlestium*," A.D. 1543; and afterwards more fully elucidated by the investigations of Kepler, Newton, and others.

Its first assertion is, that the apparent daily motion of the stars from east to west, to which we have already referred, as well as a corresponding part of the motion of all the other heavenly bodies, is to be explained by the rotation of the Earth upon its axis in the opposite direction, viz., from west to east, once in every 23 hours 56 minutes; the deficit of about 4 minutes, by which this period falls short of 24 hours, being caused (as we shall presently explain in Lecture IX.) by the motion of the Earth in its own orbit.

The Sun is taken as the controlling centre of the movements of all the planets.

In order from it,—see Fig. XXXIII., in which the size of the various orbits is indicated as nearly as possible upon a correct scale,—there revolve

		Millions of Miles.		Days.		Years.		Miles per Hour.
MERCURY	at a mean distance of about	36	in a period of about	88	or about	$\frac{1}{4}$	with a mean velocity of about	107,000
VENUS	,,	67	,,	$224\frac{7}{10}$	,,	$\frac{8}{13}$	,,	78,000
THE EARTH	,,	93	,,	$365\frac{1}{4}$	,,	1	,,	67,000
MARS	,,	142	,,	687	,,	$1\frac{7}{8}$	,,	54,000
JUPITER	,,	484	,,	4,333	,,	$11\frac{6}{7}$	,,	29,000
SATURN	,,	887	,,	10,759	,,	$29\frac{1}{2}$	,,	22,000
URANUS	,,	1,785	,,	30,688	,,	84	,,	15,000
NEPTUNE	,,	2,796	,,	60,181	,,	$164\frac{3}{4}$	,,	12,000

After a most laborious comparison of numerous positions of the planets, and more especially of those of Mars, it was discovered by Kepler, early in the 17th century, that their orbits round the Sun are not circles, but slightly oval curves,

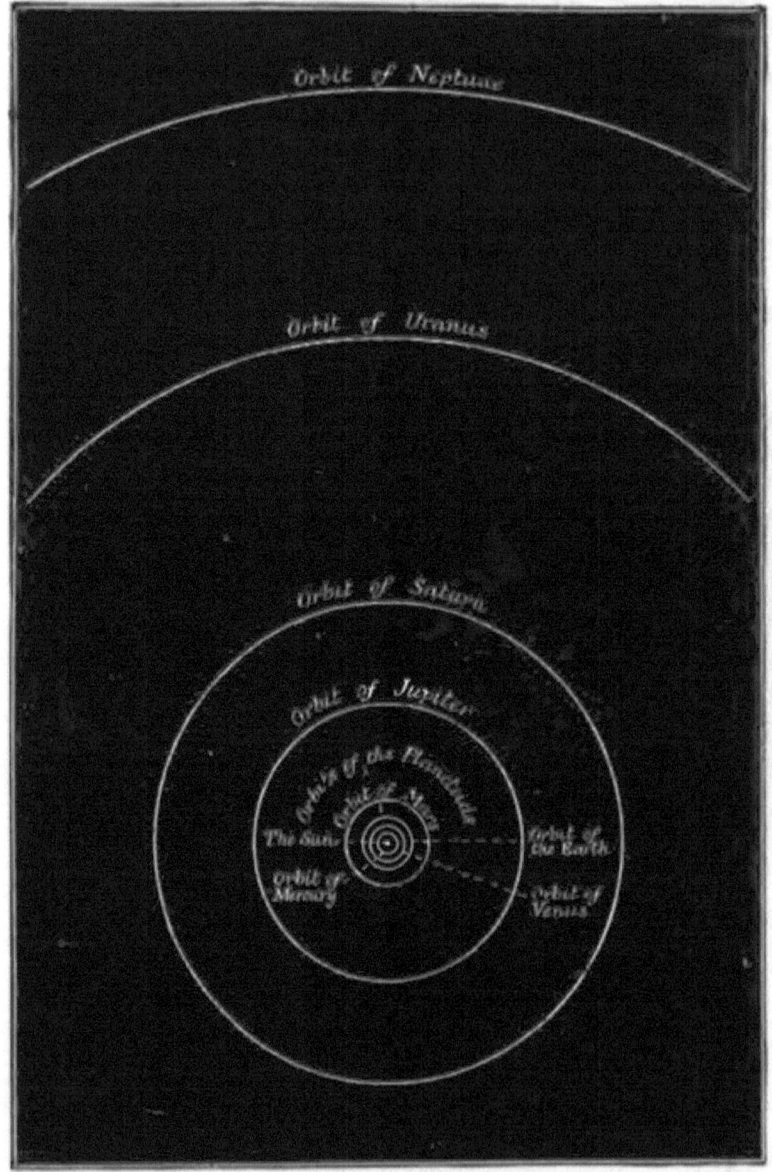

Fig. XXXIII.—The Copernican Theory of the Solar System.

called Ellipses, two specimens of which, of different degrees of ovalness, are shown in the following diagram, Fig. XXXIV.

An Ellipse is the curve described by a point P, which moves so that the sum of its distances from two other given points, s and H, remains constant. It may in practice be most easily drawn by fastening a thread, or string, at s and H; the point of a pencil which is carefully moved so as to keep the string stretched will then describe the curve.

s and H are called the Foci of the Ellipse, c half-way between them is its centre. The ovalness of the curve depends upon the proportion which the distance sH bears to the longest diameter ASHZ, which always equals in length the sum of the two lines sp and Hp. This diameter is generally termed the Major Axis; another at right angles to it through c being called the Minor Axis. If s and H are brought together to c

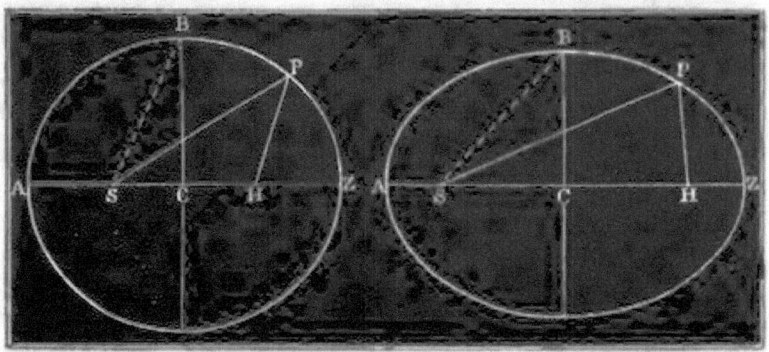

Fig. XXXIV.—Two Ellipses of different degrees of ovalness.

the ellipse then becomes a circle. The ovalness is not very apparent to the eye, unless the Foci are considerably removed from c towards A and z. It is found, for instance, that unless sc be greater than $\frac{1}{8}$th of CA, the semi-minor axis CB will not differ from the semi-major axis AC by more than $\frac{1}{50}$th part of the latter; the approximate rule for obtaining the difference in any particular case being, to square the number of times that CA contains CS, multiply by two, and take the result as the denominator of a fraction with a numerator unity, which will indicate to what extent BC will fall short of AC. For example, if CS equals $\frac{1}{60}$th of CA, then twice the square of 60 being 7200, BC will only fall short of AC by a $\frac{1}{7200}$th part. The above rule depends upon the properties of an ellipse which are proved in treatises upon Conic Sections.

The difference of AC and BC in any orbit is not, however, a very important matter. On the contrary, the difference of the greatest and least distances of a planet from the Sun, which may be represented respectively by SA and SZ, is of the utmost importance.

The only remarkable fact connected with the point B is that BS is found to be always equal to CA or CZ; *i.e.*, to one-half of the sum of SA, the greatest distance from S, and SZ, the least distance from S, in consequence of which BS is often called the Mean Distance. We may also observe, that, in such a case as that which we previously supposed, in which SC equals $\frac{1}{3}$th of AC, the difference of the greatest and least distances from S will be twice SC, or $\frac{2}{3}$ths of CA (or SB), the mean distance. It is interesting to notice that such a degree of ovalness as this corresponds very nearly to that of the orbit of the planet Mercury, and is more than double that of the orbit of Mars; while it much exceeds that of the orbits of any other of the principal planets. And yet, according to the rule previously stated, such an orbit would only vary from a circular form, so far as to make the difference of its greatest and least distances from its *centre*, C, amount to about $\frac{1}{26}$th part of the larger, which in the case of Mercury would be about 865,000 miles. The fact of the Sun being situated at a point corresponding to S and not to C, in the case of such a planetary orbit (which was Kepler's first great discovery), has therefore a much more important effect in causing the distance of a planet from it to vary, than the actual deviation of the orbit from a circular form * would produce if the Sun were situated at its centre.

When Sir Isaac Newton, near to the close of the seventeenth

---

* It is found, that if we denote the ovalness of the orbit by the ratio of CS to CA, *i.e.*, by $\frac{1}{3}$th for that of Mercury :—

        it will only be about . . . . $\frac{1}{150}$th for Venus,
    ,,   ,,   about . . . . $\frac{1}{60}$th for the Earth,
    ,,   ,,   rather more than $\frac{1}{15}$th for Mars,
    ,,   ,,   about . . . . $\frac{1}{20}$th for Jupiter,
    ,,   ,,   rather more than $\frac{1}{18}$th for Saturn,
    ,,   ,,   rather less than $\frac{1}{16}$th for Uranus,
    ,,   ,,   rather more than $\frac{1}{125}$th for Neptune;

the largest of which fractions is less than one-half of that for Mercury.

century, subsequently discovered the great law of gravity, viz., that every particle of matter in the universe attracts every other particle with a force which is proportional to the inverse square of their distance apart, he showed that two other laws of the planetary movements, which Kepler also discovered, as well as the law to which we have just referred, were necessary results of gravitation. Let us, therefore, now state all three of these important laws in as accurate and simple language as possible.

Kepler's first law is : *The planets rotate round the Sun in ellipses, each of which has the centre of the Sun as its focus.*

If we speak somewhat more accurately, we may say, that the common centre of gravity of the Sun and any particular planet is the focus about which (so far as those two bodies are concerned) they both describe an ellipse;—the planet a large ellipse, the Sun a small one. It is, however, also true, that the orbit of any planet *relatively to the Sun* (apart from any perturbations caused by the attraction of the other planets), is an ellipse as stated by Kepler, although the Sun is always moving round about the common centre of gravity of itself and all the planets, from which the average distance of its centre is about half-a-million of miles.

Kepler's second law is :—The velocity of each planet is greatest when it is nearest to the Sun, and less when further away, and in exactly such a proportion as involves the following rule :—

The area formed by the curved path of any planet, in any given time, and the straight lines joining its places at the end and the beginning of that time to the Sun, is always proportional to the time in question, or always the same, in the same length of time, for any given planet.

This is illustrated by Fig. XXXV., in which the areas sml, sfe, sxy, are supposed to be described, each in the same interval of time, around the Sun (s); and, consequently (by the above law), to be equal to one another; the actual portions of the curved path, viz., ml, fe, xy, described by the planet, decreasing in length as its velocity diminishes with its increase of distance from s.

Kepler's third law, which is of very great value, may be stated as follows :—

If we take the mean distance of any planet from the Sun, and the periodic time in which it goes once round its orbit, the ratio of the square of the time to the cube of the distance will be found to be the same for all the planets.

This may, perhaps, be best illustrated by one or two simple examples, although only approximately, and not so as to show the beautiful precision with which the law holds good. For instance, the mean distance of the Earth is 2·6 times that of Mercury, and its period is 4·2 times. The square of the latter number is 17·64, while the cube of the former is 17·576, both of which, to one place of decimals, give the same result, 17·6.

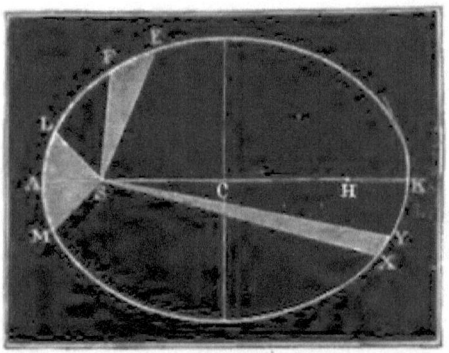

Fig. XXXV.—Kepler's Second Law: A planet describes equal areas round the Sun in equal times.

Again, the period of Uranus is close upon 84 years, and 84 squared is equal to 7056. The distance of Uranus is very nearly 19·18 times the Earth's, which number cubed produces 7055·8, the result being therefore in almost exact agreement with the law.

So, if a new planet were found, and its period were observed to be five times that of the Earth, we should be able, by squaring 5 (which gives 25), and taking the cube-root of this square, which will be rather less than 3, to say at once that its distance from the Sun must be rather less than three times the Earth's distance.

In connection with this mention of the planetary distances, we may as well allude in passing to the curious relation existing between them, first announced by Titius, in a translation of Bonnet's "Contemplation de la Nature," published at Leipzig, in 1766, which, being afterwards brought into more general notice by Bode, about A.D. 1778, is consequently usually termed Bode's Law. A series is formed by taking the number 4, and adding to it in succession 3, twice 3, four times 3, eight times 3, sixteen times 3, and so on; so that we obtain the numbers:

4, 7, 10, 16, 28, 52, 100, 196.

These numbers, if we pass over the fifth (which, however, corresponds to the gap between Mars and Jupiter, in which so many small planets have of late been discovered), are found to agree very approximately with the ratios of the distances of Mercury, Venus, the Earth, Mars, Jupiter, Saturn, and Uranus, from the Sun. For if we multiply the terms of Bode's series respectively by 10, we obtain:

40, 70, 100, 160, 280, 520, 1000, 1960;

whereas, if the Earth's distance be represented by 100, the actual distances of the various planets are respectively;—

	for Mercury	Venus	the Earth	Mars	Jupiter	Saturn	Uranus
as the numbers	39	72	100	152	520	954	1918.

The near agreement of these two sets of numbers is therefore very remarkable. The rule, however, fails for the last-discovered major planet Neptune, the next term of Bode's series being 3880, which differs greatly from the number 3005, which would represent the real value.

The truth is that Bode's arrangement hardly deserves to be termed, as it frequently is, a *law*, no theory having yet been suggested which gives a satisfactory reason for its existence, although occasional attempts have been made to connect it with the gradual formation of the planets in succession, according to Laplace's nebular hypothesis. It bears no comparison with the three laws of Kepler, which, although originally discovered without their cause being understood, have since been proved to be, in the very strictest and minutest possible degree, necessary consequences of the action of gravitation. Bode's so-called law must at present be looked upon as little more than an example of a curious coincidence.

Thus far, then, we have shown that it is principally to Copernicus that we owe the credit of the promulgation of the all-important statement that the planets revolve in orbits round the Sun; while we are indebted to Kepler and to Newton for our knowledge of the more accurate nature of those orbits involved in the laws we have just described.*

But we have not yet seen how the Copernican theory accounts for those special phenomena of the apparent planetary movements, which called forth the complicated cycles and epicycles of Hipparchus and Ptolemy, nor have we, as yet, contrasted the explanation afforded by it, with that which so long held its ground amongst the philosophers of old. In other words, we have to show how so simple an arrangement as that theory involves, of the movements of the planets with regard to the Sun, *involves all those intricate orbits* seen from the Earth, with the description and illustration of which we commenced this lecture.

To begin with, let us take the case of Mercury or Venus, whose apparent movements, as we have already seen, differ in certain important respects from those of the other planets. It is, we think, very easy to understand, that, if they move, according to the assertion of Copernicus, in orbits round the Sun as a centre, within the Earth's orbit, they must appear to oscillate from side to side of the Sun within a certain distance of it, which distance will be much less for the former than for the latter planet. It also follows that the greatest apparent distance of either of them from the Sun, as seen from the Earth, will occur, when the relative positions occupied by them are such as are indicated in Fig. XXXVI.; *i.e.*, when the

* It should also be mentioned, that calculations depending upon the law of gravitation, if accurately made, show (as careful observations also prove to be the case) that each planet's orbit lies upon a plane, or level, of its own; so that it moves just as a ball would, if, while floating in a vessel of water, it were made to travel round another larger one floating in the same vessel and representing the Sun. In the case of the Earth this plane is called the Ecliptic. The planes of the motion of the other planets are tilted at certain angles to it and to one another; but this tilt is in every case slight, except in that of some of the small planets between Mars and Jupiter, whose orbits are altogether so peculiar that they require separate consideration.

straight line joining the Earth to the planet is a tangent to the path of the latter.

Moreover, we think it is also evident, in the above diagram, that, if it were not for the tilt of the orbits to which we have recently referred, each of these two planets of which we are speaking, would, at frequent intervals, pass exactly between the Earth and the Sun, when in such a position as is marked I.C. But that tilt necessitates that there can only be two points in each of their orbits, in, or very near to which, they can be in a position to do so; viz., the two points in which those orbits cut the plane of the ecliptic in which the Earth

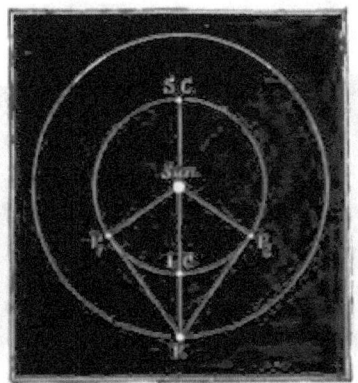

Fig. XXXVI.—The Earth being at E, an inferior planet when seen at P_1 or P_2 is at its greatest apparent distance from the Sun.*

moves. In other words, it is only possible that either of them should be exactly in a straight line between the Earth and the Sun, at two particular times in each year, when the Earth, as it would be seen from the Sun, lies in the direction of one of the two points in question; which points are, of course, different for the orbit of Mercury from those which belong to that of Venus. This may be best understood by the consideration of such a diagram as is drawn in Fig. XXXVII., in which the Earth is supposed to move in the plane of the paper.

The above figure shows that, if the Earth is in such a position as E or E', when the planet also happens to be at, or very

* If the Earth be at E, and a planet be in the position I.C., it is said to be seen in *inferior conjunction* with the Sun; if it be at S.C., it is in *superior conjunction*.

near to, such a position as P or P', the planet will either pass exactly behind the Sun, as seen from the Earth, or else it will pass in transit across the Sun's disc as a small black spot; in the former case it might be at P' when the Earth is at E, or at P when the Earth is at E'; in the latter, at P when the Earth is at E, or at P' when the Earth is at E'. For the planet Mercury the positions E and E' of the Earth are in the months of May and November; for Venus in June and December.

At all other times, the tilt of the planet's orbit carries it somewhat above or below the Sun's disc, as it passes over or under a straight line, joining the places of the Earth and the Sun; just as in Lecture IV. we explained that the tilt of the Moon's orbit prevents Solar or Lunar eclipses from occurring,

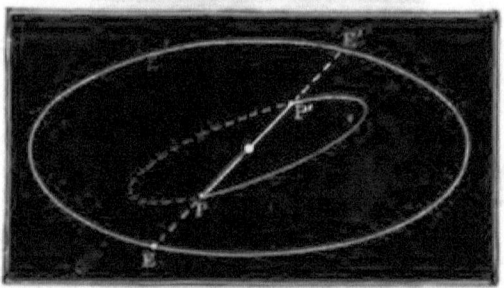

Fig. XXXVII.—The curve E E' is the Earth's orbit lying in the plane of the Ecliptic. The curve P P' is a planet's orbit in a plane inclined or tilted to the former.

as they otherwise would, upon the occasion of every new and full Moon.

We shall make further reference to the transits of Mercury in Lecture VI. At present we will only remark that Mercury, when in transit, is too small to be seen by the naked eye, so that we are not surprised that no very ancient observations of its transits are recorded. But we think it somewhat strange that the astronomers of early days do not seem to have observed any Transit of Venus; in fact, that none was noticed, as we have stated in our first lecture, until the year 1639.

Such Sun-spots, however, as are visible to the naked eye, were observed in very ancient times. It is therefore just possible that a Transit of Venus, the disc of which is quite large enough to be distinguished upon the Sun without any telescopic aid, may have been seen, and that no special notice was taken of it, the planet being supposed to be a spot.

So far, then, the Copernican theory explains the occasional occurrence of transits of Mercury and Venus, as well as their oscillation from side to side of the Sun. It also indicates that, when either of these planets is seen in transit, or passes slightly above, or below, the direction of a straight line drawn from the Earth to the Sun, it must turn its unilluminated hemisphere towards the Earth. And a little consideration will show that it further requires, not only that they should each display a fully illuminated hemisphere, when at their furthest from the Earth (in direct contradiction to the Ptolemaic theory);

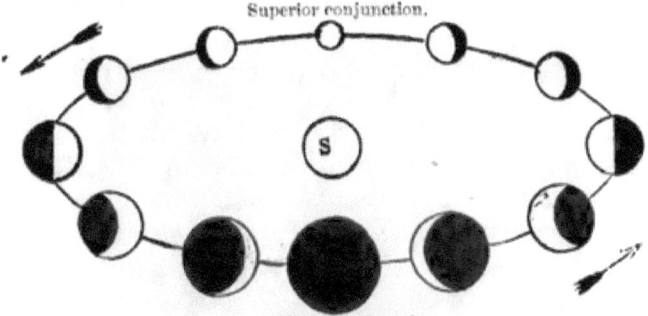

Fig. XXXVIII.—Phases of an inferior planet.

but that they should, at other times, pass through a regular succession of phases, such as are shown in Fig. XXXVIII.

It is very noticeable, however, in the above figure, that in one respect such phases decidedly differ from those exhibited by the Moon, inasmuch as the apparent diameter of the disc does not (as in the case of the Moon) remain nearly constant; but becomes much wider as the phase becomes less and less; because the planet at the same time approaches nearer and nearer to the Earth. Such phases are easily seen in a small telescope, and the change in the diameter of the disc is found to correspond precisely with the change in the planet's distance from the earth caused by its passage round the Sun.

We have dwelt at some length upon this point, because it may be considered to be a crucial test of the truth of the Copernican theory, in the case of Venus and Mercury. The

Ptolemaic system is here utterly at fault, and would have been known to be so from the first, if the phases of these planets had been visible without telescopic aid.

It may further be well to mention that certain apparent irregularities in the greatest distances to which Venus and Mercury are seen to depart from the Sun, or in their movements in general, are also found to agree exactly with those refinements in the Copernican theory, which, as we have previously stated, involve the knowledge of the tilt of their orbits and of their elliptical form about the Sun as a focus.

But to keep to more simple matters, we may remind our readers, that, in the table given in connection with Fig. XXXIII., it is said that Mercury goes round the Sun in about 88 days; while observation shows (as is also indicated in Cassini's diagrams, see Figs. XXIV. and XXVI.) that the average time between that planet being seen at its greatest distance on either side of the Sun, and reaching a similar position upon that same side of it again, is in general about 116 days. We may ask, do these facts agree with one another according to the Copernican theory? Does it offer any simple explanation of them? Yes!—It tells us that the onward movement of the Earth only allows Mercury to overtake it, upon an average, with about $\frac{3}{4}$ths its own speed of angular motion round the Sun; and that, therefore, in order to regain a similar position to that which it occupied with regard to the Sun as seen from the Earth at any given date, it takes about $\frac{1}{3}$rd longer than it occupies in circling once round its own orbit.

In like manner, the Copernican theory explains, equally as well as the old Ptolemaic system, why it is that about $1\frac{3}{5}$ years, or 584 days, elapse, between Venus being seen twice in succession in any given position relatively to the Sun, although its own orbital period is only 225 days, or $\frac{9}{15}$ths of a year. The apparent effect is, in fact, the same as would take place if the Earth were supposed to stand still in its orbit, and Mercury were to revolve round the Sun in 116 days, Venus in 584, in such paths as are shown in Fig. XXXIX.

It is also evident from the above figure, that a planet, thus revolving round the Sun, in an orbit nearer to it than that of the Earth, has a much longer path to traverse, between being at its furthest apparent distance upon one side and again upon the opposite side of the Sun, when it goes round *beyond* it between the two positions, than when it passes, in the meantime, *between* the Earth and the Sun. The two periods, in fact, last, upon an average, $43\frac{1}{2}$ days and $72\frac{1}{2}$ days respectively, for Mercury; 143 and 441 days respectively for Venus.

The whole interval of 116 days in the case of Mercury, or of 584 days in that of Venus, which elapses between its being thus

Fig. XXXIX.—Comparative positions of Mercury and Venus when seen from the Earth at their greatest elongations from the Sun.

seen twice in succession, in conjunction or opposition, or at extreme eastern or western elongation from the Sun, or in any other given position relatively to it, is called a *synodic* period, which term means that the period depends upon the joint effect of the movements of the planet and of the Earth in their respective paths.

In like manner, it may be deduced from the Copernican theory that the period of Mars in its own orbit being 687 days, it must occupy, upon an average (when certain variations in its rate of movement are allowed for) 780 days between being twice seen from the Earth in conjunction or opposition, or in coming back from any given position to the same position again, relatively to the Sun. For Mars, therefore, the orbital period is 687 days, the synodic period 780 days. For Jupiter, the two periods are respectively 4333 and 399 days; for Saturn, 10759 and 378 days. These figures, according to the Copernican theory, involve an important distinction between the cases of those planets whose orbits are exterior to that of the Earth, and that of the inferior planets Mercury and Venus, viz., that it is the Earth which has to catch them up, instead of their overtaking it. Moreover, it is the fact of the Earth's moving so much more quickly round the Sun than Jupiter or Saturn (or any planet still further away), that enables it to do so in their case, as the last two sets of numbers indicate, in a little more than a year; although it requires more than two years in order to overtake Mars, whose angular velocity is only somewhat more than one-half of its own.

It may also be easily understood, as a consequence of the same theory, that, the orbits of these planets round the Sun being outside that of the Earth, they may be seen, in the course of any such period, at any angular distance from the Sun, between it and the exactly opposite part of the heavens; and that they will not, like Mercury and Venus, simply appear to oscillate within a short distance on opposite sides of their great ruler. It is, moreover, evident that, as they alternately come nearer to and depart further from the Earth, their apparent paths seen from it must, according to the *Copernican theory*, bear a general resemblance to the series of curved loops drawn by Cassini in Fig. XXIV.; the successive portions being described in their respective synodic periods, which also agree with the times assigned for their description by the ancient astronomers as the result of the combination of the cycles and epicycles of the *Ptolemaic theory*.

Thus far, then, the hypothesis of Copernicus is found to agree in general with the chief results of observation. But we have not yet shown how exactly it explains that which we think to be the most peculiar part of all the movements involved in the apparent paths of the planets. We refer to their *stations*, and to their alternately *direct* and *retrograde* motion amongst the stars involved in the description, near to the times when they make their closest approaches to the Earth, of the smaller loops in the apparent orbits shown in Figs. XXIV.—XXVIII. The explanation, although not very obvious, may, we hope, be understood without much difficulty.

In order that we may clearly realize it, let us briefly restate, from the earlier part of this lecture, what it is that we now wish to show that the Copernican theory explains. It is this. If any planet be watched, and its place amongst the stars be noticed night after night,—although, upon the whole, it moves amongst them in a direction opposite to that in which they appear to go round the Earth day by day; or, in other words, although its path is for the most part from west to east; it will be found that, at a certain epoch, before or after the planet passes through one of its nearest approaches to the Earth, it seems to stand nearly or quite still for a while; and that between these two stationary positions it reverses the direction in which it travels amongst the stars, and appears to go from east to west.

In order to indicate how the Copernican theory explains all this, let us, first of all, take the case of a planet nearer to the Sun than the Earth, such as Venus; and let us suppose V_1, V_2, V_3, etc., to represent successive positions of Venus in its orbit; while E_1, E_2, E_3, etc., are corresponding positions of the Earth, which is moving at a slower rate round the Sun. An observer upon the Earth will see Venus amongst the stars, in the prolongation of the line joining him to it at any time, and will refer its place to an imaginary celestial sphere situated at a great distance. We will represent such places, corresponding to the positions E_1 and V_1, E_2 and V_2, E_3 and V_3, etc., respectively of the Earth and Venus, by the small letters v_1, v_2, v_3, etc. We may also notice in passing that, when the Earth is at E_5 and Venus at V_5, the latter is said to be in the position of inferior conjunction with the Sun, so that it might then, for a few hours, be seen in transit across its disc; but that in general the tilt of its orbit will carry the line of view clear of the Sun, so that it will be referred to v_5, and may be altogether invisible for a day or two owing to the glare of the surrounding Solar light. It should

9*

also be observed that in the figure a motion of rotation from west to east round the Sun is supposed to take place in a direction the reverse of that in which the hands of a watch, whose face would be represented by one of the circles, would revolve. This direction of motion is indicated by all the arrows, except those in the highest row, which are drawn to represent motion from east to west.

We hope that a little meditation upon the diagram in Fig. XL. may enable our readers to see that Venus, according to the

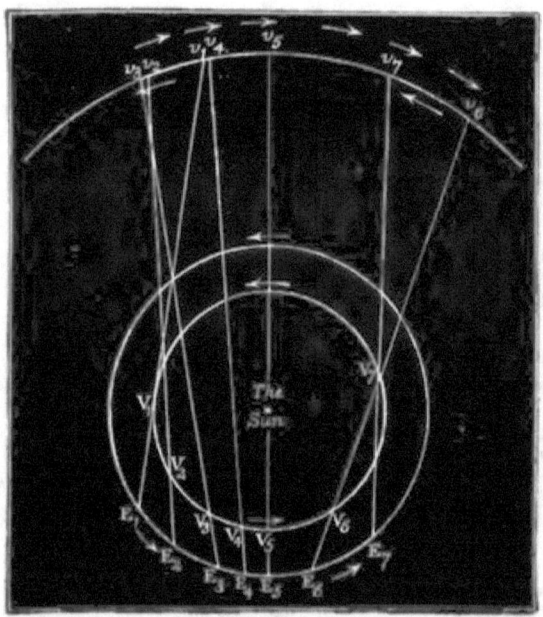

Fig. XL.—Showing how the motions of Venus and of the Earth in their orbits cause the progression and retrogression of Venus in its apparent path.

Copernican theory, would appear to move onwards, or from west to east, amongst the stars, from the position v_1 to about the position v_3, and, again, from somewhere near to v_6 on to v_7; and that it would always appear so to move except between such a position as v_2, and such an one as v_6, during which time it would appear to travel from v_3 to v_4, and then past v_5 on to v_6, in the *reverse* direction. This temporary reversal of its movement, or its retrogression, is therefore very simply accounted for, as a result of the theory in question, by the way in which its more rapid angular motion around the Sun enables it at such a time to overtake and pass by the Earth. It may also be noticed that the result is brought out by the diagram, without any mathematical calculations, simply by joining corresponding places of the Earth and of the planet, and by producing the lines so drawn to meet the imaginary celestial sphere to which our eyes refer the place of the latter.

It follows, then, that as Venus travels between, or nearly between, the positions v_3 and v_6, or for a certain distance on each side of v_5, *one of its nearest approaches to the Earth*, it will appear to *retrograde* in the sky. And we may see that,

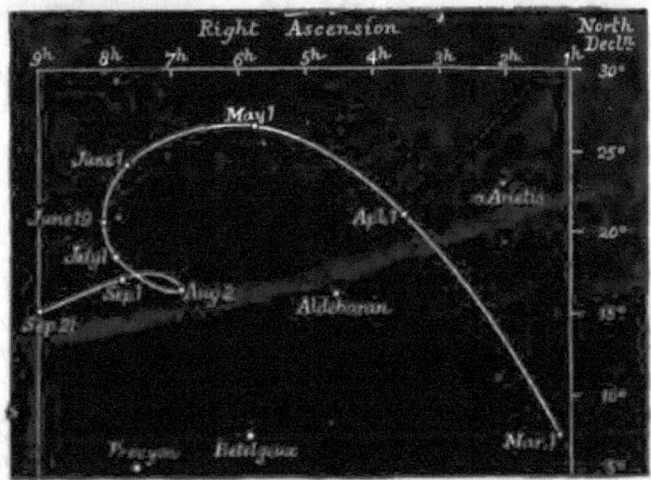

Fig. XLI.—Path of Venus during a part of the year 1884.

at such a time, as observation also proves, its motion, as it gradually overtakes and passes the Earth, will seem to be much slower than that which it will manifest when it is on the opposite side of the Sun, and moving in a direction the very opposite of that of the Earth.

Moreover, it would also result, when the planet reverses its apparent movement from progression to retrogression, and *vice versâ*, that it must make a slight pause, and for a few days move so slowly as to seem to remain very nearly still. This may be further illustrated by a diagram (see **Fig. XL.**) of a part of its apparent path during the year 1884, deduced from the places given in the Nautical Almanac, in which year it will be stationary (in what is termed Right Ascension*) on June 19th, on which date it makes a pause. This is also the case on August 2nd, when its Right Ascension, which is measured horizontally in the diagram, begins to increase again, after having diminished in the interval between the two dates, as the consequence of a retrograde motion, from about 8 hours 1½ minutes to 6 hours 49 minutes.

Without, however, considering any such refinements connected with the orbits of Venus and the Earth, as are involved in the use of the terms right ascension and declination, or are otherwise referred to in our footnote, we may ask, when would the exact moment occur, when their respective movements round the Sun shown in our previous diagrams (which are really only those which they possess in celestial longitude)

---

* Hitherto we have spoken of the stationary points of the planets without any regard to the tilt of their orbits to the plane of the ecliptic on which the Earth moves. In fact, we have only considered the way in which its motion round the Sun, and their motions, in what is termed Celestial Longitude, measured upon the plane of the ecliptic, cause them, at times, to appear to stand still. But besides this, the tilt of their paths will take them alternately to the north or south of that plane, and move them, even when they are stationary in longitude, to a certain extent (corresponding to the amount of that tilt) in what is termed North and South Celestial Latitude. It is, however, customary in the ordinary estimation of celestial positions, to determine the places of the heavenly bodies, not by their longitude on the plane of the ecliptic, and their latitude above or below it, but by their right ascension, measured on the celestial equator, and by their distance north and south of it, which is called their declination. And the planet is *usually* spoken of as *stationary* when it is so in *Right Ascension*, although its declination may still be undergoing some change; partly as an effect of the tilt of its orbit, and partly because it really is not stationary in longitude at the same dates as in right ascension, and its movement in longitude consequently still slightly alters its declination. Any apparent discrepancy in the form of the curve, or loop, in Fig. XLI., and of those in some of our previous diagrams, may be thus explained. Nevertheless Fig. XLI. sufficiently indicates how a pause must occur, when progression changes to retrogression, and *vice versâ*. We may also remark that right ascension is not simply measured (like longitude) in degrees, but is generally estimated (as in our figure) in hours, minutes, etc.; twenty-four hours being reckoned as equivalent to 360°, in agreement with the way in which the whole heavens appear to urn once round the pole, day by day, in twenty-four sidereal hours.

would cause Venus to appear to stand still? Our answer is that it will do so when the extra curvature of the smaller orbit of Venus so exactly counterbalances its extra speed round the Sun, that the straight lines joining the Earth and Venus remain for a short time parallel. Considerable calculation is required to work out the question accurately; but we may state that the dates are found to be, in the case of Venus, about twenty-one days before, and after, it is in such a position as E_5 in Fig. XL. In like manner they occur for Mercury (upon an average) about twenty-two days before and after it occupies a similar position, the limit, however, in its case, being subject to considerable irregularity.

Fig. XLII.—Venus will appear to be stationary when the line joining it to the Earth remains for a short time parallel to itself.

Our next diagram, Fig. XLII., illustrates this point, and shows that, at such a time, while Venus, in the course of a day or two, goes on from v_1 to v_2, and the Earth from E_1 to E_2, the length of the arc $v_1 v_2$ is decidedly greater than that of $E_1 E_2$; and yet the difference in the curvature of the two paths causes $E_1 v_1$ to be parallel to $E_2 v_2$; so that an observer upon the Earth, being unconscious of his own movement, and one day seeing Venus in the direction $E_1 v_1$, and the next day (or a few hours afterwards) in the *parallel* direction $E_2 v_2$, considers these two directions to be *the same*, because the supposed imaginary sphere of the heavens, to which he refers the place of Venus, is so far distant that he can detect no change in the planet's position. It therefore, for the time being, appears to remain stationary.

It follows from the above explanation, by parity of reasoning, that, at the very same time, the Earth would also seem to be stationary to an observer upon Venus. And, therefore, if in Fig. XL. v be supposed to represent the Earth, and e to be a planet, having its orbit outside that of the Earth, it is clear that such a planet would appear to stand still at a certain interval before, and after, the Earth would pass through its nearest position to it, between it and the Sun. It may thus be seen that Mars, Jupiter, Saturn, Uranus, and Neptune (equally with the inferior planets, Mercury and Venus) must all have their stationary points in their apparent paths, as seen from the Earth, on each side of their periodical nearest approaches to it. They must also, of necessity, between two such stationary positions, appear to move amongst the stars from east to west, instead of pursuing the usual direction of their movement, viz., from west to east.

We have therefore shown that the simple, but most beautiful, theory of Copernicus explains, in every respect, all the apparent complication of the movements which we observe. In its present exact and perfected form it teaches us how a steady regular movement of the planets, each in its own orbit round the central Sun, causes them to appear to us sometimes to stand still; sometimes to move quickly; sometimes slowly; sometimes to reverse their motion. It tells us that the interesting succession of the phases of Mercury and Venus is a necessary consequence of the situation of their paths; as also that the days of their greatest elongations from the Sun will not coincide with those upon which they are apparently stationary. It proves that the intervals between the consecutive nearest approaches of any planet to the Earth must exceed that in which it travels round its orbit, according to a rate which is altogether different for an inferior planet, from that which applies to one whose orbit is exterior to that of the Earth. It does away with all the confusion of cycle and eccentric, and of epicycle piled upon epicycle; and gives in its place a system perfect in its application to every observed movement, and in its power to satisfy every advance that has yet been made in theoretical astronomy ; while the old system of Ptolemy, although exceedingly clever and ingenious, and apparently able to account for most of the facts known while it held its sway, was not only fearfully complicated, but must, if it had not been previously exploded, have utterly failed to agree with observation, when the telescope revealed the phases of the inner planets, and the general improvement of astronomical instruments enabled the movements, both of those within and beyond the Earth's orbit, to be watched and recorded with increased accuracy.

We can hardly wonder that Alphonso the Wise, tenth King of Castile, is said to have remarked, more than six hundred years ago, that if he had been consulted at the creation he could have devised a better arrangement than the Ptolemaic system. But we should be greatly surprised if any of our readers should still be so far in doubt as to which is the true theory, as with any lingering uncertainty to exclaim :—

> "What if the Sun
> Be centre to the world, and other stars
> By his attractive virtue, and their own
> Incited, dance about him various rounds ?"

LECTURE VI.

THE PLANET MERCURY.

"First Mercury completes his transient year,
Glowing refulgent with reflected glare."—*Chatterton*.

"Neque a Sole longius unquam unius signi intervallo discedit, tum antevertens, tum subsequens."—*Cicero*.

In passing on to discuss the various planets individually, it will, we think, be most convenient to take them in order according to their distances from the Sun, beginning with the nearest. In attempting to do so, we must, however, at once face the question:—Do we know which is the nearest? We have already mentioned that the proximity of Mercury to the Sun makes it often impossible to observe it; may we not then ask, whether, in spite of all our recent progress, in spite of all the advance that has been made since the early days of the Chaldæan astronomy, there may not be planets between the Sun and Mercury which we have as yet failed to detect? We notice in Bode's so-called law, that the first term, which is taken to represent the distance of Mercury, is an empirical one; and for this, as well as for other reasons, we feel that we have no good grounds for affirming that there is any *à priori* improbability against the existence of such planets.

On the other hand, Le Verrier, than whom no greater authority could be quoted, has made elaborate calculations of certain peculiarities in the movement of Mercury, which indicate that a very considerable amount of matter lying between it and the Sun must constantly exercise an attractive effect upon that planet. He has said that a body of about $\frac{2}{3}$rds of the weight of Mercury, revolving at one-half of its distance from the Sun, would produce the required effect; or, that a larger amount of matter, in a more scattered form, or divided into a number of small planets, and situated partly, or wholly, still nearer to the Sun, would suffice.

It may however be asked,—May not the matter which produces the Zodiacal Light, and which (as stated in Lecture II.) extends so far beyond the orbit of Mercury as probably to reach to that of the Earth, produce the effect of which we are speaking, and so relieve us from the supposition of the existence of any additional planet or planets, especially as such matter is doubtless much denser in the near neighbourhood of the Sun than further from it? But we may answer that it would not suffice to do so, because it does not, so far as we know, fulfil another requisite condition, viz., that it should be symmetrically situated with regard to the plane of the orbit of Mercury. The *à priori* probability in favour of the existence of one or more intra-Mercurial planets consequently remains.

But, if any such planet were of a size approaching to that of Mercury itself, it seems very unlikely that it would not have often been seen. It would no doubt be so much the harder to detect it, as it might be nearer to the Sun; but the greater brightness which that proximity would confer would most likely counter-balance, or more than counter-balance, the effects of the glare through which we should view it. If, on the other hand, there be several such planets, but of much smaller size, we can easily understand that it would not only be difficult to see them in a telescope, even if their places were accurately known, but very improbable that they would be discovered by any but the most prolonged and persevering search.

In fact there are only two special occasions upon the occurrence of which we could hope to make such a discovery. It would either be when such a planet is seen in transit across the Sun's disc, or while the Sun's light is obscured by a total Solar eclipse.

We must allow that a considerable number of instances have been recorded of dark, and apparently circular, spots having been observed to pass comparatively rapidly across the Sun's disc, so as to present the appearance of the transit of an intra-Mercurial planet. At the same time, in order to make such an observation trustworthy, it is to be remembered that not only is it necessary that the spot seen be perfectly round, uniformly dark, and without any contrast between the centre and the periphery such as occurs in the penumbra surrounding the umbra of an ordinary sun-spot; but its speed of movement must be accurately determined.

It does not suffice to say that a certain appearance was seen one day and not seen a day or two afterwards, cloudy weather having intervened, inasmuch as sun-spots undoubtedly, at times, appear and disappear with most surprising rapidity. Nor must it be forgotten, that, as the Sun crosses the sky in its diurnal path, it does not describe that path round the zenith (or that point of the sky which is vertically above an observer) as a centre, but round the pole of the heavens; and that, consequently, the point on the edge of its disc which is nearest to the horizon does not remain the same during the day. An observer who is unacquainted with this fact, may therefore fancy that a sun-spot, situated near to the edge of the disc, has, in a few hours, moved upon it to some considerable extent; whereas the effect is only a result of the slight apparent rotation of the Sun's disc, produced in the way above explained, round a perpendicular to the plane of the horizon, for which, in all such cases, due allowance must be made.

Another fact also deserves very careful attention. Just as it was explained (in Lecture IV.) that Solar and Lunar eclipses must take place near to two opposite positions of the Earth in its annual orbit, and (in Lecture V.) that transits of Mercury only occur in May or November, and those of Venus only in June or December; so, unless the inclination of the orbit of an intra-Mercurial planet to the ecliptic be slight, which would render its transits more frequent, there will be two, and only two, opposite periods of the year at which they can occur.

One planet, for instance, might transit in March and October; another in December and June; and so on. Now, even if we make all reasonable allowance of every kind, we find (as we have shown in a lecture delivered at GRESHAM COLLEGE, in February 1879, and shortly afterwards published for private circulation) that the dates at which the supposed observations of which we are speaking have been made, would require the existence of at least four such planets. In the lecture in question, we have referred to nearly all the more important of such records, including one which is specially interesting to citizens of London, as it was made in the year 1847, by Mr. Benjamin Scott, now the Chamberlain of the City, whose observation was confirmed by a simultaneous and perfectly independent one by the well-known optician Mr. Wray.

The most celebrated instance of all is, however, that of Lescarbault, of the date of March 26th, 1859, in which the great Le Verrier seems to have put much faith. But Lescarbault's supposed planet has never been seen again, although often very carefully looked for, in the months of March and September, and especially in years when the calculations of Le Verrier, made from Lescarbault's original statement, show, that, if it really existed, it ought certainly to have appeared in transit.

Upon the whole, therefore, we can only say that it is very remarkable, that, if any of such observations have not been delusive, no recurrence of the transits has been detected. On the other hand, it is also surprising that so many observers should have been deceived, although it must be confessed that they have not all enjoyed the advantage of long and practised experience in observing. It is moreover certain that some apparently well-authenticated cases have been subsequently proved to have been erroneous.*

The question of the existence of intra-Mercurial planets is therefore still an open one, and of great interest. It is essential that, for some time to come, especial arrangements be made for the very careful examination of the sky, in the immediate neighbourhood of the Sun, during total eclipses. If any such planets exist, this appears to be the most likely method for their successful detection.

In connection with this remark, we may mention,† that

---

* See an article by Professor Peters in *Astronomische Nachrichten*, No. 2254; also one by M. Tisserand in the *Annuaire du Bureau des Longitudes*, 1882.

† See *Nature*, vol. xviii., p. 663.

during the total eclipse of 1869, four observers at St. Paul's Junction in Iowa (one a lady) saw *with the naked eye* what they termed " a little brilliant " at a distance about equal to the Moon's diameter from the Sun's limb; and a Mr. Vincent, with a small telescope, saw a small crescent-shaped object about three times as far from the Sun. Dr. Gould of Cordoba saw what he supposed to have been the star π^2 Cancri, but which, from its brightness, may have been the object seen by the four above-mentioned observers.

Again, in 1878, Professors Watson and Swift, in America, each announced the discovery of two intra-Mercurial planets during the total eclipse which occurred on July 29th of that year. Their observations were, however, so hurried, and the contiguity of certain stars (θ and ζ Cancri) to the supposed places of the planets so close, that further confirmation of their discoveries is desirable. Professor Watson, however, seems to have distinctly stated that one of the planets seen by him was *between* the Sun and the star θ Cancri. It is deeply to be regretted that his recent premature death has interfered with his intention to search for these planets, not only during eclipses, but, if possible, by making special arrangements for scrutinizing the neighbourhood of the Sun at other times. It will be very interesting, should one (or more), at some future time, be discovered and its orbit be determined, to compute its past movements, and to see whether they agree with any of the supposed observations which we have described.

Let us, however, now pass on to the planet MERCURY, of whose existence there is no doubt, while its proximity to the Sun is very remarkable, even if others be nearer still. It is, at any rate, so near, that we are able to discover much less than we could wish with regard to it; so near, that, if others lie between it and the Sun, we can never hope to know much more about them than their existence, their orbits, and their size.

In about three-quarters of an hour less than 88 days Mercury performs the circuit of its course, at a mean distance of very nearly 36,000,000 miles from the Sun, and with an average speed of about 107,000 miles per hour. Its diameter is believed slightly to exceed 3,000 miles ; and is therefore not quite half as large again as that of the Moon ; between 400 and 500 miles less than that of the largest of the satellites of Jupiter ; and probably not above ¾ths of that of Titan, the largest belonging to Saturn.

We need not again describe the succession of this planet's phases ; its apparent oscillation from one side of the Sun to the opposite ; its return to corresponding positions relatively to the Sun, as seen from the Earth, at a mean interval of 116 days ; and its alternate onward and retrograde motion amongst the stars ; inasmuch as we have sufficiently explained all these matters in treating of the general character of the planetary orbits and movements, in our fifth lecture. We will, however, repeat in a somewhat different form our previous statement as to the shape, or ovalness, of the orbit of Mercury, since it may the better help us to realize how decidedly excentrical the position of the Sun is with regard to it.

So far is the Sun from being in the centre of this orbit, that, if we represent its longer diameter by ten inches, the Sun's place will be one inch from its middle point; *i.e.*, four inches from one end and six inches from the other. To correspond with the actual scale of the orbit we must suppose that one inch represents about 7,000,000 miles ; 6 times, and 4 times which, respectively give about 42,000,000, and 28,000,000 miles for Mercury's greatest and least distances from the Sun ; still more exact values being about 43,250,000 and 29,500,000 miles.

According to Kepler's 2nd Law (see Lecture V.), it follows that the speed with which Mercury moves in its orbit must be very much greater when it is at its nearest to the Sun, than when it is so much farther away. In fact, it varies from about 35 miles to about 23 miles per second.

The most ancient observations of this planet that have been recorded are said to have been made in the years B.C. 265 and A.D. 118.[*] From that time onwards, it was doubtless carefully looked for, but until the invention of the telescope it was only possible to see it on the occurrence of those by no means

---

[*] Chambers' "Astronomy," p. 61.

very frequent occasions when it was particularly well situated for observation.

Those who dwelt in regions within, or near to, the tropics, were doubtless assisted in detecting it *by the shortness of the twilight*, as well as by the clearness of the air. On the other hand, we read that Copernicus, although he frequently made the attempt, never succeeded. He was, perhaps, especially hindered by the mists and fogs of the Vistula; but it is, in any case, only possible to do so with the naked eye, in such a locality as that of Cracow, where he dwelt (which is in nearly the same latitude and has consequently nearly the same duration of twilight as London), when the planet is in an especially favourable position, so that its setting or rising is separated by as long an interval as possible from that of the Sun.* It is rarely that so propitious an opportunity occurs as that which happened at the end of April, 1877, when, as may be seen by reference to such an Almanac as Whitaker's, the Sun set at 7.19 p.m., and the planet at 9.30 p.m. It was also particularly well situated in May 1882, when its close proximity to Venus was most interesting.

Notwithstanding the strong twilight and moonlight, the writer, in company with various other persons, saw it shining brilliantly night after night during the fortnight following May 22nd, 1882. When the sun set at about 8 p.m. it was clearly visible about 8.45 or 8.50 p.m. Owing to the sky being fortunately free from clouds, he thus watched it with the naked eye on seven out of nine consecutive nights. The position of Venus, which no one could fail to see, was no doubt of great help in guiding the eye to the smaller planet, but the brilliancy of Mercury was certainly surprising; while the flashing and scintillating appearance which it pre-

---

* We do not here discuss the variation in the amount of light received from Mercury caused by the extent of its illuminated phase, or its distance from the Earth at any given time. In our next lecture it will be seen that such a discussion is important in the case of Venus. If, however, Mercury is to be seen at all by the naked eye, it must be at such an apparent distance from the Sun, and at such an actual distance from the Earth, that its phase will always be nearly the same as that of the Moon when half-full.

sented recalled the appellation—ὁ Στίλβων, *the glittering*, or *sparkling one*, given to it by the ancient Greeks.

The telescope has, however, revealed its phases, and has enabled us to detect and watch it when in transit. But, otherwise, it has taught us very little of importance, or of interest, with regard to it. Mercury is a disappointing and difficult telescopic object. It is hard to get a well-defined and distinct view of its disc. Its very considerable distance from the Earth, when otherwise most favourably placed for observation; its comparatively small size; the intense brightness of its light; the solar glare that surrounds it; and some inherent peculiarity of its surface, or of its atmospheric envelope;—all combine to render our telescopic observations disappointing.

It is true that, near to the end of the last century, Schröter believed that a high mountain was periodically visible as an inequality in the boundary of the illuminated portion of its disc; and that he saw a belt probably equatoreal, and indicating an axis tilted at about 70°* to the plane of its orbit (an inclination very similar to that of the Earth's axis), during a period of 47 days in May and June 1801. But we believe that the careful observations of Sir W. Herschel did not confirm these statements; while most modern astronomers have felt it useless to devote much time to this planet. Indeed, we are almost forced to suppose that if Schröter was not deceived as to what he saw (a remark which may, in some degree, also apply to his, and some other, observations of Venus made a generation or two ago) the atmosphere, or cloudy envelope of Mercury, must

---

* We have tried in vain to meet with a copy of Schröter's " Hermographische Fragmente " in several English libraries. By the kindness of M. Niesten, of the Brussels Observatory, in which the book is to be found, we have, however, obtained a copy of Schröter's original statement, and we are surprised to notice that in almost every English book (except in Dr. Ball's recent quotation of Houzeau's astronomical constants in his " Elements of Astronomy," published in the Science Text-Book Series) the complement of Schröter's angle of about 70°,—*i.e.*, an angle of about 20°,—is given in place of the inclination of the axis really stated by him. Very erroneous deductions have consequently been made as to the seasons and climatic zones which Schröter's value would involve. They would really be, according to him, in close agreement with those of the Earth.

have, of late, undergone some change, which makes it a much more difficult telescopic object than it formerly was.

With regard, however, to Mercury, it should at any rate be noticed, that, quite apart from any changes of season caused by the inclination of its axis, its varying distance from the Sun must produce a change of temperature, which cannot but be most important. It is believed by some, who have carefully investigated the question, that the fact that the Earth's distance from the Sun varies by 3,000,000 miles in the course of each year, has a decided influence upon the comparative intensity of the seasons of its northern and southern hemispheres.

If so, let us try to conceive what would be the effect upon the light and heat received, if, instead of the variation in distance being 3,000,000 miles, it were to amount to 36,000,000 miles; in other words, if the Earth's distance from the Sun were sometimes reduced to 75,000,000 miles, and at other times increased to 111,000,000 miles; which distances are very nearly in the same ratio as the greatest and the least of the orbit of Mercury.

Let us imagine how the inhabitants of our globe would feel, if the Sun's apparent disc and the light and heat received from it were at one time more than twice as great as at another. It is true that in some northern latitudes, in the present temperate zone, something like a perpetual summer might ensue, but it is also probable that a large part of the southern hemisphere would be rendered uninhabitable by the intensity of the winter cold and of the summer heat; while, in the tropics, the difference of temperature involved in the alternation of the seasons would also be most unpleasantly great.

In the case of Mercury it is easy to calculate that the actual variation in the Solar light and heat received would be fully as great as we have just supposed. The apparent diameter of the Sun would, at different times, be just so much greater as the planet might be nearer to it, and *vice versâ*; and we have already stated that its distance is sometimes more than half as great again as at others. The Sun's apparent diameter, when greatest, would therefore exceed its least value rather more than in the ratio of three to two. Its corresponding area would vary as the squares of these numbers, or in the ratio of nine to four.

In other words, the disc of the Sun, when Mercury is at its nearest distance from it (or in *Perihelion*), would appear to be considerably more than twice as large as when it is at its farthest distance (or in *Aphelion*); and the same proportion would hold with regard to the respective amounts of light and heat received. This is illustrated in Fig. XLIII., in connection with which it is also to be observed that the corresponding increase and decrease of temperature would occur every forty-

Fig. XLIII.—Comparative sizes of the apparent disc of the Sun, as seen from Mercury, when at its least and greatest distances.

four days, the periodic time of Mercury round the Sun being eighty-eight days.

Moreover, as may be seen in Fig. XLIV., the apparent mean, or average, size of the Sun as seen from Mercury is more than $2\frac{1}{2}$ times as large in diameter, and about $6\frac{3}{5}$ times as large in area, as it is when seen from the Earth, and twice as large in area as when seen from Venus. At intervals of forty-four days the light and heat received from it by Mercury must consequently vary from about four times to about nine times that which the Earth receives. We may therefore well imagine, if there be any inhabitants upon that Sun-stricken orb, that the terms *Perihelion* and *Aphelion* are not considered by them, as they probably are by some of our readers, to involve ideas of a rather intricate or recondite nature, but that every child in the humblest village school upon Mercury (if such schools exist), as soon as it can appreciate the simplest teaching, must learn, every six weeks, by the temperature at which its own Mercurial feelings stand, whether it is in the one position or the other.

It is, however, well-nigh impossible to suppose that Mercury can be the abode of beings in any important respect resembling ourselves, the difficulty involved depending, as we opine, rather upon the effect of the Sun's *heat*, than of its *light*. Its light, we may imagine, might be moderated by eyes possessing a very slight sensitiveness to its influence; but it is not in any way easy to conceive how the intensity of its heat could be

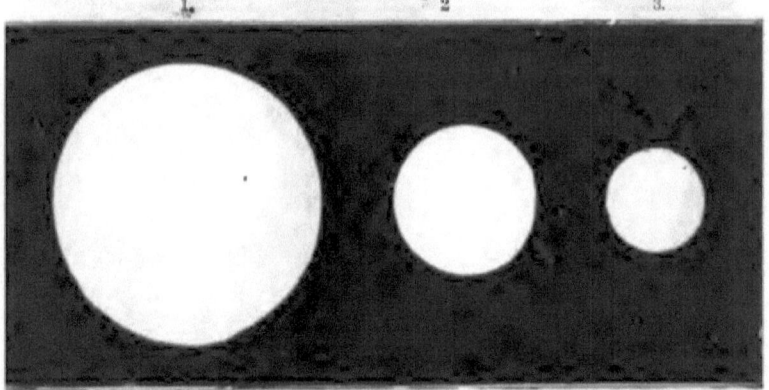

Fig. XLIV.—Comparative apparent areas of the disc of the Sun as seen (1) From Mercury, (2) From Venus, (3) From the Earth, when at their mean distances respectively from it.

endured, unless it were in some way diminished by the interposition of a dense and continuous canopy of cloud overshadowing the planet's sky.

Observation, however, does not seem to indicate that such is the case. The surface of Mercury seems, it is true, to be very much, if not entirely, hidden from us, but the brightness of its light is by no means so great as a mass of cloud might be expected to reflect. When the Sun is suitably situated we occasionally notice what an astonishing amount of light a small portion of cloud, in our own skies, reflects. From a mountain summit, or from the car of a balloon, the same effect may often be seen, upon a much larger scale, when the Sun is shining upon masses of clouds below the observer. But the amount of light received from Mercury, when its size and its nearness to the Sun are duly taken into account, is much less than would be the case if its surface (area for area) reflected even as much of the light received as does that of the planet Jupiter. This has been carefully ascertained at times when the planets have been so situated that the comparison could be easily and accurately made. The brilliancy of Mercury is also proportionally much less than that of Venus.*

But even if the existence of such a cloud canopy could be proved, it would still be hard to believe that it could make the planet an endurable, much less a pleasant place of residence.

Upon the other hand, if its skies were comparatively cloudless, so as to allow a large proportion of the Solar light and heat to pass, no imaginary supposition, either as to the density or the rarity of its atmosphere, would overcome the difficulty. If its atmosphere were dense, and contained a considerable amount of vapour, so as to render it less pervious to the passage of the heat incident upon it, it would, according to well-known laws, retain what it allowed to pass, and gradually store it up to an unbearable extent. If, being much rarer, it permitted the heat received to escape more easily, as is the case with the rarefied air in mountainous regions upon the Earth, the direct effect of the Sun's rays would be all the more terrific, as is often, in some small degree, found to be the case in noon-tide hours among Alpine snows and glaciers. This is of course a rough way of stating the dilemma, but it may suffice to indicate, that an atmosphere which would allow heat to escape would also allow it, as it came, to strike down unmercifully;

* In a letter to the *Times*, dated 28th September, 1878, Mr. Nasmyth states that he saw the two planets together in the same field of view of his telescope for some hours, and that, while Venus looked like clean silver, Mercury appeared more like lead or zinc. Yet Mercury was at the time very close to its Perihelion, as Mr. Proctor remarks in a subsequent letter to the same journal, dated October 1st, 1878. In his "Other Worlds," p. 68, Mr. Proctor also shows that, if Mercury had the same reflecting power as Jupiter, it ought to appear, when in Perihelion, and at its greatest elongation from the Sun, twice as brilliant to the naked eye as Jupiter does when at its nearest to the Earth.

while such an one as would hinder its direct effect would retain so much that the result would be equally painful.

We have already briefly alluded (in Lecture V.) to the occasional transits of Mercury across the Sun's disc; the possibility of their occurrence only in the months of May and November; and their dependence upon the tilt, or inclination of the plane of the planet's orbit, to that of the orbit of the Earth. It may here be interesting to mention that the requisite conditions for a transit are only fulfilled at intervals, which sometimes consist of seven, but more often of thirteen years, in the case of two successive November transits; while the May transits are less frequent, and occur, in general, at intervals of thirteen years, or of twice, or even three times thirteen, plus seven years; *i.e.*, after the lapse of thirty-three; or forty-six years.

Some slight irregularities may be noticed in the following table, extending from A.D. 1600 to A.D. 2000, extracted from the *Annuaire* of the Royal Observatory of Brussels for the year 1881, which we quote, with the hope that it may interest our readers; *e.g.*, there is an interval of only six, instead of seven years, between the transits of November 1776 and November 1782; while the unusually long interval of forty-six years occurs between the May transits of 1661 and 1707.

PASSAGES OF MERCURY ACROSS THE SUN.

(The asterisks denote that the Transits have been observed.)

1605.	1 November.	*1743.	5 November.	*1868.	5 November
1615.	3 May.	*1753.	6 May.	*1878.	6 May.
1618.	4 November.	*1756.	6 November.	*1881.	8 November
1628.	5 May.	*1769.	9 November.	1891.	9 May.
*1631.	7 November.	1776.	2 November.	1894.	10 November.
1644.	8 November.	*1782.	12 November.	1907.	12 November.
*1651.	2 November.	*1786.	4 May.	1914.	6 November.
*1661.	3 May.	*1789.	5 November.	1924.	7 May.
1664.	4 November.	*1799.	7 May.	1927.	8 November.
*1677.	7 November.	*1802.	9 November.	1937.	10 May
*1690.	10 November.	1815.	12 November.	1940.	12 November.
*1697.	3 November.	*1822.	5 November.	1953.	13 November.
*1707.	6 May.	*1832.	5 May.	1960.	6 November.
1710.	6 November.	*1835.	7 November.	1970.	9 May
*1723.	9 November.	*1845.	8 May.	1973.	9 November.
*1736.	11 November.	*1848.	9 November.	1986.	12 November.
*1740.	2 May.	*1861.	12 November.	1999.	24 November.

It also deserves notice that the observation of only five of the transits which have occurred since 1631 has been missed. That of 1881 was carefully watched in Australia, but was not visible in England; nor will those of 1891 and 1894 be well seen here, the former ending very soon after sunrise, and the latter beginning very near to the time of sunset. The writer of this lecture was especially fortunate in having a very good view of that of May 6th, 1878, although in most parts of England clouds were exceedingly troublesome, or altogether obscured it.

Apart, however, from the interest attaching to the transits of Mercury in connection with the computation of the dates of their occurrence and their dependence upon the comparative dimensions and position of its orbit and of that of the Earth; a great additional importance belongs to them, inasmuch as they may afford special facilities for observations of the physical characteristics of the planet. It is, for instance, very interesting to find that, at such times, a considerable number of observers have noticed indications of something like an atmosphere surrounding the planet. An appearance resembling a luminous ring was seen around it in 1736, and again in 1791; in 1832 a similar ring, somewhat tinted with violet; in 1868, Dr. Huggins, than whom no more careful and accurate observer exists, saw an aureola of light* around Mercury, of a breadth equal to about $\frac{1}{3}$rd of its diameter. He also noticed a luminous point of light not far from the centre of its disc. In the diagram (Fig. XLV., taken from Chambers's "Handbook of Descriptive Astronomy"), which is copied from that of Dr. Huggins, these appearances are shown, but are not repeated in the figure of the planet as it is seen considerably distorted in shape passing off the edge of the Sun.

Such an annulus as is here drawn was also noticed by the present Astronomer Royal (Mr. Christie) and by Mr. Dunkin, at the Royal Observatory, in the transit of May 1878; Mr. Christie perceiving an inner and narrower ring of greater brightness very close to the planet. Mr. Criswick also observed the halo with a refracting telescope of six inches aperture.

* See R. A. S. Monthly Notices, vol. xxix., p. 25.

To him it appeared wider than in the larger instruments. I find that in my own notes of the transit which I observed (as I have already mentioned) under very favourable circumstances in Suffolk, with a refractor of very nearly the same aperture as that which Mr. Criswick used, I have stated the apparent width of the halo at the same value as he gives for it, viz., about ⅔rds of the diameter of the planet's disc. An assistant who was with me also saw it. It was not, however, noticed when the observatory was darkened, and the image of the Sun was projected upon a white paper screen.

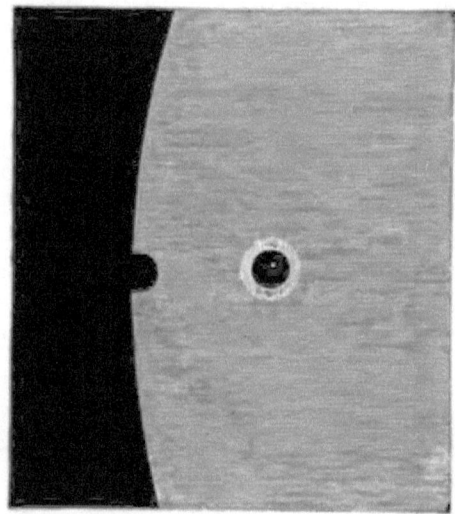

Fig. XLV.—The planet Mercury in transit, November 5th, 1868.

Some observers have described this ring as appearing darker than the neighbouring surface of the Sun, although more have seen it brighter. But it must be confessed that the majority (as may be seen by reference to an article in the *Annuaire* of the Observatory of Brussels for 1881) did not notice it in 1878; nor did they observe a bright spot which has been frequently observed upon the planet's surface at such times by others besides Dr. Huggins; and which was also seen by myself in 1878 a little to the south-west of its centre.*

* On the other hand, we find that in the recent transit of November 1881, Mr. Tebbutt, at Windsor, N. S. Wales, saw a faint whitish spot in

We are, therefore, almost forced to conclude that such appearances are in some way produced by the effect of the contrast between the great brightness of the Solar surface and the tiny black disc of Mercury, or are related to the diffraction of light, or to slight imperfections in the telescope or in the adjustment of the focus of its eye-piece, which are especially evident when a small dark circular object is viewed against a very strongly illuminated background. It is, however, very desirable that all such peculiarities should be carefully recorded and watched in future. There may be more in them than we realize at present, or it may possibly be found that Mercury possesses a gaseous, or cloudy envelope, susceptible of considerable changes from time to time. It will be specially important to notice whether any ring of light, such as an atmosphere might cause, can be detected just as Mercury is entering, or is about to enter, upon the disc of the Sun, or just as it is leaving it. At such a time the dark body of the planet may undoubtedly be seen projected upon the Sun's corona at least two minutes before the first contact of the edges of the discs occurs.*

After all, however, whether the planet be observed by the unassisted eye when at its greatest elongation from the Sun; or whether we watch, by means of a telescope, the waxing and waning of its phases; or study it upon those rarer occasions when it appears as a dark round spot in transit between the Earth and the Sun; we regret that we can discover but little of interest connected with it; in fact, we fail to detect even such indications of its physical condition as the observers of a century or more ago have recorded.

We believe that it has been stated that some indications of watery vapour have been noticed when the spectroscope has been applied to its light. But its situation is generally most unfavourable for such observations; since it must be seen, either in the midst of the overpowering glare of the Sun's rays, if

the centre of the planet's disc; and Dr. Little, at Shanghai, a darkish halo round Mercury, but no bright spot upon the disc. See R. A. S. Monthly Notices, January 1882.

* See the observation at Orwell Park Observatory, in the R. A. S. Monthly Notices, vol. xxxviii, p. 414.

the Sun be not set; or else low down in the mists of the horizon, and through an immense extent of the Earth's atmosphere. We are not surprised, therefore, that even the spectroscope, which tells us so much about most of the heavenly bodies, and, in revealing to us the secrets of their condition, finds distance no obstacle so long as their light can be duly examined, fails to give us any very certain indications with regard to Mercury.

We must then be content, in spite of every effort that we may make, to know but little of this planet. At the same time we have endeavoured to show that we know quite enough to make us very strongly doubt its habitability. If we choose to speculate upon its future, we may perhaps allow ourselves to imagine that it is waiting to be the abode of life, until the Sun be so far cooled that its proximity to it will be just such as to secure for its inhabitants a comfortable temperature; by which time it may be that the vapours which now boil and seethe around the globe of Mercury may have become condensed so as to form useful constituents of its soil, or be changed into seas and oceans, rivers and lakes, some of which may then be detected through its clearer atmosphere as dark spots similar to those which are now seen upon the disc of Mars.

This however assumes that, under such circumstances, there would be astronomers upon the Earth still unfrozen, and ready to observe them,—which is hardly probable. For, if it be true, as some suppose, that the Sun is really gradually cooling, it is also most likely, by the time that Mercury would be habitable, that the Earth would be as dead and desolate as the Moon already is. If we are to believe in the habitability of other planets, in addition to that of the Earth, it is therefore most reasonable to suppose that each has its turn as the abode of rational beings, rather than that more than one or two are habitable at the same time.

Of this we may be tolerably certain, that we should at present be very much too warm upon Mercury, although we have, of late years, often wished to feel somewhat more of the Sun's genial rays upon the Earth.

LECTURE VII.

THE PLANET VENUS.

"The star that bids the shepherd fold,
Now the top of heaven doth hold :
And the gilded car of day
His glowing axle doth allay
In the steep Atlantic stream."—MILTON.

WE must now wing our flight to a planet much larger, much more brilliant than Mercury; one that from time to time approaches nearer to the Earth than any other; but one whose very beauty is so dazzling that, bold as our eyes may be, we are forced to turn from its contemplation with baffled gaze.

We refer to Venus, admired, we may feel sure, ever since the Earth has been inhabited, when seen as the morning or evening star; but, in these days, when the habit of late rising is, we fear, becoming increasingly common, more often as the latter than as the former.

The size of Venus is nearly the same as that of the Earth; its diameter only falling short of the Earth's by about $\frac{1}{30}$th part; its surface by about $\frac{1}{15}$th part; its volume by between $\frac{1}{10}$th and $\frac{1}{11}$th part. But its orbit and the speed of its movement round the Sun differ much more considerably from those of the Earth. We will therefore repeat, in a somewhat varied form, so much of the table, which we gave in Lecture V., as may enable us to contrast its orbital motion with those both of Mercury and of the Earth.

	Days.			Miles.		
MERCURY	has a year of nearly 88	a mean distance from the sun of about	36,000,000	a mean velocity	of nearly 30 miles per sec.
VENUS	,, 225	,,	67,000,000	,,	,, of between 21 & 22 ,,
THE EARTH	,, 365	,,	93,000,000	,,	,, 18 & 19 ,,

The only point which calls for special remark in the above figures is, that the greater each planet's distance is from the

Sun, the longer is its year, and the slower its velocity of movement; while a little arithmetical calculation would show that the numbers stated above agree very satisfactorily with Kepler's 3rd Law, which was explained in the same Lecture.

There is, however, one peculiarity connected with the orbit of Venus, which distinguishes it from every other planetary orbit; viz., that it is more nearly circular than any other. We have stated, in Lecture VI., that the fraction $\frac{1}{5}$th represents the ovalness of Mercury's orbit, because the Sun is situated at one-fifth of the distance between the centre and one end of the major axis of the ellipse which that planet describes. For the ellipse in which the Earth rotates around the Sun the corresponding fraction is $\frac{1}{60}$th, which indicates an orbit much more closely approximating to a circular shape. But for that of Venus it is only about $\frac{1}{150}$th; as a consequence of which it follows, that the distance of Venus from the Sun never exceeds or falls short of its mean value of about sixty-seven millions of miles by so much as half a million. The whole change in its distance is therefore less than one million of miles; while Mercury's distance from the Sun varies from twenty-nine to forty-three millions of miles;—or, in other words, changes by as much as fourteen millions of miles in the course of its year;—and that of the Earth varies from ninety-one and a half to ninety-four and a half millions of miles; or changes by three millions of miles.

If we compare the semi-minor axis and the semi-major axis of the orbit of Venus by the rule which we gave in connection with our explanation of the form of an ellipse, in Lecture V., page 119; we find that they only differ by about $\frac{1}{15000}$th part of either; that is, by less than 1,600 miles.

It is also worthy of notice that the plane of the path of Venus is tilted to that of the Earth, or, in other words, to the plane of the ecliptic, less than one-half as much as that of Mercury. With the exception, however, of some of the minor planets, the tilt is greater than in the case of any other planetary orbit.

Venus may, of course, be seen, from time to time, at a much greater apparent distance from the Sun, and at a much less actual distance from the Earth, than Mercury, as is indicated by our previous diagram (Fig. XXXIX.), in which the comparative paths of each of these planets and their maximum elongations from the Sun are represented. We are thereby much assisted in our observations of Venus. At the same time, the excessive brilliancy of its light (to which we have already briefly referred) acts as a counterbalancing disadvantage, and to a considerable extent neutralizes the effect of the more favourable positions in which we see it. It is in general found that the best views are obtained when a telescope is directed to it during daylight. But in the middle of the day the Sun's heat often causes the atmosphere to be very much disturbed. The most favourable time of all is consequently shortly before sunset, or soon after sunrise ; and even then a slightly tinted glass may frequently be used with advantage, in order to diminish the brightness of the telescopic image.

With regard, however, to naked-eye observations of Venus, the most interesting point to be discussed is the way in which the relative distances of its orbit and of that of the Earth from the Sun, and the consequent changes in its apparent diameter and in its phase (both of which are illustrated by Figs. XXXVIII. and XXXIX., Lecture V.), cause the brilliancy with which it shines to be at times unusually great. Or, to use somewhat more technical language, we may ask,—under what circumstances is the total amount of light received from Venus a maximum?

Before, however, we proceed to answer this question, let us carefully notice that we are not speaking of the *intrinsic* brightness (as it is usually termed) of the planet's surface when seen in a telescope. This simply depends upon its distance from the Sun, and its own reflective power, and is not altered by any change in its distance from the Earth; because the brightness of any illuminated area is unaffected by its distance from an observer (if the light from it be supposed to traverse empty space), so long as the size of the area is appreciable.*

* The truth of this statement depends on the fact that, if the area be brought nearer, it appears just so much larger as the total quantity of light received from it is greater, which light consequently only suffices to give it the same intensity of illumination as before. The same argument, of course, holds good in a reversed direction, when such a surface is removed further away.

But if the surface observed is sufficiently distant to have no visible area (as in the case of Venus when viewed *by the naked eye*), so that the brightness of the object is estimated simply by the whole amount of light received from it, then its apparent brightness will be the greater the nearer it is to the observer.

In regard, therefore, to naked-eye observations of Venus, its light and its apparent brilliancy would be greatest when the planet is at its nearest approach to the Earth; and least, when it is at its farthest departure from it; *were it not that its phase* (or the ratio which the illuminated portion bears to the whole area of the disc which is turned towards us) changes from time to time.

Before, however, we begin to take this change of phase into consideration, we must remember that each degree of phase corresponds to a certain distance from the Earth. We must therefore, in the first place, calculate what amount of light we should receive from the planet, not only at its greatest or least distances from the Earth, but at any intermediate distance, supposing that its whole disc remained illuminated; and then, in the second place, estimate how much the corresponding *phase* still further alters the amount of light which reaches us.

Now apart from any change of phase, the total amount of light which we receive must vary, according to a certain well-known law of optics, inversely as the square of the distance of the planet from us. But the apparent area of its disc, as seen by means of a telescope, varies according to exactly the same law.* We may therefore take the apparent size of the *telescopic disc* of Venus at any given time to represent the total amount of light that an observer watching it, *without the aid of a telescope*, would receive from it; so that the whole areas of the first and fifth circles in Fig. XLVI., which are drawn to represent the comparative sizes of the disc, seen with the same telescopic power, when

---

* For instance, if Venus were to recede, at any given date, to twice the distance from the Earth at which it was at a previous date, the apparent area of its disc, seen by a telescope, would be reduced to one-fourth of its previous size, inasmuch as that area varies as the square of the apparent diameter, which would be reduced by one-half.

Venus is respectively at its furthest distance from the Earth, and very nearly at its nearest approach to it, will equally serve to represent the comparative total amounts of light which would be received from it on the two occasions in question, and the comparative brilliancy which it would display *to an observer viewing it without the aid of a telescope,* if in each case *the whole of its surface turned towards him* were illuminated by the Sun's light.

But, as is represented in the view of the phases of an inferior planet given in Fig. XXXVIII., the whole of the half of Venus seen from the Earth, when its position corresponds very nearly

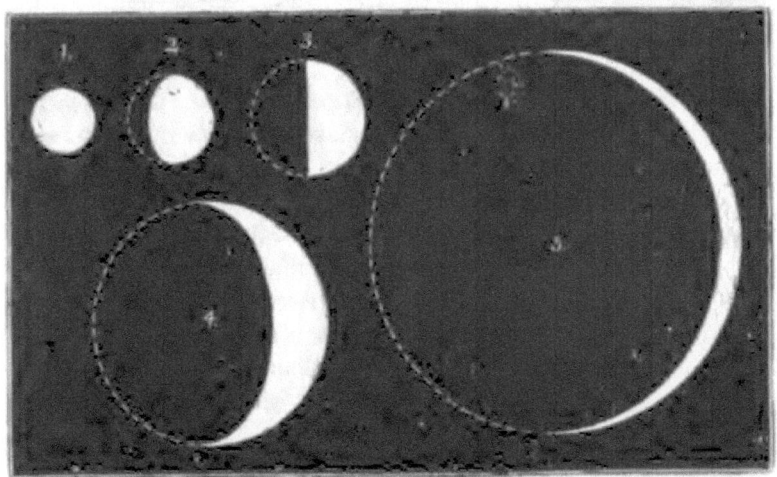

Fig. XLVI.—Comparative diameters and phases of Venus: 1. When at its greatest distance from the Earth; 2. When gibbous; 3. When seen at its greatest elongation from the Sun; 4. When at its greatest brilliancy; 5. When very nearly at its nearest approach to the Earth.

to the size of the fifth circle in Fig. XLVI., would be dark; while the whole of the smaller apparent disc represented by the first circle would be bright. It is therefore clear, that in some *intermediate* position the greatest total amount of light will be actually received from it. And it is found that this occurs when its phase is as in No. 4 of the five figures shown above, or when about one-fourth of its disc is illuminated.

In fact, all we need to do in order to determine the exact position in question, is first of all to draw a series of circles, whose diameters are greater or less according as Venus is nearer to or farther from the Earth; next, to blacken all that portion of each which, as seen from the Earth, would, according to the corresponding phase of Venus, be dark; and then to measure our various figures and find which of them has the largest area unblackened.

The final result of all this explanation is therefore as follows:—When Venus shines with its maximum brilliancy, as seen by the naked eye, its phase is (as in No. 4 of the above figures) very nearly the same as that of the moon when it is about $3\frac{1}{2}$ days old, or one-fourth full. Venus is at such a time considerably nearer to the Earth than when it is at its greatest apparent distance, or elongation (see Fig. XXXVI.) west or east of the Sun. But if it come any nearer, the diminution of its phase more than counteracts the increase in its light which its greater proximity would otherwise produce. On the other hand, if it go further away, the phase increases in width, and soon appears in a telescope to be similar to that of a Moon half-full (as in No. 3 of the above drawings), and presently similar to that of one more than half-full, or gibbous (as in No. 2); notwithstanding which, the total light received from it is less. In fact, it may be computed, that the half-moon phase, which corresponds to the greatest apparent elongation of Venus from the Sun, sends to us about three-fourths of the greatest possible amount of its light; while the planet would only appear to shine with one-fourth of its maximum brilliancy if, in spite of the solar glare, it could be seen at its furthest distance from the Earth upon the opposite side of the Sun, notwithstanding that its whole disc would then be bright.

We have explained in Lecture V. that the combined effect of the orbital motions of the Earth and of Venus is to make the latter *appear* to travel once round the Sun in 584 days, although its actual revolution in its orbit is accomplished in 225 days. An interval of 584 days consequently takes place between its being seen in any given position relatively to the Sun, and its being so seen again. Once therefore in every 584 days (*i.e.*, upon an average, since some allowance must be made for slight irregularities in its movements), the planet will attain its greatest brilliancy as an *evening* star, and once in the same time as a *morning* star. Our previous explanation (in Lecture V.) also makes it evident that these two occurrences will not divide the whole period of 584 days into two equal parts. In fact, the interval between the planet's being at its greatest brilliancy, first as a morning and then as an evening star (during which time it passes beyond the Sun), is about 512 days; leaving only 72 days between its being successively at its greatest brilliancy as an evening and then as a morning star.

Again, it should be noticed, that five times 584 days very nearly amount to eight terrestrial years, and are also equal, within less than one day, to thirteen times the actual period of Venus itself around the Sun, which is almost exactly $224\frac{7}{10}$ days. It follows, therefore, that, if at any given time, in any given year, a conjunction of circumstances causes the position of Venus, as seen in our northern latitudes, to be particularly favourable; that is to say, if the Earth happens to be close to that part of its own orbit in which it is at its nearest to the Sun (and therefore also to Venus), at the same time that the planet is in such a part of its orbit that it is seen after sunset, or before sunrise, at a considerable altitude above the horizon, —very nearly the same state of things will recur eight years afterwards. The planet was, for instance, exceptionally brilliant and well placed for observation in January 1870, and in January 1878, and we may therefore predict that it will be so in January 1886. When thus situated, its light is fully sufficient to cast a very distinct shadow, which may be well shown by opening the window of a dark room, while a rod is held in front of a sheet of white paper, upon which the shadow will be plainly visible.

At such times Venus may be seen with comparative ease with the naked eye at mid-day, by those who know where to look for it. Occasionally, when the requisite conditions are fulfilled with exceptional accuracy, it has been so conspicuous as to attract the notice of a crowd.

Arago tells us* that there is a tradition recorded by Varro, that, "in his voyage from Troy to Italy, Æneas constantly perceived this planet, notwithstanding the presence of the Sun above the horizon." He also states that Bouvard had informed

---

* Arago's "Popular Astronomy," English Edition, vol. i., p. 700.

him, how General Bonaparte, upon repairing to the Luxembourg, to attend a fête given to him by the French Directory in 1797, was much surprised at seeing the multitude in the Rue de Tournon pay more attention to the region of the heavens above the palace than to himself or his retinue. The General then learned to his astonishment that, although it was noon, they were watching the planet Venus, which they supposed to be the star of the Conqueror of Italy; and on looking in the direction indicated, he also saw it himself.*

By way of contrast to the case of Venus, whose greatest brilliancy is attained (as we have shown) when its phase is less than half-full, it is a somewhat interesting fact that the greatest brilliancy of Mercury occurs when it is more than half-full, and when it is, consequently, in the further part of its orbit from the Earth; the diminution of light, owing to the *decrease* of the apparent diameter of its disc, being more than compensated, in its case, by the *increase* of its phase. To prove that this must be so would need some calculation, but it may not be hard to understand that it arises from the fact, that the diminution of its apparent diameter, caused by the increase of its distance, is proportionally less than it is for Venus, while the proportionate increase of phase is the same. It is stated in Dr. Ball's "Elements of Astronomy" (p. 264), that a planet revolving at 41,000,000 miles would be brightest when half-full; one further from the Sun, when its phase has a crescent

---

* Although Venus is at times more conspicuous than was the case in 1881, the writer observed that, even on Easter Monday, *twenty-two days after* March 27th, the date of its maximum brilliancy, the planet could be easily distinguished at noon with the naked eye, when its position had first been discovered with an equatoreally-mounted telescope. It may also be worthy of mention, that, at the beginning of April, in the same year, his brother-in-law returned home one afternoon, and said that he had been much surprised to see, in the middle of the day, while walking along a street in the east of London, the Sun, the Moon, and a star, all comparatively near together in the heavens. It was the young Moon and Venus which he thus perceived, without knowing where to look for the planet, or even being aware that it might be possible to see it. Again, on May 23rd, in the same year, *sixteen days before* its next subsequent date of maximum brightness, it was seen with the naked eye at noon, in Suffolk, by the writer of this lecture.

form, as in the case of Venus ; one nearer, when gibbous, as in the case of Mercury.

We have said in our last lecture that Mercury does not appear so brilliant as we should expect that it would, if its light were for the most part reflected from a dense canopy of cloud. On the other hand, the brilliancy of Venus is so great, that we have little doubt that what we see is chiefly cloud, and not the true surface of the planet. Chacornac found its brightness to be ten times greater than that of the most luminous parts of the Moon ; which would assign to it a light-reflecting power fully five times as great as that of those parts, since the sunlight received by it is not quite twice as intense. But this can hardly be the case ; inasmuch as the late Professor Zöllner found the *average* reflective power of the Moon's surface to be such as to send back about $\frac{1}{5}$th of the light incident upon it. The best investigations yet made indicate that the light-reflecting power of Venus (or, as it is sometimes technically termed, its *albedo*, or whiteness) is rather more than three times as great as that just stated for the Lunar surface as a whole.

The period of the rotation of Venus upon its axis rests upon stronger evidence than that of Mercury. So long ago as 1666 the eldest Cassini, while observing it, in the clear sky of Italy, watched a bright spot upon its surface, which returned to a similar position in about 23 hours. This observation was afterwards, to some extent, confirmed by others made by Bianchini, who about A.D. 1724 detected the apparent rotation of several dark spots.* Again, in 1789, Schröter, having undertaken a careful examination of the disc with a seven-foot reflector, discovered a small luminous point in the dark hemisphere, slightly separated from the southern horn of the crescent. This point he supposed to be a high mountain, whose summit caught the Sun's rays before they reached the surrounding regions. He also tells us, that a blunted appearance

* Bianchini announced the period as 24 days 8 hours, but Cassini's son (who is generally known as Cassini II., or James Cassini) showed that Bianchini had not been able to watch the planet continuously, and that a rotation of 23 hours 20 minutes would equally satisfy his observations, by supposing that it had really gone round 25 times, while he imagined that it had only rotated once.

of the southern horn of the crescent occurred at regular intervals, which lasted for a short time, and which he considered was in all probability caused by the shadow of a peak of the same mountain range, whose illuminated summit he believed that he saw. From the interval at which the same appearances recurred, he deduced a rotation of the planet in 23 h. 21 min. 19 sec., which was in close agreement with the value previously obtained by Cassini.† We may remark that we have ourselves at times thought the above-mentioned blunting of the southern horn (the uppermost in a telescopic field of view) to be very evident. Upon one occasion it appeared to be so distinct, that, when we requested several of our friends, notwithstanding that they were unaccustomed to such observations, to look through the telescope in succession, and independently to record anything remarkable which they might see in either horn, they unanimously selected the same one, and noticed this peculiarity in it. At the same time we must accept with very much reserve Schröter's measurements of the supposed height of this, or of any other mountains, which he mentions as irregularities in the edge of the illuminated part of the disc. To the loftiest of such mountains he assigned an approximate height of twenty-seven miles, which represents a proportionate elevation about five times as great as that of the highest upon the surface of the Earth.

There are some other peculiarities connected with Venus which deserve mention; for instance, that when it is seen as a crescent there appears to be a gradual fading-away of its light towards the boundary between the illuminated and unilluminated parts; as also that the corresponding dividing line between these two portions does not appear to be exactly a straight line, so as to make the phase similar to that of the moon when half-full, at the time when calculation indicates that it ought so to do; there often being a discrepancy, one way or the other, of three or more days in the time when the half-full moon phase is seen.

---

* See Grant's "History of Astronomy," p. 234.

† Subsequent observations by De Vico and others at Rome, in and after the year 1839, have also very closely confirmed this result.

These effects may be in some way connected with the nature of its cloudy, or vaporous envelopes, but we cannot say that any thoroughly satisfactory explanation of them has as yet been given. It has sometimes been stated that the outer edge of the crescent phase appears to be its brightest part; but careful observations, in which its light has been continuously diminished by the use of a wedge of dark glass, have indicated that the brightest part is not at the outer edge, but somewhat within it, so that the light is reflected, to some extent, as it would be from a vitrified or polished spherical surface, in which one special spot would shine with surpassing brilliancy. Hence it has been recently suggested that the reflection of the light of Venus may be specular, or mirror-like; an idea which must, however, be considered to be extremely hypothetical. It may, moreover, be well that any such hypothesis should be allowed to lie by for the present, as its suggestion originated a considerable amount of rather animated controversy.

Much more certain are the indications of an atmosphere of the planet, which are afforded by the extension of the cusps, or fine terminating points of the crescent, beyond a semi-circumference, into the darkened part of the disc. This has been noticed by Sir W. Herschel, as well as by Schröter, and by many others in more recent times. Observations have also been made by Professor Lyman, of Yale College, U.S., and, we believe, by at least one other observer, of a fine thread of light seen all round the planet, at the time of its inferior conjunction, when its dark side is altogether turned towards the Earth. And certainly during its transits in 1761, 1769, and again in 1874, both just before and after it entered upon the Sun's disc, a ring of light indicating an atmosphere of considerable refractive power was seen.

We have therefore very good grounds for believing that Venus is possessed of an atmosphere; which we also suppose to be heavily laden with clouds, owing to the brilliancy of the light reflected to our gaze.

We have already discussed the transits of Venus in our first lecture upon the Sun, in which we explained their exceptional importance in reference to the determination of the Sun's distance from the Earth. In this place, therefore, we need only remark with regard to them, that we hope, that in the coming transit of 1882 very careful attention will be paid to any indications which may be seen of the planet's physical condition, or of the nature and extent of any atmospheric appendage belonging to it.

Our readers must certainly by this time (if not before) have begun to think how little that is certain, or important, has after all been discovered about this planet, so beauteous to the naked eye, so baffling in the telescope. Nor is it much less tantalizing, that we are unable to give satisfactory explanations of some appearances, which are in themselves more sure. As an instance we may refer to one, which is generally described as a greyish light occasionally noticed upon the dark part of the planet's disc. It is most often seen in observations made during daylight or bright twilight, and is termed by French observers *la lumière cendrée*. It faintly enlightens the remainder of the disc, when the planet is so situated that its illuminated phase is slight (as in No. 5 of the drawings in Fig. XLVI.) How it is caused it is hard to say, unless it be by some optical delusion. No reflected light from the Earth, such as illuminates the dark part of the disc of the young Moon, could possibly suffice to produce it. It has been suggested that it may arise from phenomena of an auroral character, or even from some kind of phosphorescence in the planet's surface; but it must, we think, be confessed, that the enigma is one which is still unsolved. It is certainly sufficiently interesting to deserve further and careful study.

Another curious fact in the history of the planet is, that for about a hundred years, viz., from 1672-1764, observations were made from time to time of a supposed satellite belonging to it. It is generally considered that such appearances were caused by minute images of the planet formed by successive reflections between the glasses of the eye-piece of the telescope used, or between it and the observer's eye,[*] which would the

---

[*] See also in R. A. S. Monthly Notices for January 1882, p. 111, how a small bright crescent was seen near to Venus by Mr. Denning, which was

more easily prove deceptive, because the phase of such a satellite (if it existed) would, when seen from the Earth, necessarily be the same as that of the planet, and cause it to look like a smaller image of the latter.

A curious incident of a somewhat similar effect may be cited in the tiny companion star, which M. Otto Struve, for some time, believed to belong to the bright star Procyon. Professor Newcomb also states, in his "Popular Astronomy" (p. 297), that one of the eye-pieces of the great Washington telescope shows, under certain circumstances, a beautiful little satellite alongside of Uranus or Neptune, which, however, disappears when the telescope is moved. Nevertheless Mr. Webb remarks (in his "Celestial Objects") that it is difficult to apply the explanation of an optical delusion, produced by instrumental effects and sometimes called a *telescopic ghost*, to Short's observations in 1740, who, by the employment of two different instruments, and four different eye-pieces, took every precaution against error.

For our own part, although we can hardly believe that a satellite of Venus has as yet really been seen, we must confess that it appears difficult to explain how observers of some time since should have been able to see, or should have thought that they saw, so much more than we can see. We are almost tempted to ask,—Did their telescopes, or did their fancy, deceive them? It may, however, be well, if further observations, with a view to the possible discovery of a satellite, be made with modern instruments of the most improved construction, either under the clear skies of Italy or of Algiers, or on some elevated site, such as that of the new Lick Observatory in the mountainous regions of North America. It will also be quite worth the while to watch carefully with very powerful telescopes during the transit of December 1882, for the appearance of any very minute satellite or satellites of Venus (similar in size to those of Mars) upon the disc of the Sun.

found to be caused by sunlight falling upon the sliding-tube of the eye-piece. In the above paper many interesting observations of the planet are recorded, accompanied by some striking drawings of spots and other markings and irregularities of its surface.

We have already mentioned that the size of Venus does not greatly differ from that of the Earth. It is also found, from calculations of its mass, that the attraction of gravity upon its surface is within one-fifth part of being the same. In this respect Venus would therefore be much better suited than Mercury (upon which the attraction of gravity is only about one-half as great) for the habitation of beings like ourselves. Nor would the intensity of the Solar heat be nearly so fierce as upon Mercury, since the Sun would only have an apparent diameter two-fifths larger than that which it presents to us; while its apparent area, although nearly double that which we see, would only be of one-fourth of its average size as seen from Mercury.* The heat and light received would, in like manner, be about twice that received by the Earth, and about one-fourth of that received by Mercury. We have, moreover, shown that there is much more reason to believe, in the case of Venus than in that of Mercury, in the existence of a dense envelope of cloud, which might possibly in some degree mitigate the Solar heat. But, if the observations of Schröter, or even the more recent observations of De Vico and of those who assisted him in 1839-1841 be correct, it would seem, that the tilt of the axis of the planet to the plane of its orbit is such, that the climatic conditions upon it would be very greatly opposed to its habitability.

Schröter considers the tilt of the axis to the plane of the orbit to be about $15°$, and De Vico about $37°$, while in the case of the Earth the corresponding inclination is $66\frac{1}{2}°$. As a consequence the arctic zones of the Earth extend to a distance of $23\frac{1}{2}°$ from its poles, i.e., to a latitude of $66\frac{1}{2}°$ from its equator, while the tropics extend to $23\frac{1}{2}°$ on each side of the equator, or $66\frac{1}{2}°$ from the poles. Upon Venus, if Schröter were correct, the arctic regions would extend to $75°$ from the poles, and the tropics to $75°$ from the equator; while about $53°$ would be the limit in either case, according to De Vico.

But even on this latter more moderate supposition the arctic circles would come down below the half-way parallel of latitude of $45°$, and the tropics would reach to $8°$ beyond it. A

---

* See Fig. XLIV., Lecture VI.

zone of 16°, in what would otherwise be the most temperate region, would consequently be the least suited for habitation; as in one half of the year it would experience a tropical summer, and in the other half an arctic winter, during which latter season the Sun would wholly disappear (according to the latitude) for a greater or less number of days. As is the case upon the Earth, so, in like manner, for places situated within the more extensive tropical regions of Venus, there would be two days in each year in which the Sun would be vertical at noon; but between those dates the tropical summer days would not (like our own) be but little longer than one-half of the number of hours in which the axial rotation of the planet is performed; on the contrary they would increase to a maximum duration of nearly twenty-three of our hours (if we suppose that to be the true rotation-period of Venus) for places situated near to 37° of N. or S. latitude, as in places upon the Earth lying but little below an arctic circle; while in localities situated between 37° and 53° of (tropical) latitude upon Venus daylight would be continuous for a certain number of days in the middle of the period in question. In either case the gradual increase of temperature caused by the excess of the heat received by day over that lost by night, would, we think, of necessity become utterly unbearable.

Upon the whole it therefore seems, that, even if other reasons might not force us to deny the habitability of Venus, any such axial inclination as the above would make it impracticable, unless perchance to a number of unsettled beings, who might, according to the season, migrate from place to place, in order to choose the most bearable climate; so that, instead of emigration being a remedy for a local intensity of over-population, as upon the Earth, a perpetual state of migration would be the normal condition of existence upon Venus. When we remember how few of those who would in winter time be glad to change the snows, gales, and fogs of England for the gentler climate of southern France, or of the Mediterranean, are able to do so, we may realize how improbable it is that there are any residents upon Venus, if it be only habitable upon some such supposition as the above.

The probable tilt of the axis of Venus is, however, so very uncertain, that it may be well to consider the above paragraph as a geometrical exercise upon the possible effects of a certain axial inclination in a planet, rather than a statement of what is really to be considered as the state of things probably existing upon Venus itself.

It is certainly the case that the observations and drawings of the spots and markings upon the surface of Venus, to which we have already referred as made by De Vico and others at Rome, in 1839-42, as well as those of some observers of more recent date, for the most part agree very fairly with those of Bianchini in the years 1726 and 1727; in fact, sufficiently well to give, as we have stated, remarkably accordant results for the supposed period of its rotation deduced from their apparent movements. And yet modern astronomers (see, for instance, the opinion of Dr. Vogel, in Part II. of the "Bothkamp Observations," published in 1873) feel great doubt as to the period in question. Much more, therefore, may we feel extreme uncertainty as to the inclination of the planet's axis, with regard to which not only do modern observations give no information, but older observations are very discordant.*

To conclude this lecture we will quote, from p. 71 of Chambers' "Handbook of Descriptive Astronomy," a translation of the curious anagram in which Galileo announced to Kepler the detection, by means of his telescope, of the phases of Venus; a discovery which afforded a most important confirmation of the truth of the great Copernican theory of the Solar System, and which, we think, it may still be interesting to read in the form in which it was originally expressed. Galileo wrote:—

"Hæc immatura a me, jam frustra, leguntur.—o. y."

"These things not ripe (for disclosure) are read, as yet in vain, by me;" or,

"These things not ripe : at present (read) in vain (by others) are read by me."

---

* For further information upon these and many other points of interest connected with the planet, we may refer our readers to the new edition of the Rev. T. W. Webb's "Celestial Objects for Common Telescopes."

This, when transposed, becomes—

"Cynthiæ figuras æmulatur Mater amorum."
"The Mother of Love (Venus) imitates the phases of Cynthia (the Moon)."

To follow the gradual change of these phases as their width continuously diminishes, while the diameter of the disc simultaneously increases, until at last the illuminated area, as is shown in Fig. XLVI., is reduced to the finest possible sickle-like line of light; or, *vice versâ*, to watch the increase of phase while the diameter diminishes;—is perhaps, even for those whose telescopes are by no means powerful, one of the most beautiful and interesting sights that the heavens afford.

LECTURE VIII.

THE EARTH.

"Age here on age
"Lies heap'd like wither'd leaves. And must it end?"

BAILEY.

IN discussing the Earth as a member of the Solar System, or in comparing it with the other planets which circle round the Sun, either within or without its orbit, it is almost impossible to say where astronomy begins and geography and geology end; or, to use a term of somewhat recent invention, it is by no means easy to define the boundary-line between astronomy and physiography.

Some of the smallest variations in the Earth's movements involve results of the deepest import in its past geological history, and in the interchange of glacial and torrid epochs. A deviation from perfect regularity in its shape, which might at first sight seem of little moment, causes the precession of the equinoxes, and changes the whole aspect of the heavens, as they are seen from any given terrestrial locality, in the course of every 26,000 years. The temperature of a mine; the density of the air at the highest altitude the aeronaut can attain; the speed with which a pendulum oscillates at the equator, or near to either pole, or in any intermediate latitude; the greater accumulation of ice at the south than at the north pole; the varying duration of twilight; the beauties of the Aurora; the Earth's ever-changing magnetic currents; the disturbances of the action of telegraphic instruments; the rainfall of any given year; the tides and currents of the ocean; the fury of a cyclone; the throbbing of an earthquake; the glowing fire of a volcano; the regulation of every watch and clock we use; the adjustment of our calendars; the determination of our standard weights and measures; the energy required to leap over a wall, to draw a train, or to propel a cannon ball,—are all closely connected with important discoveries in Astronomy; while much besides, which might at first be thought to belong to the simplest and most ordinary events of every-day life, may also be shown to depend upon, or to illustrate, some fundamental astronomical principle.

Where, then, are we to fix the limits of our discussion, amongst the fascinating lines of thought which at every point tempt our pursuit? Were we to yield to the temptation, we should soon find how many subjects there are connected with the Earth, each of which, for its worthy treatment, might well

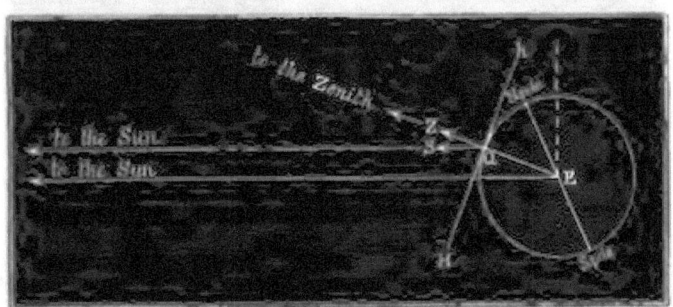

Fig. XLVII.—Incidence of the Sun's rays upon a place in the North Temperate Zone at noon at the time of the Summer Solstice.

demand a long series of lectures. It will, we think, be our wisest course simply to endeavour to put together here a few of those elementary truths connected with the Earth, and its present, or past, condition and circumstances, which are most closely related to Astronomy; although we may, in so doing, devote a little additional space to some points of special interest, which, in our opinion, frequently meet with less attention than they deserve. At the same time, we must crave the kind indulgence of our readers if we make little mention of some well-known facts which are explained in almost every elementary text-book; or omit some others because of their abstruse and difficult nature, however fully they may belong to astronomical science.

Amongst the former we may place the general explanation

of the cause of the seasons of the Earth's year—Spring, Summer, Autumn, and Winter—to which our allusion shall be comparatively brief. The diagrams in Figs. XLVII. and XLVIII. may be sufficient to show, that, in the *summer* of either hemisphere, the Sun's rays fall with a nearer approach to verticality at noon in the temperate zones, than in the *winter*, so that the heat of the summer season of the year is consequently increased. In the former figure a place, Q, is supposed to be situated in 45° of north latitude. QZ is the direction of the zenith. QS is the direction of the Sun's centre at noon, which, owing to its immense distance, is almost exactly parallel to the line drawn from E, the Earth's centre, to that of the Sun. The direction

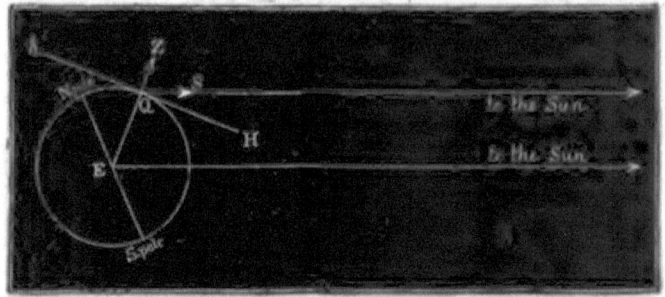

Fig. XLVIII.—Incidence of the Sun's rays upon a place in the North Temperate Zone at noon at the time of the Winter Solstice.

of the Earth's axis is drawn so as to correspond to the date of the summer solstice.

In Fig. XLVIII. the northern extremity of the Earth's axis (which axis remains parallel to its previous direction) is located just as much beyond a perpendicular to the direction of the Sun, as, in Fig. XLVII., it was on the sunward side of the same perpendicular (which in that figure is indicated by a dotted line); and it is evident that the angle ZQS is much larger than before, so that the Sun's rays will deviate much more from the vertical—*i.e.*, from the direction of the zenith—than in Fig. XLVII.

It is only another way of stating the same fact (as these diagrams also indicate), to say that, at noon on any day in summer, the Sun will be higher above Hh, the horizon of such a place as Q, than in winter; the day upon which it is highest

THE EARTH. 175

being called that of the summer solstice; a name which signifies that the Sun, after having gradually approached nearer and nearer to a pole of the heavens, stops or pauses for a brief interval before it begins to recede again. And, since the Sun appears, like all the other heavenly bodies, day by day to describe a circle round one of the celestial poles, in consequence of the rotation of the earth upon its axis, it follows, as a result of the smaller radius of the circle which its smaller distance from the pole involves, that it will be much longer

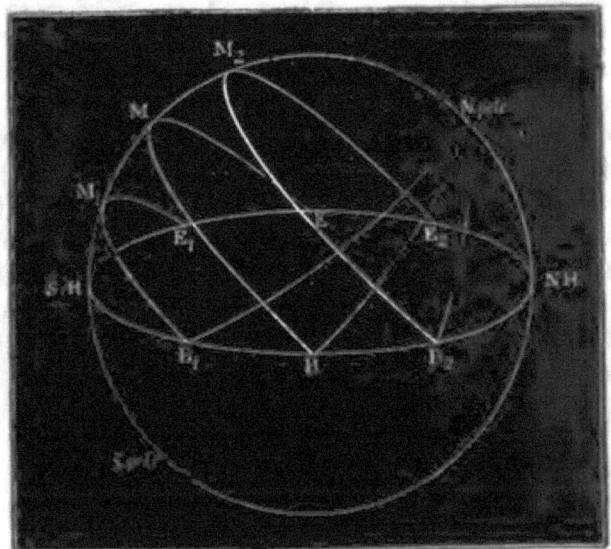

Fig. XLIX.—The Sun's diurnal paths BME, E₁M₁E₁, E₂M₂E₂, at an equinox and at the winter and summer solstices respectively, for a place in 45° of north latitude.

above the horizon of any given place in summer than in winter.* The diagram in Fig. XLIX. is intended to illustrate this last statement.

In the above figure, N. H. and S. H. represent respectively the north and south points of the horizon of any given place; the position of the north pole of the heavens being shown for one which is situated in about 45° of north latitude.

* That is to say, with the exception of a place situated upon the equator, where the days and nights respectively are of equal length all the year round.

When the Sun rises exactly in the east, and sets exactly in the west (as it always does at the time of an equinox), its apparent diurnal path will be BME, B and M being exactly half-way between N.H. and S.H. The circle BME consequently represents a day of 12 hours in length, of which B is the beginning, M the middle (at noon), and E the end. But in the height of summer, the Sun's distance from the pole will be so much less, that its path will correspond to the circle $B_2 M_2 E_2$; while in the middle of winter it will be $B_1 M_1 E_1$. It is not difficult to see, that, in the latter case, the duration of the day is shortened so as to be less than 12 hours, while in summer it is correspondingly lengthened.

From our preceding statements it therefore follows that there are two causes which conjointly produce the difference between the various seasons of the year; the first, the greater altitude of the Sun, and the consequent nearer approach of its rays to perpendicularity; and the second, the greater length of the day in summer than in winter. This latter effect also (so to say) intensifies itself, from the fact that, when the duration of the day exceeds 12 hours, and is consequently longer than that of the night, the heat received accumulates, because more is gained in the day than is lost in the night.* It therefore results that the greatest heat of summer is not attained at the time of the summer solstice (*i.e.* at the end of the third week of June), but some time afterwards, the temperature in the meanwhile increasing to some extent; just as it does, day by day, for a certain length of time after noon, the daily maximum not being reached until about 2 p.m.

We have thus briefly referred to the fact that the length of the day in either hemisphere depends upon the Sun's distance from the corresponding pole of the heavens, and we have indicated, in Fig. XLIX., that, the further it is from the pole, the sooner will it in setting meet the horizon, the later will it cross it in rising. We think, however, that a somewhat fuller consideration of its apparent *daily* path, as seen from different *localities*, and at different *seasons* of the year in any

* In the latitude of London the longest day is about 16 hours 34 minutes, and the shortest night about 7 hours 26 minutes.

given locality, may for several reasons be both interesting and profitable.

To this we will, therefore, now proceed ; but, before we actually enter upon the explanations involved, we will mention a little difficulty which we believe to be often felt in connection with the definitions usually given of those points in the heavens to which we have already several times alluded as its *North and South Poles*. They are frequently defined in astronomical treatises to be the points where a prolongation of the Earth's axis would meet the imaginary celestial sphere to which the places of all the heavenly bodies are referred ; but at other times they are said to be the points in which a line drawn through the centre of the Sun, parallel to the aforesaid axis of the Earth, would meet the same sphere; and some of our readers may have wondered that two definitions, apparently so different, should thus be given. We wish, therefore, to state that they are really equivalent.

In comparing these two definitions, we must allow that, at first sight, it might be thought that it must necessarily make some difference, whether the direction in which the north or south pole of the heavens lies be that of a straight line drawn through the Earth's centre at any given time, or of a parallel line through a point (viz., the Sun's centre) which is 93,000,000 miles distant. Still more, according to the first definition, might it seem hard to believe, that it matters not, whether the parallel direction be taken through the Earth's centre at any given date, or six months afterwards, when its position has changed by a distance equal to 186,000,000 of miles. But, as we have stated, it really makes no appreciable difference, and for this reason,—that the celestial sphere, or spherical surface upon which the poles are located, is imagined to be so immensely distant, that the whole area of the Earth's orbit, stretching across a space whose diameter is not far short of two hundred millions of miles, is *as a point*, or *as nothing*, in comparison with the distance to that sphere.

It is therefore impossible to distinguish between the intersections of the prolongation of the Earth's axis with the far-distant celestial sphere, at any two given dates, while the

Earth keeps travelling onwards in its orbit, and at the same time maintains the direction of that axis (except so far as some slight effects of precession and nutation are concerned) constantly parallel to itself. It is for a similar reason that the two rows of trees in a very long straight avenue seem to meet in the extreme distance; or, to use the language of mathematics, that parallel lines may equally well be said, either never to meet, or only to meet at infinity.

It may consequently, we think, be understood, that all the year round the pole of the heavens would appear to a spectator at the pole of the Earth to be immediately above him, unaltered in position by the orbital motion of the Earth. If, therefore, the pole star were exactly at the pole, instead of being about $1\frac{1}{3}°$ from it, its place would always be exactly in his *zenith*. In

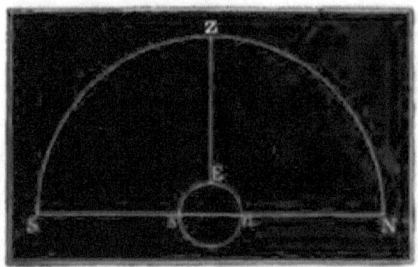

Fig. L.—The poles of the heavens are seen in the horizon by an observer at the Earth's equator.

the same way an observer upon the Earth's equator, whose zenith lies in a direction perpendicular to the Earth's axis (or at an angle of 90° from it), sees the poles of the heavens, all the year through, 90° from the point of the celestial sphere immediately above him; *i.e.* he sees them (approximately) in his *horizon*, their positions respectively corresponding to its north and south points. This is illustrated in Fig. L., in which z is the zenith, E the observer's place upon the equator, N and S the north and south points of the horizon, n and s the poles of the Earth.

The only difficulty which the above figure really involves is this;—that the horizon-plane shown in it is supposed to pass through the Earth's centre, instead of being drawn (in accordance with the most usual acceptation of the word) as a parallel plane touching the Earth at E, the place of observation. We are glad, therefore, of the opportunity which this diagram thus affords us, to mention that the former plane is technically termed that of the *rational* horizon, the latter that of the *sensible* horizon. Although, however, this difference exists, we must remember that we have shown in our previous explanation of the position of the poles of the heavens, that we may consider them to be so almost infinitely distant, that the whole area of the Earth's orbit may be considered to be a mere point. We are, therefore, still more justified in treating the Earth itself as if it were a point in comparison with the radius of the celestial sphere. If we do so, the two planes in question at once become coincident, and the difficulty, which might otherwise be seen in the diagram, disappears.

In fact, the difference of the rational and sensible horizons need only be regarded when we are concerned with the observation of those few of the heavenly bodies, such as the Moon, the Planets, or the Sun, which are sufficiently near to the Earth for its size to be in some degree comparable with their distances from it. We do not propose to explain the way in which a slight allowance is made in such cases, according to which of the two above-mentioned horizons is used in any given observation, although the explanation really involves no practical difficulty. It suffices for our present purpose that we are justified in making the statement, that either pole of the heavens would be seen vertically overhead *all the year round*, by an observer at the corresponding pole of the Earth; and that it would in like manner be approximately seen in the horizon by an observer situated upon the Earth's equator.

The next point to be noticed with regard to the position in which the pole of the heavens appears, when seen from other places upon either hemisphere of the Earth, is,—that, from any intermediate locality between the equator and one of the Earth's poles, the celestial pole is seen in a corresponding and intermediate position between the observer's horizon and his zenith. If we travel northwards from the equator, the north pole of the sky gradually rises higher above the horizon as we increase our north latitude; if southwards, the same is the case with the celestial south pole as we increase our south latitude. And the rate at which the elevation of the pole takes place exactly corresponds with the change of an observer's latitude, so that, in either hemisphere, the rule to find its position is this:—Look at a point whose altitude above the north point of the horizon in the northern hemisphere, or above its south point in the southern hemisphere, is measured by the number of degrees in the latitude of the place of observation; you will then look in the direction of the corresponding celestial pole.

This rule is generally expressed in the following brief form: *The apparent altitude of the pole of the heavens is equal to*

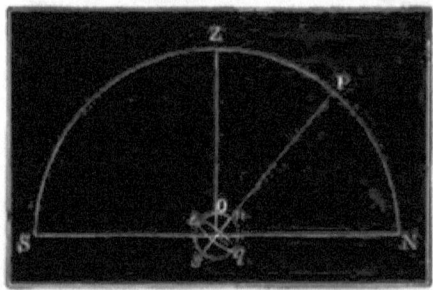

Fig. LI.—The altitude of the Celestial Pole above the horizon is equal to the latitude of the place of observation.

the terrestrial latitude of any given place. It may also easily be understood that this law would bring the pole to the observer's zenith, or 90° above his horizon (as we have previously shown ought to be the case), by the time that he would reach one of the poles of the Earth, or a latitude of 90°.

Fig. LI. illustrates the application of the above rule to a place situated in about 50° of north latitude.

In the above figure, o is the observer's position; z his zenith; *n* and *s* the poles of the Earth, and *eq* its equator; N and S the north and south points of the horizon; P the north pole of the heavens. It is not difficult to see that the angle between the two straight lines drawn from N and P to the centre of the Earth (which is measured by the arc NP); *i.e.* the altitude of the celestial pole is equal to that between two others respectively perpendicular to these two,* viz., those drawn from o and e to the Earth's centre, the angle between which is measured upon the Earth by the arc eo, *i.e.* by the observer's terrestrial latitude;—in other words, the rule given above holds good, that the altitude of the pole of the heavens, as seen from any place, is equal to the latitude of that place.

Having premised thus much, we may now be able to consider what the apparent daily motion of the Sun will be, as seen from any given place, at any given time. Owing to the diurnal rotation of the Earth upon its axis, which is from west to east, it will, of course, appear to describe a circle, during each day, around the pole of the heavens from east to west. And as we have shown where that pole will be seen, the only remaining question that we now need to ask is, how far from it will the Sun be on any given day? That distance will then decide the size and position of the diurnal circle which the Sun will appear to describe.

But the tilt, or inclination of the Earth's axis, is at an angle of $23\frac{1}{2}°$ from a direction perpendicular to the plane of the ecliptic in which the Earth performs its journey round the Sun. It is not, therefore, difficult to perceive, from the diagram in Fig. XLVII., that, at the time of the summer solstice of either hemisphere, when the pole of that hemisphere is tilted towards the Sun, the angular distance from it at which the Sun will be seen will be $23\frac{1}{2}°$ less than a right angle, *i.e.* $66\frac{1}{2}°$ (a right angle being measured by 90° out of the 360° into which the whole circumference of the heavens is usually divided). In like manner, at the winter solstice of either hemisphere, it may be equally

---

* This statement, that the angle between any two straight lines is equal to that between two others respectively perpendicular to them, is, of course, always true; and not only in the above special instance. The unmathematical reader may easily convince himself that it is so, by drawing diagrams of a few particular cases, and measuring the angles involved; or he may take two straight lines inclined at any angle, and first suppose one of them, and then the other, to revolve through a right angle, in the same direction; when it will be evident that the angle between them must be the same as before, although each occupies a position perpendicular to that in which it was originally placed.

well perceived, in Fig. XLVIII., that the angular distance of the Sun from the corresponding pole will be $23\frac{1}{2}°$ added to $90°$, or $113\frac{1}{2}°$. It also follows, that the polar distance of the Sun at the equinoxes, which occur just half-way between the solstices, will be half-way between these two values, *i.e.*, exactly $90°$.

Remembering this, let us first of all take the case of an observer supposed to be at the north pole of the Earth, and see what will be the effect during the year of these varying distances of the Sun from the north pole of his heavens. That pole (as we have previously explained) will always be vertically over his head, $90°$ from the horizon. On the day of the summer solstice the Sun, being only $66\frac{1}{2}°$ from the pole, must attain to an elevation of $23\frac{1}{2}°$ above the horizon. It would therefore go round, during the day, *exactly in a horizontal circle*, were it not that, directly it has thus attained its shortest apparent distance from the pole, it begins to increase it again, in consequence of the Earth's continuous orbital motion. It will therefore really appear, day after day, to describe a spiral, the successive loops of which will be very nearly horizontal, until, at the end of six months, on one particular day (viz., that of the equinox) when its apparent distance from the pole becomes equal to $90°$, its centre will travel almost exactly round the horizon. After this, the Sun will very soon altogether disappear below the horizon, and remain unseen during the next half-year.

To an observer travelling southwards from the north pole, the Sun's daily path would become more and more tilted from a horizontal direction, owing to the position of the pole, round which it daily appears to revolve, being (as we have explained) gradually lowered; until at the Arctic Circle there would only be *one day in the year*, viz., that of the summer solstice, in which it would just escape passing below the horizon. On that one day, as is shown in Fig. LII., the distance of the Sun from the pole being $66\frac{1}{2}°$, and the distance of the pole above the north point of the horizon being also $66\frac{1}{2}°$, the Sun at midnight would just graze the horizon.

On the other hand, at the equator, the poles of the heavens being in the horizon, and the polar axis being horizontal, the

Sun's daily path around either pole, as is shown in Fig. LIII., will always be vertical. The Sun will rise at E (due east), and set at W (due west), on the days of the two equinoxes, owing to its distance of 90° from either pole being then the

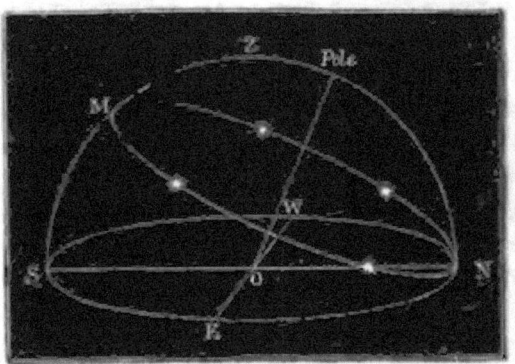

Fig. LII.—Diurnal path of the Sun on the day of the summer solstice, seen from a place upon the Arctic Circle in 66½° of north latitude.

same as the distances of the points E and W, which are half-way between the poles, or 90° from either of them. It will also, on the same days, be vertically overhead at noon, since the zenith, Z, is also 90° from the poles. In fact, its path will be

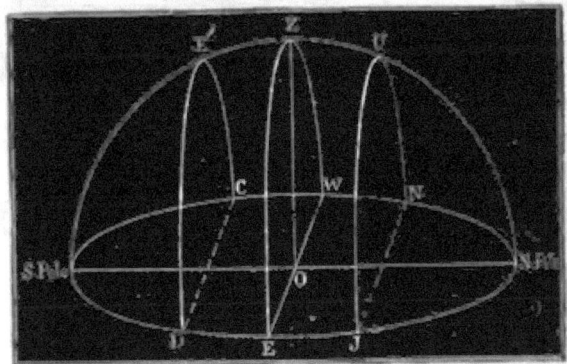

Fig. LIII.—Diurnal paths of the Sun EZW, DE'C, JUN, as seen from the equator of the Earth, at an equinox, and upon the days of the December and June solstices respectively.

in the circle EZW. But before, or after, those dates, it will cross the horizon, in rising and setting, to the north or south of east and west respectively; and it will cross the meridian at noon, north or south of the zenith, owing to its diminished, or increased, distance from the celestial pole; its extreme positions being attained on the days of the solstices; upon one of which (in December) its distance from the north pole of the heavens is (as we have previously mentioned) $113\frac{1}{2}°$, while upon the other (in June) * its distance is only $66\frac{1}{2}°$. Its diurnal path upon the day of the December solstice is therefore represented by the curve DE′C; and upon the day of the June solstice by the curve JUN.

If, instead of the poles being supposed to be in the horizon, as in the above figure, the *north* pole be supposed to be elevated to a moderate extent above it, and the south pole to be equally depressed below it, as would be the case for a place situated between the equator and the Tropic of Cancer (for which the elevation of the pole, being equal to the latitude of the place, would be less than $23\frac{1}{2}°$), it is not difficult to understand that a somewhat similar figure would apply. The Sun's path would not be quite vertical, as in Fig. LIII.; but it would pass through the zenith, so that the Sun would be *vertically overhead* at noon, on *two days* in each year, when its angular distance from the north pole would be equal to the distance of that pole from z. Indeed, for any such place this latter distance must be between $66\frac{1}{2}°$ and $90°$; and the Sun's north polar distance diminishing from $90°$ at the equinox to $66\frac{1}{2}°$ on the day of the June solstice, it is a necessity of the case, that on some day *before* that solstice, and at an *equal interval after* it, the two must be equal to one another; from which it would result, that the Sun would pass through z at noon. In the interval between these two dates it is also evident, since the Sun's angular distance from the north pole would be less than that of z, that it would cross the meridian (E′ZU in the above figure) to the right of z, or on the opposite side of it to that upon which it would pass during the rest of the year. It may in like manner be seen that a similar effect, relatively to the south pole, would occur for places situated between the equator and the Tropic of Capricorn, by supposing the *south* pole to be elevated above

---

* The respective dates at present are June 21st or 22nd, and December 21st or 22nd.

the horizon to a distance not exceeding $23\frac{1}{2}°$ in a figure similar to Fig. LIII.*

In neither case must it, however, be supposed that the temperature of the season would necessarily be lowered between the two dates in question. It is no doubt true that, during the first half of the interval between them, the Sun would fall increasingly short of being vertically overhead at noon, until, after the occurrence of the solstice, it would begin to draw nearer to the zenith again. But it must be remembered that, at the same time, the duration of daylight would increase up to the date of the solstice, in consequence of which the heat might upon the whole become greater.

The above explanation, at any rate, shows that it is not quite correct to say (as is sometimes done) that, in such intertropical localities, there are two summers in each year, but rather that there are two epochs at which the Sun is *vertically overhead at noon*. It will also be evident, if such figures as we have suggested be drawn, that, at those dates, the Sun rises and sets at points of the horizon somewhat to the north of east and west respectively, as seen from a place between the equator and the Tropic of Cancer; and somewhat to the south of east and west respectively, as seen from one which is between the equator and the Tropic of Capricorn; while on the days of the equinoxes, when it rises and sets due east and west, it crosses the meridians of two such places respectively south and north of the zenith, because its equinoctial distance of $90°$ from the pole is then greater than the distance from the pole to the zenith.

Such are the principal peculiarities in the Sun's diurnal path produced by a change in an observer's latitude, or by the effect of the inclination of the Earth's axis in different positions of the Earth in its annual orbit. We have endeavoured to present, in as clear a form as possible, the explanation of them which is usually given. But it must be confessed, that,

* If the place of observation be exactly upon either tropic, the distance of the corresponding pole from z will of course be exactly $66\frac{1}{2}°$, and there will be only one day, viz., the day of the corresponding solstice, upon which the Sun will pass through the zenith.

in order to the thorough understanding of the matter, no diagrams or words can be so effective as some simple model or apparatus,* by which the movements in question may be graphically represented.

Many other very interesting phenomena, to which we can only allude in passing, are closely connected with that recurrence of the seasons which we have thus described. Amongst these, the trade winds deserve especial mention: as also the existence of the neutral zone, between them, commonly called the Doldrums; and the shifting, to a certain extent, of the trade winds, which corresponds to the Sun's alternate passage to the north or south of the equator. Then again the monsoons, induced, when the Sun thus passes northwards, by the highly heated condition of the African deserts:—the descending currents, north and south of the regions of the trade winds, of the air which has risen with an augmented temperature over the equator, and has been cooled at higher altitudes:—the east winds which vex England in spring, and arise from the excess of the heat of the Atlantic Ocean over that of the central parts of the continent of Europe, which causes the air above that ocean to ascend, and to be replaced by a colder indraft from Russia:—are all related to the recurrence of the seasons.

To a similar cause many great ocean currents may also be attributed; amongst which we may name the mighty volume of the great Gulf Stream, to whose far-reaching warmth our own climate is doubtless much indebted; and, by way of contrast, the frigid current which gives an almost arctic character to the climate of Labrador;—a current which, flowing southwards in the depths of the ocean from the neighbourhood of the north pole, gradually obtains, after the manner of the trade winds, a certain amount of relative motion towards the west (owing to the more rapid eastward motion of the parts to which it travels), so that it impinges vigorously against that rocky coast, and, as it surges up, cools the surface of the surrounding sea.

But a more interesting question than any other which is

---

* When this lecture was delivered at Gresham College it was illustrated by models specially constructed for the purpose.

connected with the seasons and climates of the Earth, is that which has become so well known through its elaborate discussion in Dr. Croll's learned work, "Climate and Time."

When we go back in thought to ages long past; when we investigate the stratification of the Earth's rocks, and study the revelations of geology; we find the Earth's climatic zones have been far from constant. At one time, for instance, there is every proof, in the fauna and fossil remains that have been found, that England had a climate similar to that of the torrid zone. If so, the first idea suggested by such a discovery might perhaps be, that, at such a time, the Earth had not cooled down to its present condition, but was altogether hotter than it now is; in fact, that the temperature of *every* climatic zone was intensified by the Earth's internal heat.

But such a hypothesis cannot for a moment be maintained. For we have the strongest evidence that such periods of greatly increased heat *alternated*, from time to time, with what are termed *Glacial Epochs*, in which the whole of the British Isles were covered to a great depth with solid ice; so that they were, in fact, in much the same condition as that in which Greenland is at present. We are, therefore, at once tempted to inquire,—What other cause or causes could have brought about so remarkable a result?

Our Sun being doubtless a periodical star to a slight extent, *i.e.*, so far as the occurrence of sun-spots is concerned, we may ask,—Can its periodicity ever have been sufficient to have produced such epochs? But we are forced to answer that there is no evidence to justify such a supposition.

We must consequently look to some cause connected with the Earth itself, with its movements, or with its orbit, or with the inclination of its axis to the plane of the ecliptic, for the origin of the phenomena in question.

Now, it is undoubtedly the case, that the united effect of the attraction of the other planets changes the inclination of the Earth's axis to the plane of the ecliptic by about $2\frac{3}{4}°$ in the course of every ten thousand years; its present value of $66\frac{1}{2}°$ being very nearly the mean of its greatest and least values. It is also true, that the smaller this value the greater would be the distance of the arctic circles from the poles. The above change of inclination would, therefore, in all probability, produce an effect upon the temperature of the Earth, as a whole, and especially upon the climate of the polar regions, which would be by no means inconsiderable.*

But the climatic effects to which we refer are much greater than any which could thus arise. It has, therefore, been suggested that the alternation of glacial and torrid epochs may have in some way arisen from the elevation, or depression, of large districts, which may, from time to time, have produced a great change in the comparative distribution of land and water, and in the great ocean currents of the globe, and consequently in the distribution of heat upon the Earth. We prefer, however, to leave to geologists the discussion of any such hypothesis as this, and the results to be deduced from it; only remarking, that, if we understand the question rightly, changes of climate, involving the occurrence of glacial epochs, have not only taken place in England in very remote ages, but also in what, speaking geologically, may be termed comparatively recent times, during which no extraordinary alteration has occurred in the elevation or depression of neighbouring continents and seas.

It is purely from an astronomical point of view that we wish to refer to Dr. Croll's theory, especially as we believe that its importance may be decidedly increased, if it should be possible absolutely to prove that glacial epochs have not only alternated with tropical epochs, but that the occurrence of the one epoch in a given hemisphere has corresponded to the occurrence of the other in the *opposite* half of the globe.

Dr. Croll suggests that the cause of the occurrence of these epochs is to be found in an accumulative result, produced by a small change in the *ovalness*, or *eccentricity*, of the Earth's orbit; which change, although very slow, continues in the same direction for very long intervals of time, and arises from the united effect of the attractions of all the other planets. At present, as we have explained in Lecture V., the Earth's orbit is an ellipse which differs but little from a circle, its major axis

* See "Climate and Time," pp. 398 *et seq.*

being about 186,000,000 miles in length. The eccentricity, or ovalness, of this ellipse is represented by ·016, or very nearly by the fraction $\frac{1}{60}$; which means that the Sun is $\frac{1}{60}$th part of the way from its centre to its circumference, measured along its principal diameter; in other words, that the Sun is about 1,500,000 miles from the centre of the curve.

The Earth is, therefore, at present, at the time of *mid-winter* in our hemisphere, very nearly 3,000,000 miles nearer to the Sun than it is when in the opposite part of its orbit. Moreover, it is generally considered that, at a given latitude, winter is colder in the southern than in the northern hemisphere, and summer cooler in the northern than in the southern hemisphere; *i.e.*, it seems as if the greater proximity of the Sun mitigates our winters, and intensifies the southern summers. There may be some who will at once say: "Of course, this result might naturally be expected." But the connection of cause and effect is not in this instance as obvious as it might at first appear to be. For it also happens (see Lecture V., page 121) that the Earth travels at its highest speed when it is at its nearest to the Sun. It therefore gets through the most favourable part of its orbit for receiving heat most rapidly; in fact, it moves, upón the whole, just so much the more slowly at one time than at another, that it receives altogether an exactly equal amount of solar heat in each half of its yearly course.

Nevertheless, as we have stated, observation appears to indicate that the more intense heat of the southern summer in some way fails to make up, in its effect upon the climate, for the more intense cold of the southern winter. And we believe the principal reason of this is, that, when snow is deposited and formed into ice, as it is in an arctic winter, the quantity of Solar heat which, under the actual climatic conditions involved, is required to remove it, is much greater than would be the case if it could be applied under more favourable circumstances. For the heat of the summer, as it shines upon the frozen surface, raises fogs which hinder its own action, while it is engaged in dispersing the ice by changing it from a solid to a liquid, or vaporous, state. Over the equatoreal regions the same heat also raises an excessive quantity of vapour and cloud, which is carried by upper aerial currents towards the polar regions, there again to increase the fall of snow.

We have not space in which to discuss this question at any length (and we are aware that a good deal may be said against, as well as in favour of Dr. Croll's theory), but we must confess that it may be very fairly maintained, that the Earth's greater distance from the Sun during the southern winter does really cause the cold of the southern arctic zone, and of the southern winter as a whole, to exceed that of the northern hemisphere. It is also strongly urged in support of this opinion, that, even if all possible allowance be made for the configuration of land and water, or for the action of ocean currents, and for the effects of the existence of a great southern antarctic continent, there nevertheless appears to be no other sufficient explanation of the decidedly greater accumulation of snow and ice which is met with by mariners who have sailed towards the south pole; so that, although it has been found possible to reach a latitude of more than $83\frac{1}{3}°$ north,* the nearest approach to the south pole as yet attained is but little beyond 78°† of south latitude.

Assuming this to be the result of an eccentricity of $\frac{1}{60}$ in the Earth's orbit, we may easily understand that a considerably increased eccentricity would, in all probability, produce a greatly enhanced effect of a similar kind. And it has been calculated that some 210,000 years ago the eccentricity equalled about $3\frac{1}{2}$ times its present value, and that it has varied according to the subjoined table:—

The eccentricity of the Earth's orbit‡ is at present	.	·0168
20,000 years ago it was	.	·0188
40,000 ,, it had a minimum value	.	·0109
100,000 ,, it was	.	·0473
200,000 ,, ,, ...	.	·0569
300,000 ,, ,, ...	.	·0424
400,000 it was almost exactly the same as now .	.	·0170

* By Captain Markham (second in command of Sir G. Nares' expedition), and Lieutenant Parr, in 1876.
† By Sir J. C. Ross, in the expedition of 1839-43.
‡ Some additional values are given in Beckett's "Astronomy without

Much fuller tables, extending to 3,000,000 years of past time and 1,000,000 years of future time, with diagrams of illustrative curves, are to be found in Dr. Croll's treatise. It is there shown that there have been various intermediate and minor oscillations of value, and that more than a million years ago the eccentricity was occasionally reduced to so small a value, that it was only equal to $\frac{1}{2}$, or $\frac{1}{3}$, or even $\frac{1}{5}$ of the smallest value, ·0109, mentioned above. But the most interesting point to be noticed is, that during a period of some 200,000 years, from about 100,000 to about 300,000 years ago, it maintained an exceptionally large value; and that about the middle of that period it reached a maximum value of ·0575.

It is also of great importance to remember that, if the eccentricity remained unaltered (whatever its value might be), the northern and southern hemispheres would interchange any effect produced by it every 10,500 years. We cannot go into the proofs of this last statement, which indicate that the precession of the equinoxes, in consequence of its period being 26,000 years, would produce such an alternation of effect once in each 13,000 years, were it not for a modification caused by a movement of the perihelion of the Earth's orbit, which reduces the 13,000 years to 10,500. The result, however, is, that about 10,000 years ago, the summer of the southern hemisphere (instead of, as at the present time, that of the northern hemisphere) must have occurred when the Earth was in that part of its orbit in which it is at its nearest to the Sun, and that the same will be the case about 10,000 years hence. As far as this precessional effect is concerned, there might consequently result at such intervals, even with the present value of the eccentricity, an increased accumulation of snow and ice in the northern hemisphere, similar to that which at present exists in the southern hemisphere.

But during the period of great eccentricity, which, as we have stated, appears to have lasted from about 100,000 to 300,000

Mathematics," p. 51, where the whole question is very ably discussed. There is also an interesting table of the past and future eccentricities of the orbits of Mars, Mercury, and Venus, as well as of that of the Earth, in Lardner's "Astronomy," edited by Dunkin, article 785a.

years ago, we may suppose that the same alternate effect in the two hemispheres must have been of very much greater intensity. It has, indeed, been estimated by Dr. Croll, that the consequent diminution of the winter temperature, when the eccentricity was at a maximum, about 200,000 years ago, must have amounted to 73 degrees Fahrenheit below its present average value in our temperate regions.* Instead of the Earth's distance from the Sun varying at opposite seasons of the year by 3,000,000 miles, it would then have varied by 10,500,000; and the winter of one hemisphere, instead of being, as at present, about one week longer than that of the other, or than its own summer, would have been nearly a whole month longer.

It may therefore be considered very probable that, under such circumstances, glacial epochs would occur, first in one half of the world, and then in the other; with intervals of about 10,500 years between their respective maxima in either hemisphere. And during a period of 200,000 years, about ten may have occurred in each hemisphere, some more intense than others, according to the value of the eccentricity of the orbit for the time being; whilst, in alternation with them, a climate of a tropical nature may have extended, first in one and then in the other hemisphere, to a much greater distance from the equator than at present.

Although it must be allowed that some of the arguments and conclusions, which we have thus quoted, have been more or less disputed, we believe that no one can deny that they deserve the fullest consideration, or that they afford a most excellent example of the importance of the results which may follow in all astronomical investigations, from causes apparently most trivial; the whole question depending upon an eccentricity which, at the highest value,—viz., ·0575,—involved in our discussion, would only produce a difference of less than 300,000 miles between the major and minor axes of the earth's orbit.

It may be some comfort to our readers to know that no glacial epoch of such a description is likely to occur again for

* The outline of the calculation involved is quoted and explained in "Astronomy without Mathematics," p. 52.

a very long time to come. The ovalness of the Earth's orbit, upon the increase of which the effect in question depends, is now steadily *diminishing*, and will continue to do so until about 25,000 years hence. It will then begin to increase again, until in 50,000 years' time it will have about the same value as at present; and it will not exceed that value, except very slightly, for 100,000 years to come.

Next to the shape of the Earth's orbit, and its effects, in combination with the inclination of the Earth's axis, upon the climate of the various regions of its surface, perhaps the most important question to be discussed is,—What is the Earth's size and shape? We need not here descant upon the proofs ordinarily given that its shape is very nearly spherical, as it may be supposed that almost every child at school in these days is taught; that the way in which the distance of the horizon is increased as the mast of a ship at sea is ascended, while it retains a circular boundary line; the disappearance, as one ship sails away from another, first of its hull, then of the lower masts, and finally of the highest rigging; the shape of the shadow of the Earth upon the Moon in a Lunar eclipse;—all prove the Earth to be approximately a sphere. Experiments made with a spirit-level upon the surface of an extended expanse of still water, which show a fall of about eight inches in a mile, indicate the same fact. Nor is it difficult by actual measurement also to obtain a somewhat rough approximation to the Earth's size.

But if the astronomer is to make reliable observations, he needs to find out much more than this. It is always fundamentally important to know, not only the exact instant at which any observation is made, but also the exact locality of the observer who makes it; and this involves an *extremely* accurate acquaintance with the size and shape of the Earth upon which he is placed. We may know where the Earth is in its orbit at the moment in question, but this is not enough; we also need to know the precise point of space to which the observer is brought by his position upon it. In order to ascertain this, long and elaborate measurements of the distances between selected points upon the Earth's surface must be made, conjointly with determinations of the directions in which given stars are seen from them. Observations of the attractive force of the Earth upon bodies on its surface are also very useful and important. We will endeavour to give some slight idea of the way in which these various kinds of observations are utilized.

As the first step in the operation, it is not very difficult to discover that the Earth's form, although not precisely spherical, is very nearly that of a surface of revolution, *i.e.*, that it is symmetrical with respect to an axis passing through its north and south poles. In travelling along the equator, it is found, by observations made with a pendulum, or by some

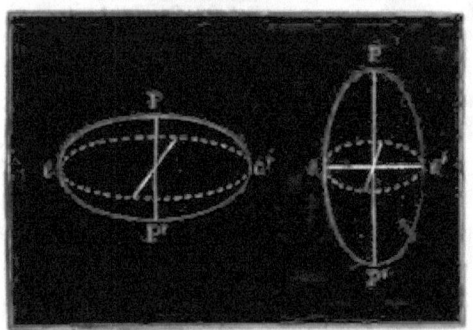

Fig. LIV.—Showing how a spheroid may be produced by the revolution of an ellipse about its shorter and longer diameters respectively.

other means, that the attraction of the Earth does not vary. It consequently follows that every point of the equator must be very nearly at exactly the same distance from the Earth's centre, or, in other words, that the equator is a circle. In like manner it may be proved that other parallels of latitude are also circles. This indicates that the Earth's form is such as would be generated by some curved figure revolving round an axis. But the next question is:—What must be the shape of this curve? If it be a circle, the Earth will be a sphere. If an ellipse, the Earth will be what is called a *spheroid*. If the diameter (P P′) about which the ellipse revolves be the longer, the shape of the spheroid will be that of a sphere with matter added near the poles, as in the right-hand drawing in Fig. LIV.; if PP′, as in the left-hand drawing, be the shorter diameter, the shape will be that of a sphere with matter added near to the equator.

How are we to decide which of these two cases corresponds with the reality? We can do so by measurements made upon a line drawn upon the Earth due north or south from the equator to a pole. If the earth were a sphere, this line would be a quarter of a circle, and would subtend at the centre o, an angle of 90° of latitude. And if it were divided into equal parts, such as AB, BC, CD, etc. (*see* Fig. LV.), each of

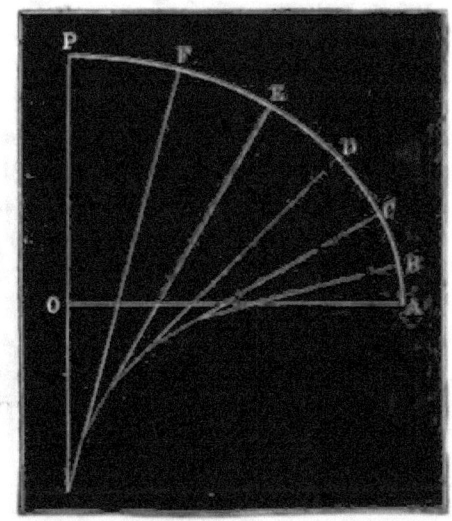

Fig. LV.—Showing that upon a spheroid of the shape of the Earth, an arc of a given number of degrees of latitude increases in length as we pass from the equator to a pole, and that the plumb-line does not in general pass through the Earth's centre.

them would subtend at o, an equal angle; while the straight lines joining o to A, B, C, D, etc., would all be perpendicular to the surface at the points where they would meet it; in other words, the direction of a plumb-line, or of the vertical to the surface, at each of them, would pass through the centre of the Earth.

Now it is easy to know when we have gone a distance corresponding to any given number of degrees of latitude, *e.g.* 15°, by noticing the successive positions A, B, C, D, etc., at which the direction of some particular star is found to have changed by 15° when compared with that of a plumb-line. When we do so, we do not find that the distances AB, BC, CD, etc., are equal. On the contrary, we are obliged, each successive time, to go somewhat further before the direction of the star has changed by the angle in question. It therefore follows that the curve which we traverse is not a circle; and it also results that the direction of the plumb-line does not continually pass through the centre of the Earth. Accurate observations prove that the curve is really an ellipse, and that the plumb-line successively passes as in the above Fig. LV.; the actual result of careful calculations being, that the length of an arc, corresponding to one degree of latitude, gradually increases from about $68\frac{1}{2}$ miles close to the equator, to about $69\frac{2}{3}$ miles close to the poles.

Some exceedingly slight tendency to an elliptic shape has also been suspected in the sections of the Earth, perpendicular to its axis, *i.e.*, parallel to the equator, which we have previously described as circles; but it is so slight, and so far doubtful, that we need take no notice of it. The Earth to all intents and purposes may be described as a spheroid, such as would be formed by the revolution of an ellipse round the shorter of its two principal axes. This ellipse differs from a circle to an appreciable, but by no means to any great extent; in fact, only so far as to make the polar diameter of the Earth about $13\frac{1}{4}$ miles shorter than the equatoreal diameter. In other words, a person at the pole is about $6\frac{2}{3}$ miles nearer to the Earth's centre than he would be if he were at the equator.

We reserve for explanation in our next succeeding lecture some interesting experimental proofs of the Earth's rotation, together with several methods by which the difficult problem of the determination of its weight has been very skilfully solved. At the same time we shall also discuss the very important question—whether its period of rotation upon its axis is constant, or slowly changing;—or, to express the same problem somewhat differently, how far the length of the day at the present time differs from what it has been in the past, or will be in the future.

LECTURE IX.

THE EARTH *(continued)*.

> " That spinning sleeps
> On her soft axle as she paces even,
> And bears us swiftly with the smooth air along."—MILTON.

WE have already several times spoken of the Earth's polar axis as that about which it rotates. It may perhaps be thought that it is hardly necessary to offer to a contemplative mind any proof that the Earth really possesses such a rotation, in addition to that weighty testimony which is afforded by the apparent daily motion of the stars, and of all the other heavenly bodies, around an axis passing through the north and south poles of the sky. When we consider the enormous distances of the stars, and the consequent enormity of the supposition of their all going round the Earth once in every twenty-four hours, as the ancient astronomers (who really knew nothing about those distances) supposed; and when we contrast with this the beautiful simplicity of the supposition that all these immense movements are only apparent, and are caused by the one simple fact that the Earth rotates day by day upon its axis, we can hardly escape the conviction (as we have more fully explained in Lecture V.) that this is the true explanation.

But proofs actual and positive may be asked for; and such can be given. For instance, if the Earth be revolving on its axis, the top of a high tower, being further from that axis than its base, must describe a longer circumference as it is whirled round from west to east; and must consequently possess a more rapid eastward motion. If a body be let fall from the top, it will therefore have the greater velocity of the point which it leaves, and will not fall exactly vertically, but a little to the *east* of the foot of the tower.

The above experiment, originally suggested by Sir Isaac Newton, A.D. 1679, was first made in that year by Dr. Hooke, of GRESHAM COLLEGE. The balls were, however, dropped from so small a height—27 feet—that any apparent deviation noticed was probably produced by other causes. In A.D. 1791 and 1792 the Italian philosopher Guglielmini let fall a series of balls from the Asinelli Tower at Bologna. They descended through 257 English feet, and exhibited a deviation, which was, upon the whole, in the right direction; but various precautions, which had not been taken, were found to be necessary in order to ensure a more accurate result. In 1802 Dr. Benzenberg therefore repeated the experiment at St. Michael's tower in Hamburg, from a height of 250 feet, when the observed deviation towards the east proved to be ·355 of an inch. This was in remarkably close agreement with ·342 inch, the value calculated by theory.

For a similar reason a body dropped over the centre of a very deep and vertical shaft will not be found to fall quite vertically, but will show a slight tendency towards the *east*; so that, if the shaft be deep enough, it will at last come in contact with its side. This has been satisfactorily tested both by the above-mentioned Dr. Benzenberg in the shaft of a mine 279 feet deep at Schlebusch in Westphalia, and still more successfully by Professor Reich in the year 1831 in the shaft of a mine near Freiberg. In the latter case the distance fallen was 520 feet, and a deviation of 1·1 inch was observed, the calculated value being 1·086 inch.*

Such experiments are, however, hardly quite so conclusive as might be wished, owing to their very delicate character, the smallness of the deviations which have to be measured, and the difficulty of making accurate allowance for the dis-

---

* See "The Earth and its Mechanism," by Henry Worms, F.R.A.S., published in 1862, chap. v.; in which, however, it is to be noticed that the distances are given in French feet, each of which is equal to 12·78933 English inches.

turbing action of currents in the surrounding air, which more or less affect the motion of the falling bodies.

More interest is consequently attached to an experiment with a pendulum, which was, for the first time, performed by Foucault in 1851. The principle involved in it is as follows:—If a heavy pendulum, or weight, could be suspended over the pole of the Earth, by a long fine wire, and be carefully set swinging, it may easily be understood that, once in every twenty-four hours, if the motion could go on undisturbed for so long, the ground would describe a complete rotation beneath it. To a spectator, unconscious of this movement, it would therefore seem as if the plane of oscillation of the weight were gradually rotated from east to west, as is indicated in Fig. LVI.; in which it may be observed that the pendulum is started by the burning of a thread by which it is previously held back, so as to prevent, as far as possible, the slightest lateral twist or tension from being given to it. The upper end of the wire is also so suspended, that no movement of the point of suspension, caused by the Earth's rotation, can affect the plane of motion of the heavy bob at the other end. Consequently, at the pole of the Earth, as above stated, the ground would appear to turn round beneath the bob with the whole angular velocity of the Earth's rotation. In three hours, as is partly indicated by the figures 1, 2, 3, etc., in the diagram, the direction of the oscillation would seem to have travelled one-eighth of the way round the whole circumference of the platform shown in the figure; in six hours round one-fourth; in twelve hours the pendulum (if it still continued swinging) would oscillate across the same line as at first, but in exactly the reverse direction, and so on.

It is not, however, difficult to perceive, that, if the same experiment were made at a place situated upon the equator, no such effect would be produced. At the pole, the point of suspension, being in the prolongation of the Earth's axis, is not carried round by the Earth's rotation; but, at the equator, it would be carried round the Earth's axis equally with the surface of the ground beneath it, and the pendulum would therefore continue to oscillate in any given vertical plane in

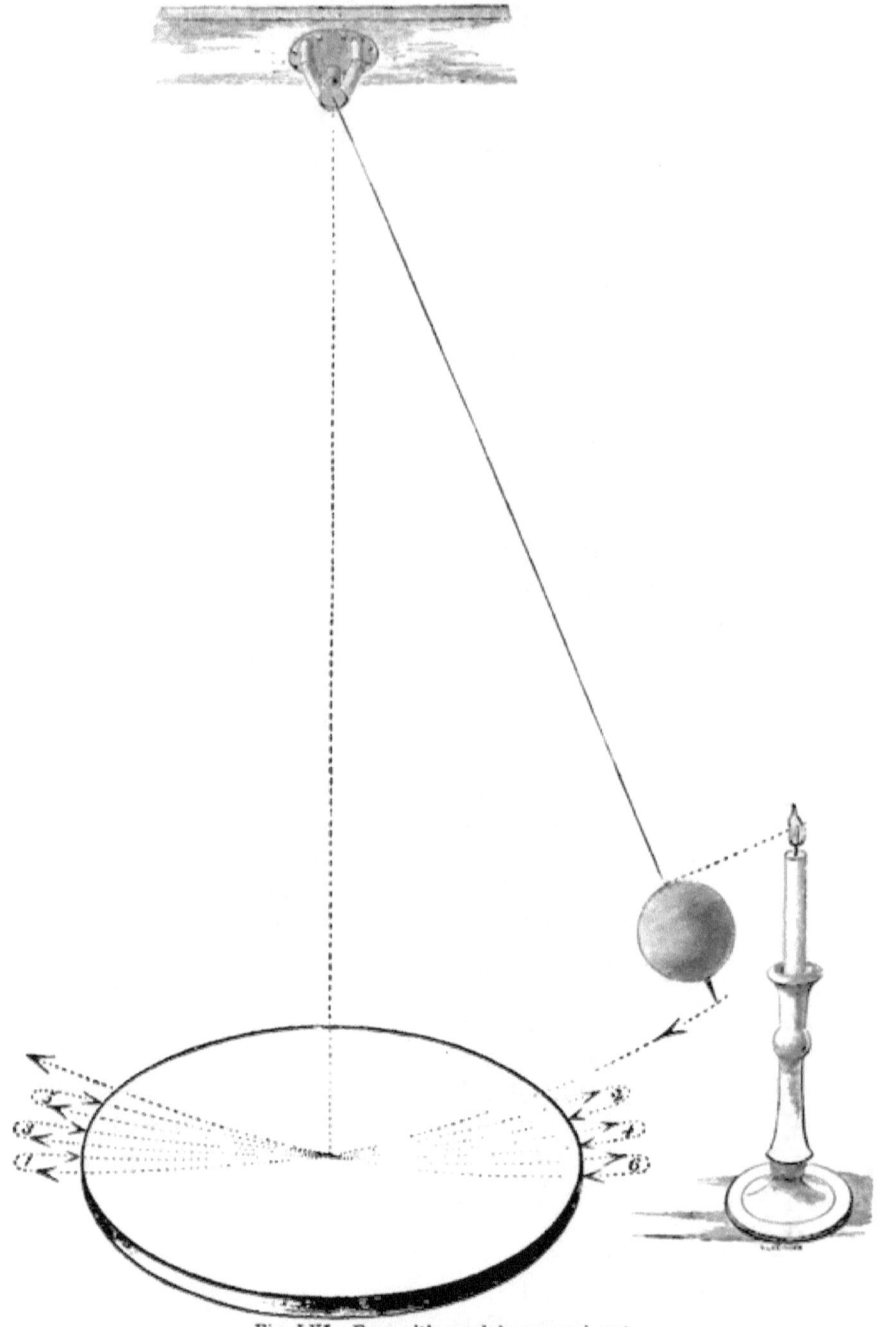

Fig. LVI.—Foucault's pendulum experiment.

which it might be started; or, to speak more simply, the bob would continue to move backwards and forwards over any given line, through the centre of the platform, over which it might at first begin to move. An attempt is made in the following little diagram to indicate by arrows the contrast between the two cases, the left-hand drawing in Fig. LVII. representing what would happen at the north or south pole; while the right hand drawing shows (as we have stated) that, at the equator, the platform would not appear to turn round beneath s, the point of suspension, at all. In the right hand drawing it may be noticed that the pendulum is supposed to be actually swinging in the plane of the equator itself; in

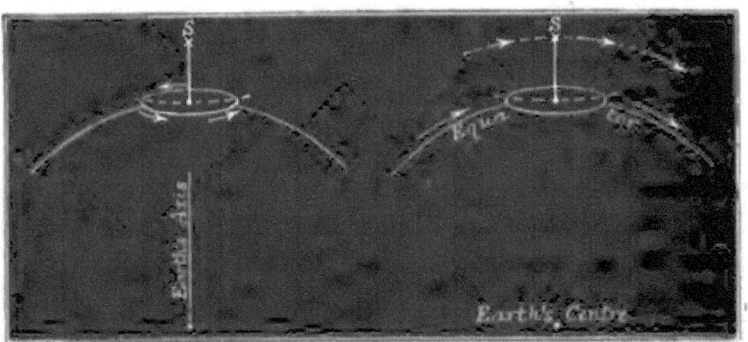

Fig. LVII.—The action of Foucault's pendulum at a Pole of the Earth, and at the Equator respectively.

which case it is perhaps most easy to realize the apparent immobility of its direction of oscillation. But it may be shown that the result will be the same for any plane, in which it may be started, at a place situated upon the equator.

In any intermediate locality of north or south latitude, it is more difficult to realize what will take place, inasmuch as the problem involves a somewhat clear appreciation of the effect of the resolution of a force into its components. It seems, however, very reasonable to suppose that the effect produced will be of an intermediate character; that the ground will, in fact, appear to turn round under the pendulum to some extent, but at a rate which is less than once in twenty-four hours. And it may be calculated that the apparent speed of its rotation will depend, according to a certain formula, upon the latitude of the place of experiment. When the experiment is properly conducted, as it was at the Pantheon in Paris in 1851, and subsequently at King's College, London, in 1859,* the calculated value is found to be in exact agreement with that observed, and a very satisfactory proof of the Earth's rotation is obtained.

Another experimental demonstration of the diurnal rotation of the Earth is afforded by a remarkable instrument which is called a gyroscope. The action of this instrument depends upon the fact that a heavy metal disc, made to revolve with extreme rapidity in its own plane, is with difficulty disturbed from that plane of motion. Such a disc, thus revolving, if suitably supported upon an axis, which is connected with a series of supports and pivots arranged in such a manner that it can, without any appreciable friction, easily move in every direction, will therefore maintain a *fixed position in space;* or, in other words, will keep its own plane of rotation unaltered, notwithstanding any change of position which the rotation of the earth may communicate to the place of observation at which it may be.

This being the case, if we set our disc in movement, and then look at a star along a telescope which is arranged so as to be perpendicular to the revolving disc, the star will appear to remain still so long as a sufficiently rapid revolution is maintained; but as the revolution of the disc slackens the star will begin to move across the field of view of the telescope, in consequence of the movement of the telescope caused by the rotation of the Earth. A similar effect may be observed by watching a long pointer connected with the disc. If a graduated scale be suitably placed, the rotation of the Earth will move the scale beneath the pointer; the pointer remaining fixed in space so long as the revolution of the gyroscope is kept up. By either of these methods the rotation of the Earth may be made evident, in contrast with the fixity in space of the plane of the rapidly revolving disc.

Another and a somewhat curious effect of the Earth's

* See "The Earth and its Mechanism," by Henry Worms, p. 112.

rotation is, that it affects the weights of bodies according to the latitude in which they are weighed. This result depends upon what is popularly, but not very accurately, termed centrifugal force ; a name which indicates a tendency to fly away from a centre, and which is used to express the fact that, if a body be revolving in a circle round a centre, its path must be bent into a circular form by some force tending towards that centre. Moreover, the faster it goes, the greater it is found that this force must be. Apart from the continued effect of some such action, it would, at any moment, when at such a point as P, in Fig. LVIII., run away from its curved path, along a tangent to it ; *i.e.* it would proceed along the straight line PT.

Fig. LVIII.—Showing how the tendency of a body to proceed along the tangent at any point of a curved orbit is counteracted by a central force.

In such a case the path might be curved either by the body being made to move in a circular groove, so that it would press against the groove, and consequently be pressed by it in turn towards C; or it might be tied to C by a string, which string, being strained by it, would pull it inwards ; or it might be attracted by some force existing at C. The force with which it needs to be attracted towards C, or pulled inwards by a string, or pushed inwards by a groove, is said to measure its centrifugal force ; which, however, is not (as the literal meaning of the word might suggest) any actual force which drives the body directly away from C, but simply the reverse of that tendency towards C, which must be given at each moment to it.

It may also be proved that the attraction, or force, employed for this purpose must be just so much greater as the radius of the path is greater (supposing that the time in which a revolution is performed remains the same), owing to the greater velocity with which the body will, in that case, describe it.

Now, if a body be at the equator, it evidently describes, in consequence of the Earth's rotation, a much longer path round its axis than it would describe in the same time in a higher latitude; and the centrifugal tendency will therefore be so much the greater. But this tendency (according to the laws of mechanics) is equivalent to two; the one along the meridian of any point upon the Earth (*i.e.* due south or north) towards the equator; the other in exactly, or almost exactly, the opposite direction to that of the well-known force of gravity, which acts very nearly towards the centre of the Earth. The effect of gravity is consequently in some slight degree neutralized, and the body appears to *weigh less* than it would if this centrifugal action did not take place.

It is evident that the only locality in which this effect will not in some degree be felt is at either of the poles; since a body so situated does not rotate round the Earth's axis at all, but simply remains *in it*. As, however, an observer increases his distance from the pole, it will be felt in a gradually increasing degree; and it will be greatest at the equator; but even there it is found that it only amounts to $\frac{1}{289}$th part of the effect of the force of gravity. This amount is, however, quite appreciable; and it is an undoubted fact that, from this cause alone, a weight of 289 lbs. at one of the Earth's poles would, if carried to the equator, weigh only 288 lbs.

The above statement is quite independent of another cause by which the weight would also be reduced, viz., that the Earth's shape is not that of a sphere, but, as we have previously stated, that of a spheroid. Owing to this, the protuberant matter at the equator removes a body placed upon it to a greater distance from its centre; and an additional diminution of apparent weight results, inasmuch as the Earth attracts anything upon its surface very nearly as if its whole mass were collected at its centre, and consequently with a power which is diminished when the distance of the attracted body from its centre is increased.

Moreover, it may be proved that the difference in the attraction thus caused will really be larger than might at first be supposed, owing to the increase of density which we have every reason to feel sure must exist in the central, as compared with the superficial, parts of the Earth. The actual result is found to be that a body weighs somewhat more than $\frac{1}{500}$th part less at the equator than at the poles, in consequence of the spheroidal form of the Earth.

In intermediate positions each of these effects will, of course, have an intermediate value. Taken together, they are sufficient to cause 194lbs. at the poles to weigh only 193lbs. at the equator; or 333 lbs. in London to weigh only 332 lbs. at Quito. This statement may be very satisfactorily tested by the use of an accurate spring balance; or it may be proved by observations of the oscillations of a pendulum, since these depend upon the effective weight of its bob, and if the bob become lighter the pendulum will oscillate more slowly. It is, in fact, found that a difference of about $2\frac{1}{4}$ minutes per diem is thus produced in the time of a clock governed by a pendulum, if it be removed from London to the equator.

Here, as previously in the probable effect of small changes of the eccentricity of the Earth's orbit upon the production of glacial epochs, we have another instance of the remote and indirect connection which astronomy proves to exist between causes and effects otherwise apparently dissociated. It is surely a useful, and we would hope a pleasing, intellectual exercise, to follow the steps by which it is demonstrated, that a clock will gain or lose, or the weight of a given volume of any given substance will be increased or diminished, according to the latitude of the locality in which it may be placed; and to such an extent as may suffice, if it be carefully observed, to demonstrate the rotation of the Earth upon its axis.

The rotation of the Earth may also be detected by artillery experiments. If a cannon be fired, due north or south, at a very long range (which with modern ordnance may amount to five or six miles), a considerable deviation of its aim will arise, from the difference of the velocity with which the Earth's rotation carries the gun towards the east (the whole of which velocity is imparted to the projectile as it leaves it), and that with which the point aimed at is carried in the same direction ; the effect being exactly similar to that met with in the case of the trade winds, which do not blow north and south, but from N.E. and S.E. respectively.

Since, then, a point on the Earth nearer to the equator is carried round towards the east more rapidly than one farther north, it follows, that the projectile from a gun in the *northern* hemisphere, which is fired when pointed due south, will fall somewhat to the *west*, or to the *right*, of the point aimed at, just as the trade wind, in approaching the equator, tends to blow towards the west. If, on the other hand, the gun be turned round to point due north and be again fired, the projectile will fall towards the *east*, or once more to the *right*.

Now let the same gun be carried into the *southern* hemisphere. It will be found that when it is pointed due north (or towards the equator) the projectile will, as before, fall to the west of the point aimed at ; and when the gun is pointed due south, it will fall, as before, towards the east ; but in either case this will, in the *southern* hemisphere, be to the *left* of the target. We are not aware that the *same* gun has thus been tried in the two hemispheres, so as positively to demonstrate that the effect, being always towards the *right* in the *one*, and towards the *left* in the *other*, is not produced by any defect in the gun. We understand, however, that, in firing at long ranges *north* or *south*, the deviation observed is very appreciable, and must be allowed for in accurate practice.* In firing due *east*, or *west*, of course no such effect is produced.

The rotation and shape of the Earth are, however, not only interesting in connection with its present condition, and with the accurate knowledge of an observer's position upon its surface. They take us back in thought to the time when, in ages

---

* A range of 10,000 yards, with a time of flight of thirty seconds, is by no means unusual. At a place having a latitude of 45° it may easily be calculated that the Earth's rotation would in such a case produce an easterly or westerly deviation of about forty-six feet upon the aim of a projectile fired due north or south. In latitudes nearer to the poles the effect would be greater.

long, long past, there are good reasons for believing that the Earth was in a fluid, or semi-fluid, condition. Under such circumstances that centrifugal tendency, which we have already explained, would give to a rotating sphere a *spheroidal* shape. It would cause it to swell out in the neighbourhood of the equator, and to become flattened at the poles; and it would do so to an extent exactly dependent upon its speed of rotation and upon its internal density. The faster its rotation, the greater would be the protuberance in its shape round about the equator; and very careful calculations have been made in the case of the Earth, which show that its shape is, as nearly as we can judge, such as a velocity, not differing much from that which it now possesses, would produce in a viscous or slightly-fluid mass, constituted according to that which we believe to be its internal law of density. We therefore find in the Earth's existing shape an important confirmation of the hypothesis that, at some epoch long past, it was in a fluid or semi-fluid condition. The same hypothesis is also supported by the much greater degree of polar flattening and equatoreal bulging which is found in the huge and less dense globes of Jupiter and Saturn, in accordance with their much more rapid axial rotation.

We have thus shown how we measure the Earth, and prove its rotation. To weigh it is a much more difficult matter; nor have the most carefully conducted experiments given results altogether accordant. One method, which has been adopted in two or three instances, has been to estimate the exact size and weight of an isolated mountain, and to notice to what extent its attraction influences the direction of a plumb-line on either side of it. Apart from any such influence, the plumb-line would be exactly perpendicular to the surface of still water, or to what is generally spoken of as the level of the Earth's surface at any given place; but, near to such a mountain, it is found that the attraction of the matter in it draws the plumb-line slightly towards it. Such an effect being in addition to that of the rest of the Earth, it may be compared with the attraction of the Earth as a whole; and if the mountain be very carefully surveyed and measured, and the weight of its component materials be estimated as accurately as possible, a very fair approximation to the value of the Earth's weight may be obtained.

This experiment is generally termed the Schehallien* experiment, inasmuch as it was first carefully made (in the summer and autumn of A.D. 1774) in connection with the Scotch mountain of that name in Perthshire, under the superintendence of Dr. Maskelyne, who was then Astronomer Royal. For further information with regard to it, in books easily accessible, we may refer our readers to Lardner's "Handbook of Astronomy," edited by Mr. Dunkin; and to Sir Edmund Beckett's "Astronomy without Mathematics," published by the Society for Promoting Christian Knowledge.

We may also remark that the first hint of the possible advantage of such an experiment is to be found in Sir Isaac Newton's "System of the World," and that some French academicians had already proved, before Maskelyne measured the attraction of the Scotch mountain, that the effect produced by that of Chimborazo was very sensible, although they did not accurately determine its value. Maskelyne tells us that he chose Schehallien because it was not only isolated, but in a direction lying east and west. The two stations for observation could therefore be chosen north and south of one another, so that it was especially easy to determine with accuracy, by means of the instrument employed (a zenith sector), the amount of deviation produced; inasmuch as the Earth's attraction, and the disturbing attractions of the mountain upon the plumb-lines used, were all in a plane running nearly due north and south.

Another very interesting method by which the problem has been attacked is generally known as the Cavendish Experiment, because it was first performed, between August 1797 and May 1798, by the celebrated chemist and natural philosopher of that name,—a grandson of the second Duke of Devonshire,—amongst whose other many researches none is, perhaps, more memorable than that which led him to the

---

* This is Maskelyne's spelling. More correctly, Schichallion, or Schihallion; or, as in the Ordnance Survey, Schichallion. Maskelyne states that in the Erse language the word means "constant storm."

discovery of the composition of water, in the year 1784. In order to weigh the Earth he employed an apparatus called a torsion balance, which was originally made (although in a less perfect form) by the Rev. John Michell, who, unfortunately, died before he had an opportunity of using it.

In performing the experiment the attraction of two large leaden balls (whose size is carefully measured)* upon two much smaller balls, is first of all calculated in the following manner. The little balls are fastened one at each end of a thin rod, which hangs in a horizontal position, and is supported by a fine wire at its middle point. The apparatus is so arranged that the two large balls may be respectively brought near to the two small ones upon opposite sides of them, and in such a way that the centre of each of the large ones may be at an exactly equal distance from that of the neighbouring small one; as also that the direction of the centre of each large one, measured in a straight line from the centre of the adjacent small one, is very nearly perpendicular to that of the length of the suspended rod, while the centres of all the four balls always remain in the same horizontal plane. The large balls being thus brought near to the small ones, it results that the whole effect produced by the attraction of the former upon the latter acts horizontally. Consequently the rod carrying the little balls is twisted round in a horizontal plane, until the resistance of the wire which supports it to further torsion produces equilibrium. But, in practice, this equilibrium is a long time in taking place, so that it is found better to determine the position in which the rod would finally remain, by observing the mean position about which it oscillates, rather than to wait for it actually to come to rest.

The experiment is then repeated, the only variation being, that the large balls are respectively brought by suitable means to the opposite sides of the small ones to those occupied in their previous positions, so that the rod is twisted in the reverse direction. Without drawing the suspensory arrangement employed in the earlier apparatus, or the platform

* The two large balls are made as nearly as possible exactly equal to one another, as also are the two smaller ones.

subsequently adopted as a better means of carrying the large balls, we have endeavoured in the following diagram,—Fig. LIX.,—to show them, first, in the positions A_1 and B_1; and secondly to indicate that A_1 is moved horizontally to A_2, and B_1 to B_2. The corresponding positions of the small balls are, for the one, a_1 and a_2; for the other, b_1 and b_2. The normal position of the rod which carries these latter, when unaffected by the attraction of the large balls, would be half-way between the directions $a_1 b_1$ and $a_2 b_2$. One-half, therefore, of the angle between these two directions is that through which the large balls twist the rod from its original position when brought

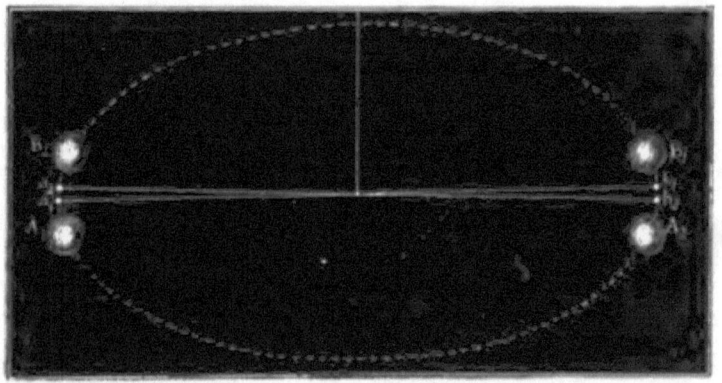

Fig. LIX.—Relative positions of the large and small balls in the Cavendish experiment.

within such a distance from the small ones as is indicated in the diagram.

By taking the mean result of a number of such experiments, in which the rod, which bears the little balls at its extremities, is first twisted in one direction and then in the opposite, it is possible to determine very precisely how far the attraction of the large balls, when at a given distance from the small ones, is able to move the latter. Very accurate observations are also made of the *time* occupied by the rod, which carries the small balls, in swinging horizontally between the extreme positions to which it moves, when it is disturbed by the attraction of the large balls upon the little ones.

We need not go into the details of the various methods which were actually adopted in practice to determine the time of this oscillation. Sometimes the large balls were suddenly moved away until the line joining them was at right angles to the original direction of the rod, so that their effect ceased, or neutralized itself, while the rod continued to oscillate from side to side of its original direction. Sometimes the large balls were very rapidly brought round to the opposite sides of the little ones, so as to encourage the oscillation still further. The essential point, however, connected with such oscillations (provided they are of comparatively small extent), is, that, whether they be caused by either of the above methods, or simply by moving the rod round in any other manner, and then letting it go, they will occupy the same length of *time*, so long as the instrument remains unaltered, whether the rod actually swings through a greater or less angle.*

In the Cavendish experiment we therefore make two distinct observations : (1) how far the big balls move the small ones by their attraction ; (2) how long the rod occupies in performing oscillations when it is set vibrating horizontally.

It is true that these vibrations actually depend upon three things : (1) the weight of the rod and little balls ; (2) the nature of the suspending wire, which actually effects the oscillatory movement by its resistance to being twisted, and its tendency, when twisted, to untwist itself and then to retwist itself in an opposite direction, which together constitute what is termed its force of torsion; (3) the length of the rod, which, as it is longer or shorter, gives a greater leverage to the balls at its extremities, and to the matter of which it is itself composed, to resist the twisting and untwisting power of the wire.

But it is most fortunately found, as the result of mathematical and experimental investigations, that the actual force of attraction upon the small balls necessary to move them a certain distance, bears a ratio to their *weight*,† which is quite

---

* This is just what also happens in an ordinary vertical pendulum. If such a pendulum swings twice as far, it gets through the distance at twice as quick a rate, so that the time of oscillation remains unaltered, or, as it is often termed, *isochronous.*

† We will here, and henceforward, neglect the weight of the *rod*, as it can easily be allowed for, and would only complicate our explanation.

independent of any *separate* consideration of what that weight may be, or of the length of the rod, or of the twisting power of the wire; provided that the *joint* effect of these three last-mentioned causes is such as to make the rod, when set oscillating, perform its oscillations in a certain observed *time*.

In fact it is found that the proportion which the force in question must bear to the weight of the little balls, in order to move them one inch from their original positions, must be rather less than $\frac{1}{50}$th of that weight, if each oscillation of the rod in the instrument occupies one second of time.* Moreover, it can also be proved, that, if the rod should oscillate in any other number of seconds, the requisite force would be less in the proportion of the square of that number. If, for instance, the duration of an oscillation were 400 seconds, the requisite force to move the small balls one inch would be nearly $\frac{1}{160000}$ of a $\frac{1}{50}$th part of their weight. And it is also a property of the apparatus in question, that the force required to move the balls any other number of inches would be almost exactly proportional to that number. In order to move them six inches, very nearly six times as great a force would be required as that which would move them one inch; the slight difference from an exact proportionality of the force being easily calculated and allowed for.

If, therefore, we find that the large balls move the little ones a certain number of inches, then, without weighing these latter, or measuring and observing anything else except the *time* in which the rod oscillates (which we can do with very great accuracy), we know exactly what proportion the attraction exercised by the big balls bears to the weight of the little ones.

Let us next see how this enables us to determine the weight of the Earth. It is a law of the attraction of spheres upon

---

* The actual proportion varies slightly with the latitude of the place of observation. The accurate value of the denominator of the fraction in the latitude of London is 39·139; and is the same, in any given locality, as the number of inches in the length of an ordinary vertical pendulum that will oscillate in a second. It is, therefore, always known with great precision.

spheres, that the comparative attractions of two, both made of the same material, upon any other, simply depend upon the distances of their respective centres from that other, and upon their respective sizes, or volumes. We can, therefore, easily calculate how many times less than the attraction of the Earth upon the two little balls, would be that of two spheres of the same size as our big lead balls, if they were made of a material of the same average density as the Earth. It would be less as they are smaller; it would be greater in proportion to the square of the smaller distance of their centres as compared with the square of the greater distance of the Earth's centre, which is 4,000 miles away. But the attraction of the Earth upon the small balls is only another name for their weight. Therefore, without actually weighing the little balls, we can calculate with what *fraction of their weight* two balls of the exact size of the big ones, and at their known distance from the little ones, would attract them, if the supposed big balls were for the time being made of material of the same mean density as the Earth. It also follows, in such a supposed case, that the rod which carries the little balls would be twisted round so as to move these latter through a certain calculable number of inches, which distance will otherwise only depend upon the time in which (as previously explained) it would swing backwards and forwards if set oscillating horizontally.

Consequently, without weighing the little balls, or determining anything else about our apparatus than the duration of its oscillations, we arrive at the following results. (1) That the attraction of the large lead balls moves the little balls a certain *observed* distance. (2) That it would move them a certain *calculable* distance, if the large balls were unaltered in size, but made of material of the same mean density as the Earth. (3) It follows, from the above (1) and (2), and from our previous statement as to the comparative distances to which any two given forces would move the small balls;* that, the latter distance (2) having proved, in the actual experiment, to be very nearly one-half of the former distance (1), the density

* Viz., that the distance through which they would be moved would vary in the same proportion as the attraction exercised.

of a material corresponding to the mean density of the Earth must be very nearly one-half of that of lead; or, that the weight of the Earth is very nearly one-half of that of a mass of lead of the same size as itself. It is surely an exceedingly interesting and important fact that so small and delicate a little instrument as a torsion-balance, by that very delicacy, enables us to solve the problem of weighing so huge a mass as the Earth.

The final result of the best determination yet made by the above method is, that the Earth's mean density is almost exactly one-half of that of lead, or about $5\frac{2}{3}$ times that of water.

The original investigation by Cavendish, made in A.D. 1797 and 1798, is recorded in the volume of "Philosophical Transactions" for the year 1798. It gave a value less than the above by about $\frac{1}{30}$th part, which was again somewhat decreased when it was repeated, with certain modifications, in 1836, by Professor Reich, of Freiburg. The value which we have stated is, however, more trustworthy, and was obtained by Mr. Baily, by an exceedingly careful series of experiments which he performed between A.D. 1838 and 1842. Diagrams of the apparatus used by Cavendish may be found in Lardner's "Handbook of Astronomy," and in other well-known books. A full description of Mr. Baily's improved apparatus (with diagrams), and of the elaborate precautions used by him to avoid the disturbing effects of currents of air, or of the heat of the observer's body, is given in the fourteenth volume of the "Memoirs of the Royal Astronomical Society," the whole of which is occupied with an account of his proceedings. His torsion-balance was very carefully enclosed in a glass case, while the large and small balls were separated by partitions, impervious to heat, the necessary movements being effected by ropes worked from a distance, and the observations made by telescopes through holes in a wall.

The especial value of the Cavendish experiment consists in the simple character of the observations involved, and in its being easy to make a torsion-balance which will oscillate so slowly (as, for instance, in 200, or more, seconds), that we are able by means of it to measure with accuracy so small an attraction as that of the large leaden balls,* which, in Mr. Baily's case, were about one foot in diameter,—an attraction so small in comparison with that of the Earth, that, in order for it to produce a measurable displacement in the direction of a plumb-line, similar to that caused by the mountain in the Schehallien experiment, the plumb-line used would need to be of the impracticable length of many miles. It is, moreover, a great advantage that the whole effect to be measured is produced horizontally, so that the movements of the apparatus involve no alteration of the Earth's attraction upon any of its parts.

It might perchance be just possible to obtain a relation between the attraction of a large ball upon a small one and the small one's weight (which is equal to the attraction of the Earth upon it), by weighing the latter in a *very* delicate balance, while the big one might alternately be placed a short distance above it, and then below it, and noting the difference caused in its apparent weight; but there would be many more difficulties involved, and much less accuracy would be attainable, than in the Cavendish experiment, the slowness of the oscillations in which not only increases the delicacy of the results deducible, but very much conduces to their own accurate measurement.

Our readers will, we hope, understand, that, in any such investigation the point to be arrived at is, to obtain a relation between the actual attractive effect of the Earth and that of a given weight of some known substance, such as lead. When we weigh two bodies against one another in any ordinary balance, although the weight of each is produced by the Earth's attraction, we only *compare its attractions upon them;* we cannot in any case determine what that attraction absolutely is. But in the Cavendish experiment we obtain a direct relation between the actual attractive effect of the big balls, so far as it depends upon the known density of the material of which they are made, and the effect which they would produce

* The balls were made of lead because of its weightiness, so that as large an attraction as possible might be obtained with balls of a moderate size.

if their density were equal to the mean density of the Earth. Consequently we obtain the value of this latter density.

Having thus found the mean density of the Earth, and having by other methods already determined its size, we of course at once know its weight; but it is practically useless to express that weight, as is sometimes done, in tons or pounds. It is much better simply to say (as we have) that it is $5\frac{2}{3}$ times as great as it would be if the Earth were composed throughout of water; and half as great as it would be if it were composed of lead.

To recur to what is perhaps the most important point of all, we may be allowed once more to remark, with regard to the Cavendish experiment, that, although we need to know the material and the size of the big balls (which it is easy to determine), we do not need to effect any independent determination of the size or weight of the small balls, or of the torsion-force of the wire by which the rod is suspended, or even of the length of the rod itself,—which, or some of which, it would be very difficult to estimate accurately. We need only to determine the *time* in which the rod swings when set oscillating, and the distance through which the centres of the small balls are moved.

In practice that distance may perhaps be most conveniently found by noting the movement of a pointer fixed upon a prolongation of the rod, a little beyond the position of the small balls upon it; in which case, in order to get the actual motion of the centres of the balls from that of the pointer, it is necessary to compare their distances from the middle of the rod with that of the end of the pointer. For this particular purpose the length of the rod may consequently be measured. We can still, however, affirm that the length in question is not involved primarily in the experiment, and is not, of itself, necessary for the calculation of the Earth's weight from the observed oscillations and displacements. The mention of its value in the general explanation of the experiment is therefore best avoided.

We trust that the importance of an accurate knowledge of the Earth's weight, and the comparative obscurity of some of the explanations usually given of the above valuable method of investigating it, will be accepted as our apology for having discussed it at so great a length.

We must not, however, pass by unmentioned another method by which the same great problem may be attacked—viz., that to which Sir G. B. Airy, while Astronomer Royal, devoted extraordinary skill and pains, in the year 1854, by means of experiments made at Harton Colliery, near South Shields. It depends upon the fact that it is found, that a pendulum oscillates *more quickly* at the bottom of a deep shaft than at its mouth. If the Earth were of uniform constitution throughout, this would not be the case, but the pendulum would, on the contrary, oscillate *more slowly;* because it can be shown by elaborate mathematical calculations, that all the portion of the Earth which lies at a greater distance from its centre than that of the pendulum, attracts the latter equally in all directions, and has consequently no effect upon its oscillation. The pendulum when at the bottom of a shaft is therefore attracted, as if it were on the surface of a sphere just so much reduced in diameter, as the bottom of the shaft is nearer than the surface of the Earth to its centre; or, in other words, as if all the portion of the Earth further away from its centre than the bottom of the shaft did not exist. The attraction would consequently, as we have stated, be less, and the oscillation of the pendulum would be slower, at the bottom of the shaft, than at the top, if the Earth were of an uniform density. The observed *increase* in the speed of oscillation proves, on the contrary, that the inner portion of the Earth is denser than the outer parts. And from a long-continued series of careful observations, it is possible to compute what the mean value of that inner density must be. The result thus obtained, although somewhat greater, decidedly confirms the values obtained for the weight of the Earth by the other methods which we have described.*

Upon the whole, then, we may conclude that the Earth is probably about $5\frac{1}{2}$ times as heavy as an equal volume of water

* See Lardner's "Handbook of Astronomy," edited by Dunkin: or Appendix III. to Airy's "Popular Astronomy."

would be, or twice as heavy as if made throughout of solid rocks similar to those of the mountain Schehallien; a statement which, as we have just explained, involves the very natural supposition that the substance of the Earth is much compressed internally.

But granting this, as we may without any demur, there still remain other very important questions for consideration—*e.g.*, Is the Earth solid throughout; or, if not, to what depth; or, does it consist simply of a thin, hollow crust, which is filled with such molten matter as ever and anon bursts through the outlets of volcanoes? What is the temperature of its internal heat? Does that heat liquefy the regions within, or does the enormous superincumbent pressure retain them in a solid condition?

Of one fact there is no doubt, viz. that, as we descend below the surface of the Earth, the temperature in general increases, and, in most localities where it has been tested, at a comparatively rapid and uniform rate of about 1° Fahrenheit for every sixty feet of depth; although the rate of increase varies to some extent with the nature and arrangement of the strata passed through, and occasionally has somewhat diminished after a depth of about 2,000 feet has been attained. If, however, it be allowed that the increase in general continues, it would follow, that, at a depth of about thirty-five miles, a temperature sufficient under ordinary circumstances to fuse the most refractory metals, must exist. Now, if any one will draw a circle of about $11\frac{1}{2}$ inches diameter, and take care that the line by which it is described be only $\frac{1}{26}$th of an inch in thickness (which would correspond to the ratio which thirty-five miles bears to the Earth's diameter of 8,000 miles), he may have some idea of the pleasant condition of things which such a circle would represent.

It would to many persons, we think, be excessively alarming to imagine, that, with so comparatively thin a crust, they were dwelling over so enormous a mass of liquid fire.

Such a crust would probably be constantly cracking and falling in, so that volcanic and other catastrophes would be most fearful in magnitude, intensity, and frequency. Moreover Sir William Thomson has shown, that the attractions and varying distances of the Sun and Moon would cause such tidal disturbances in the molten mass within, that so thin a crust would bend and yield as the internal tides passed under it; the consequence of which would be that the corresponding external tides, now seen day by day, would not be apparent upon a surface which would rise and fall equally with themselves. And other difficulties (which we cannot enter into here) connected with precession and nutation would also be involved. For all these reasons, the hypothesis that the Earth is for the most part liquid or fluid can hardly be maintained.

Nor is there any necessity to suppose that the heat, which under ordinary circumstances suffices to fuse any given substance, would do so in the interior of the Earth. In fact, it is found that the power of heat to fuse, or melt, may be counteracted by enormous pressure. It is true that we are not able to arrange experiments really equivalent to the actual state of things existing at great depths below the Earth's surface; but we can perform enough to show that the above statement in general holds good, without testing it fully.

We may therefore say that it is very likely that the Earth is solid at far greater depths than used to be supposed, and that its enormous internal pressures may neutralize the liquefying power of the internal heat. Perhaps at some great depth, although we are unable to decide how great, the heat gains the victory, and fluidity results. We cannot, however, go further into this question without involving ourselves in abstruse mathematical and geological considerations.* But

* Since the above was written, the very interesting work of the Rev. O. Fisher upon the Physics of the Earth's Crust has appeared. Its discussion of the way in which the necessary support for the weight of mountain ranges might be provided, upon the supposition of a fluid interior in the Earth, is especially important; as also are the remarks made upon the relation of volcanoes to vapour retained by pressure (like the gas of aerated waters) in a viscous subterranean ocean from which it may be slowly set free. We also refer our readers for further information to Professor Judd's very instructive treatise upon Volcanoes in the International Scientific series.

we hope that it may afford some comfort to our readers to believe, as we do, that day by day, while they go to their business or their pleasure, they are not walking upon a horribly thin crust of solid matter over a caldron of molten fire.

Indeed we conceive that there is very good reason for supposing that such volcanic eruptions as from time to time occur, are not caused by the pouring forth of any central molten matter, but that in some way the pressure existing within a portion of the crust at a *moderate* depth is relieved; whereupon the solid matter there situated immediately becomes liquid, and rises up to the aperture of the crater of a volcano.

Such a diminution of pressure may possibly be caused by the explosive generation of steam; or by chemical processes originating in the percolation of water to a depth where great heat exists; or by the escape of gases occluded in the rocks of the Earth;* or by some other cause which may lift up the weight of the superincumbent strata. In connection with any of these suppositions the remarkable proximity of most great volcanoes to the sea should by no means be overlooked.

One or two other interesting facts relating to the Earth still remain to be noticed. For instance,—that the centrifugal effect of the Earth's rotation, although it is at present very slight, is proportional to the square of its speed. If, therefore, that speed were sixteen times as great as it is, the centrifugal effect would be increased 256 times. The result would follow that anything placed upon the equator would become almost entirely weightless, instead of, as at present, according to our previous statement in this lecture, having its weight reduced by only $\frac{1}{289}$th part. A slightly greater speed would cause a person so located to feel absolutely without weight; while a rotation-period of the Earth upon its axis of less than one hour twenty-five minutes, would hurl everything placed upon the equator away into infinite space.

On the other hand, another centrifugal effect is most useful.

* Compare Shakespeare, *Henry IV.*, Part I., Act 3, Scene i. :—
   "The teeming Earth
   Is with a kind of colic pinch'd and vex'd
   By the imprisoning of unruly wind."

We refer to that produced by the Earth's velocity *in its orbit*, which is sufficient to prevent it from being drawn into the Sun, an unfortunate catastrophe which would gradually ensue, if its present speed of about $18\frac{1}{2}$ miles per second were reduced to 12 miles per second. In like manner, it may be calculated, that, if its speed were augmented so as to exceed 26 miles per second, the Earth would fly away from its present orbit altogether. The happy mean keeps everything as it should be.

It may also be worthy of mention, with regard to the speed of the Earth's rotation upon its axis, that it is owing to its very *regular* and *uniform character*, that we are unconscious of it, notwithstanding its magnitude. If it were suddenly changed, even by a very small amount, the most fearful catastrophes would be at once experienced. Many persons, we think, hardly realize how vast the Earth's orbital velocity of about $18\frac{1}{2}$ miles per second, or about 1,100 miles per minute, really is. It is about 1,200 times as great as that of our fastest express trains. If it were suddenly diminished by only $\frac{1}{1200}$th part, the universal effect would be equivalent to that produced by the instantaneous stoppage of such a train in a most frightful collision.

The speed with which the Earth rotates is, then, so uniform and regular, that, until very recently, it was believed to have been absolutely without any change, for at least some thousands of years past. But of late it has been suspected that the speed of its rotation may be very, very slowly diminishing; or, which is only another way of saying the same thing, that the length of the day may be very slowly increasing. In fact the whole question of the variability or invariability of the Earth's day, and of the possible causes which may slowly change it, has recently been the subject of much discussion. It can easily be understood that this is a subject not only full of interest in itself, but of great importance in connection with all astronomical observations, and especially in the comparison of those of ancient date with those of modern times.

We will therefore conclude our discussion of the Earth with

a brief explanation of the way in which the length of its day *
is determined, and of the most important cause by which it
may be changed. This explanation will also involve a reference,
which we should regret to leave unnoticed, to the difference
between an ordinary or *solar day*, and that which is termed
a *sidereal day*, and is more usually employed in astronomical
calculations.

As the Earth's rotation from west to east causes the heavenly
bodies to appear to travel round the heavens in the opposite
direction from east to west once during each of its revolutions
upon its axis, the duration of its rotation-period may easily be
determined with great exactness, by watching how long some
heavenly body (such as a fixed star) which has no appreciable
proper motion of its own, occupies in making one such apparent
circuit of the sky. The period thus obtained is that which is
called a *sidereal day*. It is found to be very nearly four
minutes shorter than an ordinary day of twenty-four hours in
length. Nor is it difficult to understand how this difference of
nearly four minutes arises. During one revolution upon its
axis (which we may suppose to begin at the moment of noon,
at a given place, when the Sun is on the meridian), the Earth
also traverses about $\frac{1}{365}$th of its annual path round the Sun.
This movement, apart from the Earth's rotation, would cause
any place to look at the Sun in a direction which would be
changed by the angle which $\frac{1}{365}$th of the whole circumference
of any circle measures at its centre.

When one rotation of the Earth is completed, the meridian
of the place being turned to this extent in a slightly different
direction from that which we supposed it at first to occupy,
will not have the Sun again exactly over it; and, since the
rotation of the Earth upon its axis is in the same direction
as that of the Earth in its orbit, it results, that the Earth
must turn very nearly $\frac{1}{365}$th part of another rotation in order

---

* In the following paragraphs we shall not refer to any difference
between day and night. The word *day* will be used in its more extended
meaning, in which it also embraces the duration of a corresponding
night. Thus we shall in every case speak of a day which will only differ
slightly from twenty-four hours in length.

to cause the plane of the meridian in question to pass through the Sun again. That this is really the case may perhaps be most easily understood by the simple experiment of carrying a globe slowly round a lamp, which may be supposed to represent the Sun, while the globe is at the same time revolved round its own axis. The same fact, put into astronomical language, may be stated as follows:—An ordinary day between two successive noons is about $\frac{1}{363}$th part longer than a sidereal day, and a sidereal day is consequently about $\frac{1}{366}$th part shorter than an ordinary day, the difference in either case being not quite four minutes.

We say an *ordinary* day, since the actual fact is, that our days, if measured by the interval between two successive transits of the Sun over the meridian of any given place, would exhibit very considerable deviations from uniformity; (1) because the elliptic form of the Earth's orbit causes the speed of its motion round the Sun (upon which we have seen that the length of the day, as above estimated, must partly depend) to be slightly irregular; and (2) because the plane of that orbit, viz., the ecliptic, is not perpendicular to the Earth's axis of rotation, but is inclined to it at an angle of $23\frac{1}{2}°$. We cannot, without the use of some rather difficult propositions, explain the exact results that thus follow at different epochs in each year. It must suffice to state, that it is found best to take a mean, or average, value of the solar day (or day that would be determined by the Sun) deduced from its actual values in the course of a whole year, for use as our ordinary day. This day, thus obtained, is called a *mean solar day*, and is divided into twenty-four hours of civil time. But it will be understood from our previous statement that this arrangement will sometimes cause the Sun to be on the meridian rather before, and at other times rather after, what our timepieces call twelve o'clock. In fact the difference may amount to about $16\frac{1}{3}$ minutes in one direction, and about $14\frac{1}{2}$ minutes in the opposite direction; which is expressed in almanacs by the phrase that the Sun is fast, or slow, compared with the clock.

We might, if space permitted, prove that the Sun and the clock exactly agree on four different days in each year, although the dates of those days do not occur at equal intervals apart. Our present purpose, however, only requires that we should clearly understand that the time of the Earth's rotation upon its axis is termed a sidereal day; and that the difference between a *sidereal* day and a *mean solar*, or ordinary day, arises from the Earth's motion round the Sun, in exactly the same way that a traveller, who should sail at an uniform speed, so as to journey round the world in 365 days, would find that he would gain or lose a day in the whole journey, or $\frac{1}{365}$th part of a day (*i.e.* about four minutes) in each day of his progress, according as his voyage might be made from west to east, or from east to west. Moreover, the Earth's journey round the Sun being made from west to east, it does not *lose* the one day, but *gains* it in each year. Only $365\frac{1}{4}$ ordinary days are therefore reckoned while it revolves $366\frac{1}{4}$ times upon its axis; in other words, $365\frac{1}{4}$ ordinary, or mean solar days, are equal to $366\frac{1}{4}$ sidereal days.

We have especially referred to this fact in order to be able the more clearly to state that it is a supposed gradual lengthening of the *sidereal day*, or (which is really the same thing) an extremely slow diminution of the speed of the Earth's rotation upon its axis, that has of late years been suspected. The suspicion has arisen in consequence of a peculiarity which has been detected in the Moon's apparent motion. Some observations of the Moon's place recorded more than 2,000 years ago have come down to us with great accuracy, owing to their connection with ancient eclipses. Now, if the Moon had continued, since those times, to move with the same average velocity which it at present possesses, it may be calculated that it would have attained to a position in advance of that in which we now find it. The most natural explanation of this fact might seem to be, that, in some way, the Moon must have slowly and gradually increased its speed since the date of those ancient records. But it is evident that the effect would be exactly the same if the increase in speed were not real, but only apparent; which would be the case, if the Earth's rate of rotation, by which we measure our time, and consequently estimate the Moon's speed, has been getting slower.

And it has been found that the results of the most careful theoretical investigations show, that only about one-half of the observed effect is actually due to a real, but very minute, quickening, or acceleration of the Moon's movement, caused by a very slow change in the form of its orbit around the Sun. It has therefore been asked :—Can any cause be suggested which, by affecting the length of the Earth's day, would account for the other half of the Moon's apparent acceleration as the result of the way in which we measure our time? Is there, for instance, any friction acting upon the Earth's surface which would very slightly check its rotation?

It has been replied that the Tidal Wave, as it sweeps round the Earth twice in every twenty-four hours, must by its friction produce some such result. The effect day by day, it is true, would be excessively minute, and very careful theoretical calculations indicate that it does not exceed $\frac{1}{57000000}$ of a second in a day. Nevertheless, this would be enough to cause a difference of twelve seconds, in the course of a single century, in the time at which the Moon would occupy any given position. It is shown in our foot-note that this result may be deduced by the summation of an arithmetical progression,* each term of which exceeds the preceding by the above-mentioned minute fraction of a second; the

* An arithmetical progression is one in which each term exceeds, or falls short of, the preceding by the same quantity. The sum of such a progression is found by multiplying the sum of its first and last terms by a number equal to one-half of the whole number of terms in the series— e.g., if we take the arithmetical progression 1, 3, 5, 7, 9, 11, in which there are 6 terms, then, in order to get their sum, we must multiply the sum of 1 and 11 by one-half of 6, the result being 3 times 12, or 36. If the first term, as in the case referred to in the text, be the fraction $\frac{1}{57000000}$ of a second of time, it is so small that we may neglect it in forming the sum which is to be multiplied by one-half of the number of terms in the series. If, also, each term exceed the preceding by the same small fraction of a second, as we have explained that it will in the above case, we shall have for the sum of 36,525 terms (which is the number of days in a century) very nearly one-half of 36,525 times the last term, which last term will be $\frac{36525}{57000000}$ of a second. The result (in seconds) is therefore 36525 multiplied by itself and divided by twice 57,000,000; or 1,334,075,625 divided by 114,000,000; the quotient of which evidently amounts to rather more than 12 seconds.

form of the progression in question arising from the fact that a clock, which, in consequence of such a cause, might become a certain fraction of a second wrong in one day, would the next day show an error of double the magnitude, of treble the magnitude the third day, and so on.

It may also be calculated that in 2,500 years the error produced would be about 625 times as great in its total magnitude as in one century; so that the exceedingly minute retardation of $\frac{1}{57000000}$ of a second in one day would involve a displacement of two hours of time in the place of the Moon in 2,500 years, which is equivalent to a space of one degree, or to about twice the Moon's diameter as seen upon the celestial sphere. When this apparent effect is added to the very nearly equal amount by which the Moon's place is changed by the real acceleration of its movement, the most ancient observations are brought into excellent accordance with those of the present day.

Here, then, once more, we have an interesting example of the important results which astronomy teaches us may arise from causes not only apparently trivial, but which are such as might be supposed to be utterly unconnected with one another.

Since the Moon travels across the apparent width of its own diameter in about an hour, we may put the above statement into a slightly different form, and say;—that a diminution of the length of the day caused by tidal friction only amounting day by day to $\frac{1}{57000000}$ of a second, and which, in 2,500 years, or 913,125 days, would only make the actual length of each day less by $\frac{913125}{57000000}$ of a second (i.e., by less than $\frac{1}{62}$nd part of a second), would by its *accumulative effect* upon the time reckoned by a clock, during those twenty-five centuries, cause the Moon to be seen in a given position two hours sooner than would otherwise be the case. It would also, in the same length of time, alter the bounding longitudes upon the Earth from which an eclipse would be seen by 30°, inasmuch as a difference of one hour by the clock corresponds to a difference of 15° of longitude.*

* In the above explanation we have closely followed that given by Sir E. Beckett in "Astronomy without Mathematics," pp. 177 and 182,

To some of our readers it might possibly be interesting (although we think not very profitable) to discuss the appearance which the Earth would present if viewed from those of the planets which are its nearest neighbours; what details of its larger continents and seas might be seen from Mars in comparison with the amount of Martial detail which is visible to us; or how far both the configuration of the Earth's surface and the ever-varying forms of its cloud-formations might be not only interesting, but very puzzling, to an observer upon Venus, or even to one upon Mercury.

But we must, for want of space, refrain from any such discussion. For the same reason we are forced to pass by many other questions which belong to the astronomy of the Earth, —some more, and some less, abstruse,—but all important and instructive. For further information upon these, larger and more complete treatises must be consulted. The only addition which we will at present make to our previous statements is, to insert a few numbers and statistics, not as yet given in detail, in a list, which is appended below, of some of the principal elements of the Earth's size, shape, and movements. Although the information contained in it is by no means exhaustive, but on the contrary only consists for the most part of rough approximations, expressed in very elementary language, such as may be easily remembered, we hope that it may be of some use to our readers for the purpose of occasional reference.

The Earth's sidereal year—*i.e.*, the interval of time in which it makes one complete circuit of the Sun—is $365 \cdot 256$ days, or 365 days 6 hours 9 minutes 9·6 seconds.

The eccentricity of its orbit is ·01677, or about $\frac{1}{60}$th.

Its mean distance from the sun is most probably very nearly 93,000,000 miles.

Its greatest and least distances are respectively obtained by adding to, and subtracting from, its mean distance very nearly a $\frac{1}{60}$th part, and are about $94\frac{1}{2}$ and $91\frac{1}{2}$ millions of miles.

The corresponding amounts of light and heat received from

to whose lucid exposition we beg to acknowledge that we are very greatly indebted.

the Sun by the Earth vary inversely as the squares of the above numbers, and are approximately as the numbers 16 and 15—*i.e.*, they differ by about $\frac{1}{15}$th or $\frac{1}{16}$th part of either.

The equatoreal and polar diameters of the Earth are given
          by Airy   and by Bessel
     as 7,899·170 miles   as 7,899·114 miles
     and 7,925·648 miles   and 7,925·604 miles
               the difference being
       26·478 miles   26·490 miles.

More recent investigations make the difference slightly to exceed 27 miles; or give as the values 41,852,404 feet and 41,709,790 feet respectively.

A sidereal day is 3 minutes 55·91 seconds, or very nearly 3 min. 56 sec., less than a mean solar day.

The Earth's mean density is about 4 times that of the Sun, and about $5\frac{1}{2}$ times that of water.

Its weight is between $\frac{1}{320000}$ and $\frac{1}{325000}$ of that of the Sun.

The Earth's volume is only about $\frac{1}{1300000}$ of that of the Sun.

The Sun's volume is so huge, that, if its centre were at the centre of the Earth, its external surface would be nearly twice as far away as the Moon.

The Earth's axis is inclined at about 23° 27′ to a perpendicular to the plane of the ecliptic, from which inclination the principal phenomena of the seasons result.

At present the spring and summer quarters of the year each last, in the northern hemisphere of the Earth, for $186\frac{1}{2}$ days, the winter and autumn for $178\frac{3}{4}$ days respectively. In the southern hemisphere the above durations are interchanged.

The Earth is now in the nearest point of its orbit to the Sun about the 1st January, and at its furthest away about the 2nd July; the dates varying by a day or two either way, partly from the irregularity in the length of the civil year involved in the interposition of leap years, and partly from the perturbations of its movement caused by the attractions of the other planets. After the lapse of about 10,500 years, the above dates will be reversed.

LECTURE X.

THE PLANET MARS.

> "The planets of each system represent
> Kind neighbours."
> YOUNG.

> "It timor, et major Martis jam apparet imago."
> VIRG., Æn., 8, 557.

WE must wing our flight beyond the orbit of the Earth, to a distance which may vary from about 36,000,000 to about 62,000,000 miles, in order to reach the orbit of Mars, our nearest neighbour as we journey away from the Sun. This may seem to be a rather indefinite statement, but we shall presently see, that it simply affirms the orbit of Mars to be so considerably oval in shape, that the distance between it and the nearest portion of the approximately circular orbit of the Earth is in one part nearly $1\frac{4}{5}$ times as great as in the opposite part. And this peculiarity of the path of Mars has so much to do with the occurrence of those occasions when we can best observe the planet, and with what we shall have to discuss with regard to it, that we desire, at the very commencement of this Lecture, to draw special attention to it.

Let us then see, by the quotation of a few numbers from the tables given in Lecture V., how far we are justified in the statement we have just made.

We there indicated that the Earth revolves at a *mean distance* from the Sun of about 93,000,000 miles, and that the planet Mars has a *mean distance* of about 142,000,000 miles. It follows that, if the Earth and Mars should be at any time in the same straight line, or very nearly so, and on the *same* side of the Sun, they would, if at their respective *mean distances* from it, be about 142 *minus* 93, or 49,000,000 miles apart. If, on the other hand, they were on opposite sides of the Sun, their distance apart (upon the same supposition) would be 93 *plus* 142, or 235,000,000 miles.

Now we have already seen, that the Earth's distance from the Sun only varies from its mean value, in different parts of its annual orbit, to the extent of about $1\frac{1}{2}$ millions of miles. We have therefore next to consider how far that of Mars will vary, owing to the much greater ovalness of its path.

The Sun is situated at about $\frac{1}{11}$th of the distance between the centre and one end of the principal axis of the orbit of Mars, which is expressed by saying that the eccentricity of the orbit is $\frac{1}{11}$th. The distance of Mars, when at its nearest to the Sun, is therefore only 129,000,000 miles; but at its furthest from it, it is 155,000,000 miles; $\frac{1}{11}$th of 142,000,000 (or about 13,000,000 miles) being in the one case subtracted, and in the other case added to the mean value of 142,000,000 miles.

In comparison with the difference of 3,000,000 miles between the greatest and least distances of the Earth from the Sun, this difference of 26,000,000 miles in the case of Mars is so much more important, that, if we only wish to make a rough investigation (which is all that we are at present attempting) of the extent to which the distance of the two planets may vary, when in, or very nearly in, a straight line passing through the Sun, we need only subtract 93,000,000 miles (the *mean* distance of the Earth) from 129,000,000 miles, and from 155,000,000 miles, respectively. The resulting differences are 36,000,000 miles and 62,000,000 miles. Our previous statement is therefore true, that the distance between various points of the orbit of Mars and the corresponding points of the orbit of the Earth which lie between them and the Sun, varies from about 36,000,000 to about 62,000,000 of miles.

This next leads us to the very interesting question:—How often will it happen that Mars and the Earth will be so situated in their respective paths that they will approach within, or nearly within, the smaller of these two distances?

As we have already stated, we may, in an approximate calculation, reject the variation of the Earth's distance as comparatively trivial. It will, in fact, suffice, if we calculate when Mars will be in that part of its orbit in which it most closely approaches the Sun, and then notice whether the Earth will at the same time be nearly between the two.* Such a position of Mars relatively to the Earth and the Sun, when it is as nearly as possible in the straight line joining them, but upon the opposite side of the Earth to the Sun, we shall denote by the technical term *Opposition*.

Having premised thus much, let us investigate how often *especially* near approaches of the Earth and Mars will recur, when the latter is in Opposition to the Sun, supposing the Earth's orbit to be circular, and that it is only the ovalness of that of Mars which need be considered.

We first of all notice, that, the period of Mars round the Sun being 687 terrestrial days, or about 43 days less than two of our years, the planet will be at its nearest to the Sun, or, in more technical language, in its *Perihelion*, once in every 687 days. Let us imagine the Earth to pass exactly between the Sun and Mars very nearly at a date when Mars is so situated. Then the two will make an *especially* close approach to one another. This was the case, for instance, in the autumn of the year 1877, as is indicated in Fig. LX., in which the point, P, is the Perihelion of the orbit of Mars, while the line marked 1877 joins the places of Mars and of the Earth early in the month of September of that year.

Mars having therefore approached particularly near to the Earth, when seen in Opposition to the Sun, in 1877, what we have next to determine is :—How soon can it and the Earth again occupy similar, or nearly similar, relative positions?

To do so, we must next observe that, while the Earth

---

* If the Earth could also at the same time be very nearly at its *greatest distance* from the Sun, the above-mentioned 36,000,000 of miles might become more nearly 34,000,000. In these days, the distance is occasionally less than 36,000,000 ; as in 1877, when it was about 35,000,000 of miles; but it would take an almost inconceivably long course of ages, before the exceedingly slow movements of the Perihelia of the two orbits (*P* and *p* in Fig. LX.) would allow the Earth to be in its Aphelion, and Mars in its Perihelion precisely together ; so that the nearest approach of the two ever possible might exactly occur.

would travel once round the Sun, Mars would go somewhat more than half-way round. By the time that the Earth would have completed a second circuit—that is, by the end of $730\frac{1}{2}$ days—Mars would have performed about $\frac{1}{18}$ more than one whole circumference of its orbit, of which it would have finished one complete circuit about 43 days previously. Now the Earth does not journey quite so quickly as Mars in its angular motion round the Sun; it would therefore, before overtaking Mars, describe rather more than $\frac{1}{5}$th of a *third*

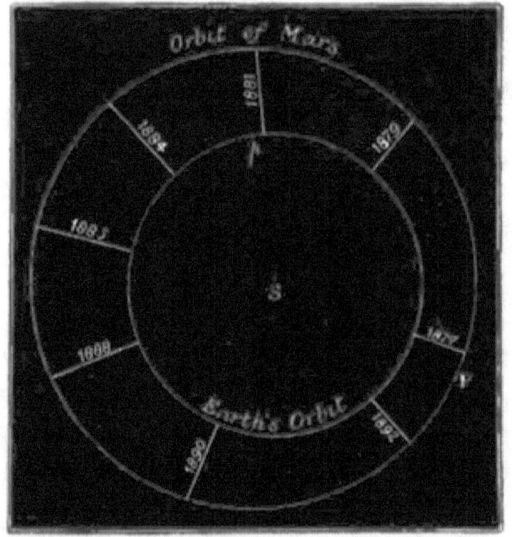

Fig. LX.—The positions of Mars in its orbit, when seen from the Earth in Opposition to the Sun, between the years 1877 and 1892.

circumference; in doing which it would occupy upon an average 50 additional days. Consequently, at the end of 780 days, or 93 more than one whole revolution of Mars, the Earth would once more be between the Sun and that planet, or, in other words, Mars would be seen in Opposition again; it being, however, by this time about 93 days' journey further away from P, the point of its orbit in which it is at its nearest to the Sun. The line marked 1879 in Fig. LX. indicates the actual occurrence of this event in November 1879.

The next subsequent occasion when Mars would be in Opposition, as seen from the Earth, would be at the end of about 780 days more, when it would consequently be about twice 93, or 186, days from its position of nearest approach to the Sun. This is shown by the line marked 1881 in the figure. In like manner there would be Oppositions in 1884, 1886, etc. But the time of one whole revolution of Mars in its orbit being 687 days, it follows that, at the date of the seventh succeeding Opposition after that with which we began, Mars would have advanced (apart from any irregularity in its velocity) $\frac{651}{687}$ths of the whole way round its orbit; and at the date of the eighth Opposition $\frac{744}{687}$ths of the whole way round (651 being 7 times 93, and 744 being 8 times 93). That is to say, it would on the seventh occasion be rather short of having again attained the position from which we supposed it to start; while on the eighth it would have passed somewhat beyond it.

In either case it would not be very far from its nearest approach to the Sun, and consequently to the Earth; and, if allowance be made for the variations in its speed as it travels in its orbit, as also, if it be not quite so close to its Perihelion to begin with as we have supposed, it may happen that *either* the seventh, *or* the eighth Opposition, may be the next at which an especially close approach to the Earth will again occur.

After this it will be necessary to wait for the next occurrence of such an event until seven or eight more Oppositions shall have taken place; and the interval between two successive Oppositions being, as we have shown, about 780 days, it therefore results, that unusually near approaches of the Earth and Mars occur at intervals of seven or eight times this number of days—*i.e.*, after about 5,460 or 6,240 days;—in other words, either after the lapse of very nearly 15 years, or after rather more than 17 years. Fig. LX. shows that the positions of the two planets on the occurrence of the Opposition of 1877 were such, that the seventh succeeding Opposition in 1892 is that which will involve their next especially close approach; inasmuch as that of 1894* would be situated con-

* The place of the line representing that of 1894 would be nearly two-thirds of the way from the line marked 1877 towards that marked 1879.

siderably further beyond the Perihelion, P, than that of 1892 falls short of it.

If the time of the year at which Mars is thus especially near to the Earth be also such as to give the planet an altitude above the horizon of any given place which enables it to be well observed, its brilliancy will be exceedingly striking; and details of its surface may be detected, which it may be very difficult to see, or which may be altogether invisible for fifteen, or seventeen, years to come. It is mentioned, for instance, in Chambers' "Handbook of Descriptive Astronomy," that De Zach states that in 1719 its brightness was so extraordinary as to cause a panic. The ruddy colour of its light, for which, according to the most ancient records which we possess, it has always been celebrated, also becomes at such times especially remarkable in naked-eye observations. This was very evident in the remarkably near approach of the year 1877; a year which, as we shall presently show, will be memorable for many reasons in the history of this planet.

We have already pointed out that it seems, upon the whole, more easy to conceive that beings somewhat akin to ourselves may exist upon a planet at a considerably greater distance from the Sun than the Earth, such as Mars, than upon one which is situated nearer to it, such as Venus or Mercury. It is therefore very interesting to find, that the telescope reveals many more points of resemblance to the Earth in the case of Mars than in that of either of these two last-mentioned planets.

As we have stated in Lectures VI. and VII., we know practically nothing of the physical characteristics of Venus and Mercury, owing to their becoming more and more involved in the glare of the Sun as they approach their nearest positions to the Earth; and also, in the case of Venus, in consequence of the cloudy envelope with which it appears to be surrounded. On the contrary, there is not the slightest doubt as to our being able to distinguish many physical features and details of the highest interest upon the surface of Mars. When Mars is at its nearest to the Earth, it is true that it is about eleven millions of miles further from it than the nearest distance within which Venus may approach; but this comparative disadvantage is far more than compensated by the fact that Mars, being in Opposition, is at the same time in its very best position for observation; that is to say, it is then seen upon the meridian at midnight at its greatest altitude in the heavens, and it is high above the horizon during all the dark hours.

What then, let us ask, may be observed at such a time in a powerful telescope? Certain darker markings or shadings upon the surface of the planet are at once noticed; which, however, occasionally appear to be of a slightly greenish hue, in contradistinction to the ruddier tint of the surrounding parts. These markings are found to be constant, or nearly constant, in form, although subject at times to some amount of temporary obscuration. As observation has become more painstaking, and telescopes have been improved, it has been found possible to recognize an increasing amount of detail, and to construct more or less accurate maps of the surface of the planet. In these maps the darker markings are named oceans, seas, gulfs, bays, or inlets, on the supposition that they are portions of water; while the intermediate regions are termed capes, continents, or islands.

Each portion of the surface is, of course, best seen when it is turned directly towards an observer, or nearly so; *i.e.*, when it is on, or nearly on, the centre of the planet's apparent disc. At other times it appears foreshortened, and may be so far distorted in its apparent shape, that it is difficult even for a practised observer to recognize it.

An approximate value of the rotation-period of Mars upon its axis may be very easily obtained by noticing the interval which any small and decided mark upon the disc, or the central point of any larger one, occupies in passing once round, before it returns to a similar position again. For a more accurate determination it is well to take a much longer interval, in which a large number of revolutions have occurred, and to divide the interval by that number; a procedure which proportionally lessens the effect of any slight errors of observation. For this purpose the ancient drawings of the marks seen upon the planet by the celebrated Professor of Geometry in

Gresham College, Dr. Hooke, made more than two hundred years ago, and others of not much less antiquity, have proved very valuable. They have incontestably demonstrated that the principal markings are of a permanent nature, so that we can compare, with great exactness, the positions into which they were then brought by the rotation of the planet, with those in which we now see them at any given moment.

From such comparisons (which have been made with special care by Mr. Proctor) it is found that *twenty-four hours thirty-seven minutes twenty-two and three-quarter seconds* is an exceedingly close approximation to the true rotation-period of Mars. The year of Mars being nearly double our own, it is, we think, all the more remarkable to find that its day and our day are so nearly of the same duration. As the larger planets, Jupiter and Saturn, rotate much more rapidly than the Earth (the one in just under ten hours, the other in about ten and a half), it might, perhaps, have been imagined that Mars, being so much smaller in size, would rotate so much the more slowly; but this is not the case. While we can say very little about the rotation-period of Venus, and still less as to that of Mercury, we can affirm with the utmost certainty that the excess of that of Mars over that of the Earth falls short of thirty-eight minutes.

It may, however, be very fairly asked, whether we are equally justified in asserting that the darker markings upon the planet are assuredly caused by water, so that they really indicate the localities of seas and oceans?

The analogy of the Earth, and the general impression conveyed to the eye, may suggest this interpretation of them; but some more positive testimony may be demanded.

If so, we reply, in the first place, that we are frequently able to detect temporary obscurations of considerable portions of the surface, which we may, with much confidence, ascribe to the effect of clouds overshadowing them; and that, if there are clouds, there must be water.

Again, we find that the markings near to the edge of the disc are always comparatively indistinct, or obscured, owing, almost undoubtedly, to the greater thickness of atmosphere through which they are seen. The non-existence of any appreciable atmosphere would (as in the case of the Moon) be opposed to the existence of water; its presence confirms it. We also find that the hemisphere of the planet which is enjoying summer, in which the weather might be expected to be drier and less overcast, is much more clearly seen, as though it were more free from aqueous vapours and cloud, than that which is passing through the season of winter.

Moreover, if we test the light received from the planet by means of the spectroscope, and compare it with that received from the disc of the Moon,* when the two are seen nearly side by side; the former shows dark lines, representative of watery vapour, which are wanting in the case of the latter.

We have, therefore, abundant indications of the existence of water upon Mars; and, as there is no doubt that water would reflect much less light than land, we are amply justified in believing that the darker markings are caused by its presence. Their occasionally greenish tint may afford some farther confirmation of this opinion.

But, in addition to all the above reasons, there is one more convincing still. It has to do with certain brilliant white spots which involve, and surround to a certain distance, the North and South Poles of the planet. As the summer progresses, in either hemisphere, the spot belonging to the corresponding pole rapidly diminishes in size. On the other hand, as the winter progresses, it undergoes a great increase and extension. It is also noticed, that the variations of size in the southern spot are greater than those of the northern one (as is also the case upon the Earth, *see* Lecture VIII., p. 190); a coincidence especially important when we remember that the summers of the southern hemispheres, both of Mars and of the Earth, occur in those halves of their respective years in which they at present pass through their nearest approaches to the Sun. We are, moreover, able to watch the increase and diminution of the southern polar spot with peculiar facility, because the south pole is that which is inclined towards us whenever the proximity of

* Such a comparison was made by Dr. Huggins; see *Monthly Notices* R. A. S. for March 1867 (vol. xxvii., p. 178).

Mars to the Earth is unusually close, as it was in the years 1862 and 1877, and as it will be in 1892.* And, when we thus watch this southern spot, we often notice, at such times, a considerable amount of irregularity in its contour; as is shown in the drawings of the planet in Plate VII.

Upon the hypothesis which we may presume has already suggested itself to our readers,—viz., that these polar spots are composed of snow,—such irregularities may be ascribed to differences of elevation; the hills and mountains of greater altitude remaining covered with snow, while those of less elevation are denuded. Moreover, a few smaller outlying white spots may be seen, which probably indicate isolated mountain peaks somewhat further removed from the poles.

It is also worthy of mention, in this connection, that one other remarkable white spot, to which the name of Hall's Snow Island has been given, has been observed from time to time when the planet has been favourably situated. This island, which is far removed from the polar regions (its situation being in about 40° of south latitude), may easily be found in the chart of Mars which forms our frontispiece. There seems to be very little doubt that there is upon that island some very lofty mountain which is always covered with snow.

Apart, however, from any such additional details as these, it must, we think, be conceded, that the alternation of *increase* and *decrease* in *winter* and *summer*, of both polar spots, combined with their extremely white and brilliant appearance—so brilliant that one of them has occasionally been visible with a telescope when a fog in the neighbourhood of the observer has hidden all the rest of the planet from his view—affords almost indubitable proof that they are caused by *snow*. If so, there can, we think, be little doubt that the darker tracts upon the planet's disc are really covered by water.

---

* Another remarkable peculiarity of the southern polar spot also deserves special attention; viz., that it does not appear to be concentric with the axis of the planet, while the northern one is very nearly so. Why this discrepancy should exist we cannot say. It may possibly arise from some local effect, dependent upon the distribution of land and water, or it may be due to oceanic and atmospheric currents.

We may notice in the chart of Mars (see frontispiece) that there appears to be a polar sea surrounding, or approximately surrounding, each polar snow-cap. It is, however, very difficult to determine the details of the planet's surface so certainly as we could wish in its higher latitudes; possibly owing to the prevalence of polar mists and fogs upon it. This is the case even in the southern hemisphere, although, as we have already mentioned, we are acquainted with that hemisphere as a whole much better than we are with the other half of the planet, in consequence of the south pole being inclined towards the Earth whenever Mars most nearly approaches it.

Very interesting drawings have been made from time to time of both hemispheres by many skilful observers, by the careful comparison of which a really surprising number of details, extending to high southern latitudes and to a more moderate distance north of the planet's equator, have been authenticated. Amongst the most successful sketches of a date previous to the near approach of 1877, we may mention those of Messrs. Dawes and Lockyer. From the examination of a large number, many of which were by Mr. Dawes, while others were by earlier observers, a most excellent, and now well known, map of the planet was some years since published by Mr. Proctor, copies of which may be found in several of his popular and instructive works upon astronomy.

More recently, in 1877, and again in 1879, very many additional drawings have been made. From the careful attention which has been given to these by M. Terby, of Louvain, we hope that conclusions of much importance may be deduced. In our own opinion, the most beautiful of all the drawings yet published are those of Mr. Green, a Fellow of the Royal Astronomical Society, and a most ardent amateur of astronomy. In order to obtain the best possible view of the planet, he was so enterprising as to take a large reflecting telescope to the island of Madeira in the summer of 1877, where, night after night, he made his observations; and, happily possessing great artistic power, depicted what he saw with the utmost skill and accuracy.

By his kindness and that of the Council of the Royal Astronomical Society, three of his sketches (the darker markings in which represent water) are appended in Plate VII., as well as the full-size copy of his Chart of the Planet upon Mercator's projection, deduced from his own and from many other drawings, which forms our frontispiece. In this no detail is inserted which has not been confirmed by at least three independent observers. The chart and twelve views of the planet, tinted by chromo-lithography, were originally published in vol. xliv. of the *Memoirs* of the Royal Astronomical Society. The copy of the chart in black and white, which we have been permitted to use, is from the *original* lithographic stone, which was fortunately found to be intact, and was most kindly placed at our disposal by Mr. Green. For this great and unexpected favour the author of this lecture desires to express in the warmest terms the gratitude which he feels. He is sure that the more this very beautiful map is known by the public, the more will its merits be appreciated.*

It may be well to compare some features of special interest in the three views in Plate VII. with their surroundings in the chart. The longitude of the centre of the first view is 7°. In it Dawes' Forked Bay is seen rather to the left of the centre; above which bay, extending still further to the left, is Phillips' Island. Mädler continent occupies a considerable portion of the right-hand part of the lower half of the drawing, and above it is

---

* It may be noticed that in Mr. Green's chart there are only a few occasional and faint indications of certain numerous and very remarkable so-called canals, which Professor Schiaparelli, at Milan, in the latter part of 1877, and again in 1879 and 1881, has seen apparently running in such directions that they split up into many smaller divisions the line of continents, which otherwise would occupy the greater part of the equatoreal regions of the planet. The most remarkable point connected with these peculiar features of the surface (if they really exist, and are not, more or less, the effects of an optical delusion, or of currents in the atmosphere of Mars) is, that they were less evident when Mars was at its nearest to the Earth in 1877, than when it had moved away to a somewhat greater distance. They were also even more clearly seen by Schiaparelli in 1879, when the planet was still further from the Earth; while in 1881 (as is shown in a somewhat rough but very interesting drawing by that learned Professor, which is reproduced in *Nature* of

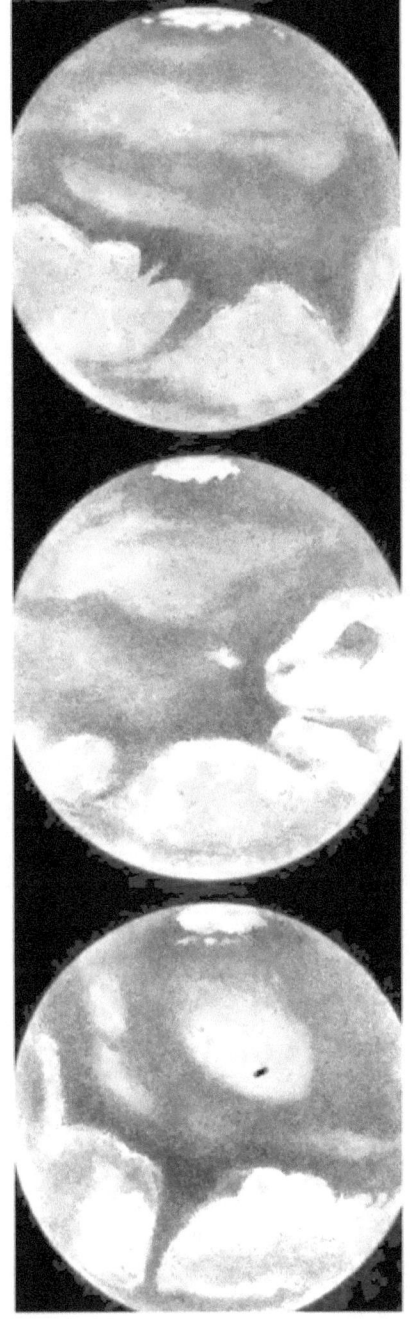

Christie Bay, somewhat foreshortened. Hall's Snow Island is hardly noticeable, but it comes out very distinctly in the next view (in which the central longitude is 43°). In this it may be also observed that the dark Schiaparelli Lake, situated nearly in the same latitude as the above-mentioned island, and to the right of it, together with Terby Sea still further to the right, form with the surrounding land a most peculiar resemblance to the head, eye, and ear, of a dog, or other animal.

The third drawing (in which the central longitude is 243°), represents by far the most perfect of all the views of the planet which Mr. Green obtained. The most prominent feature in it is the Kaiser, or (as it is sometimes termed) the Hour-glass Sea, from the lower part of which runs off the narrow Nasmyth inlet. Lockyer Land, and some other lands, are well seen in the upper portion of the view. Many very important half-tints also prove how excellent the state of the atmosphere was at the time when the drawing was made. Dawes' Forked Bay may be noticed near to the right-hand side of the view; thus showing that, in our brief description, we have completed the circuit of the planet. For an interesting discussion of many other details we refer our readers to Mr. Green's original memoirs.

It will be noticed in the views in Plate VII. that the illuminated disc of the planet is of a circular form, since they

May 4th, 1882, and in the French periodical *L'Astronomie* for August 1882) they appeared to be in many cases *double;* so as to consist of two narrow lines running parallel and very near to one another. What should have produced this duplication, or why it should be more visible in 1881, even in the clear air of Italy, when the planet was so much further away than in 1877, it is difficult to conceive. It must, however, be allowed that in the drawings of other observers there are occasional confirmations of some of these supposed canals. We think, however, that Mr. Green was quite right to omit them when he drew his chart, nor should we be as yet inclined to insert them, although they seem to have been, not only once or twice, but often, so very distinctly seen by Professor Schiaparelli that they are deserving of the most careful attention. It has, we believe, been recently suggested, that possibly temporary inundations of water may have had something to do with some of these apparent variations in the geographical configuration of the planet.

refer to dates very near to the time of its Opposition. It may, however, be well to mention that this would not always be its shape. Such a planet as Jupiter is so distant from the Earth that the change of phase by which its disc occasionally appears to fall very slightly short of a complete circle is almost too small to be detected; while that of Saturn, or of any more distant planet, is quite inappreciable; but Mars is, comparatively speaking, so near to us, that its change of phase is at times very noticeable. It must, however, be remembered that, its orbit being exterior to our own, its phase does not vary from that which we see in a Full Moon to that which belongs to a New Moon, but only diminishes to some extent from a circular form, and then begins to increase again. When the planet is in Opposition to the Sun and at its nearest to the Earth, its illuminated disc is circular. When it is upon the opposite side of the Sun, in Conjunction with it, and at its greatest distance from the Earth, it would (if visible) be also circular, although very much smaller in diameter. In intermediate positions one-half of its illuminated disc falls somewhat short of a circular form, the diminution being greatest when the planet is in quadrature—*i.e.*, when it is seen in a direction at right angles to that of the Sun. It then presents an appearance similar to that of the Moon when it is three days from being full; one diameter being about $\frac{9}{10}$ths of the length of the other.

We must, however, hasten on to speak of various other very interesting questions connected with the physical condition of Mars, although the space at our disposal will only allow us to allude to them with the utmost possible brevity. Why, it may be asked, is its light so red, that from ancient times it has been associated with wars, and stratagems, and toils; that Propertius calls it "the savage star of rapacious Mars;" and that Dante, as he describes the heaven of this planet, declares:—

> "With him shalt thou see,
> That mortal who was at his birth imprest
> So strongly with this star, that of his deeds
> The nations shall take note"?

We must confess that we can give no satisfactory answer to such a query. Some who have thought upon this subject have suggested, that a peculiar and prevailing tinge of the vegetation upon the surface of Mars may produce its ruddy hue. Others, that it may be due to a large amount of red sandstone among its rocks, or of red earth in its soil. But no explanation yet given really meets the difficulties of the case.

The inclination of the equator of Mars to the plane of its orbit round the Sun (a matter of some doubt) has until recently been considered to be about $27\frac{1}{4}°$. If so, the peculiarities of its seasons would not be very different from those of the Earth, with the exception that they would be slightly intensified; its arctic circles extending only some four degrees further from the poles (or rather less), and its tropics the same distance further from the equator, than our own limit of $23\frac{1}{2}°$.*

At any rate, so far as the question of the habitability of Mars is concerned, we may say:—Its day is only a little longer than our own: its surface is varied by land and sea, by snow-clad mountain and arctic ice: its temperate and tropical zones are not very different in extent from those of the Earth. But, at the same time (as may be seen from Mr. Green's chart), the proportion of land compared with that of water is probably much greater; so that there is about an equal extent of each, instead of one part of land to three parts of water, as upon the Earth. The chart also indicates that the water frequently winds amongst the land in long inlets, and that it may be in general possible to journey the greater part of the distance between any two given localities by water.†

On the other hand, the year of Mars is very much longer than our own. Moreover, as a consequence of the ovalness of its orbit, the number of days in its successive seasons must greatly vary. This is the case to such an extent, that in the

---

* We understand that Professor Schiaparelli has recently deduced an inclination of rather less than $25°$, which would cause the climatic zones to correspond almost exactly with those of the Earth.

† Some modifications and additions must be admitted into the above statement, if the existence of Schiaparelli's canals be confirmed, and if it be allowed that they are formed by water.

northern hemisphere there are 372 Martial days in the *summer* half of the year, while in the *winter* half there are only 296; the lengths of the four quarters being respectively;—of the spring, 191 days; of the summer, 181 days; of the autumn, 149 days; and of the winter, 147 days. In the southern hemisphere the durations are of course interchanged.

Such a variation in the length of the seasons, although considerable, would, however, not be very troublesome. And the smaller amounts of heat and light which are received by Mars might coincide with a slower growth of vegetation and of the crops, so that a year and seasons of about double the length of our own might be very convenient. There are, nevertheless, other reasons connected with the diminished amount of solar light and heat, to which we have just referred, which, in our opinion, involve the question of the habitability of Mars in very considerable difficulties.

Upon an average this light and heat would only be of an intensity equal to $\frac{43}{100}$ths, or rather more than $\frac{2}{5}$ths, of that which the Earth receives, which may be shown as follows:— The Sun would present to an observer upon Mars a diameter about two-thirds of the size of that which it exhibits to us, because Mars is at a distance from it equal to about $1\frac{1}{2}$ times that of the Earth; and the light and heat received would vary as the *surface* of the apparent disc, *i.e.* as the square of $\frac{2}{3}$rds, or nearly as $\frac{4}{9}$ths; the more accurate value being, as above stated, $\frac{43}{100}$ths.

In the preceding paragraph we have used the expression "upon an average," because it must be remembered, that a considerable variation in the above value would also be caused by the ovalness of the orbit of Mars. Its minimum distance from the Sun being 129,000,000 miles, its mean distance 142,000,000 miles, and its maximum distance 155,000,000 miles, there would be a corresponding variation in the Sun's apparent diameter of about $\frac{1}{11}$th part of its mean value. In other words, the diameter of the Sun's apparent disc, at different times in each Martial year, would be as the numbers 10, 11, and 12. The light or heat received would be as the squares of these numbers, *i.e.* as 100, 121, and 144; or nearly as the numbers 5, 6, and 7; *i.e.* it would sometimes be increased, and sometimes diminished, by a sixth part of its mean intensity.

But such a variation is not excessive. We need, therefore, only take into consideration the average amount of light or heat which the planet receives. And the habitability of Mars, as far as this point is concerned, really depends upon the answer that may be given to the following query:— Can we suppose that beings such as ourselves, if there be no other obstacle, could exist with somewhat less than one-half of the solar light or heat that we enjoy? We think that the answer must certainly be, Yes!

When we consider how readily man can adapt himself to an arctic or to a tropical climate; when we compare the average temperature in which a Greenlander, or an Esquimaux, can be contented, with that in which the Papuan, or Central African negro delights; we are inclined to think, that it would need no very great modification of bodily constitution to enable a man to live on Mars; especially if there were something in its atmosphere, or in its physical conditions, which might prevent the cold from being so great as it otherwise would be.

And it has been suggested that this must be so, because the snow-caps do not extend by any means so far from the poles as we should have expected would be the case, when we remember that the greater distance of Mars from the Sun, if in other respects it resembled the Earth, is such as would produce an average temperature upon its equator no greater than that which is found in about 62° of north or south terrestrial latitude.

At the same time we must confess, that we perceive no signs upon Mars of any specially dense atmosphere, heavily laden with vapours, such as might retain, and (so to say) hoard up the heat received. The mists, or clouds, which we see are in general thin and transitory; and other observations of various kinds all combine to indicate, that the apparent absence of frost and snow is accompanied by a rarer, rather than by a denser, atmosphere than that of the Earth. Moreover, it would be according to all analogy to suppose, that a globe so much smaller than the Earth should be enveloped by a proportionately less extent of atmosphere. At the same time, the much smaller attraction of gravity upon Mars, quite apart from any relation between the whole mass of the atmosphere and that of the planet, would cause any given quantity of air to extend to a much greater height, and to possess a much less density in its lower layers. Upon each square inch of the Earth's surface the pressure of the atmosphere amounts to about fifteen pounds avoirdupois; upon Mars (if the same amount of air, instead of a proportionately less amount, were superimposed upon each square inch of its surface) the pressure would only be about $5\tfrac{17}{20}$ pounds, so that the surface density of the atmosphere would be barely $\tfrac{2}{5}$ths of that to which we are accustomed. It would, in fact, be less in the same proportion as the attraction of gravity is less upon Mars than upon the Earth, and would be nearly equal to that of the air which is met with upon the tops of our highest mountains. If, in addition, the total amount of atmosphere were reduced in the proportion of the mass of Mars to that of the Earth, the density would be much further diminished.*

But it may be asked:—Is it not because of the rapid radiation of heat due to the rarity of the air upon mountain tops that the moisture precipitated upon them takes the form of perpetual snow? although, the rarer the air, the more freely do the incident rays of the Sun pass through it. In fact, if the atmosphere upon Mars is very rare, would not the result be an intensification of cold, and a consequent great extension of the polar snows, with an occasional deposition beyond their ordinary limits, of such large quantities of snow as would be very easily seen by means of our telescopes. Some such effects are doubtless occasionally observed, but they are far less important or extensive than we might reasonably expect.

* Some of the above reasons for the probable rarity of the density of any atmosphere of Mars are more fully discussed in a series of very able and interesting articles, published in the January, February, and March numbers of the *Sunday Magazine* for 1882, by Mr. Maunder, F.R.A.S., of the Royal Observatory, Greenwich, to whose charming explanations we beg to acknowledge that we are much indebted.

How then is this discrepancy to be reconciled with the supposition of the rarity of the planet's atmosphere?

It may be replied that, the cold must undoubtedly be very great if the atmosphere is rare, but that, owing to other causes,* there is not anything like a similar amount of cloud or vapour, rain or snow, upon Mars to that which there is upon the Earth. In a rarer atmosphere vapour would not rise or be diffused so rapidly, and atmospheric currents would be much weaker; so that the vapours formed over the seas and oceans would have a tendency to hang over the regions where they were generated, instead of being blown away as clouds, to fall upon the continents after a while, in a re-condensed form, as snow.

Possibly, in some such way, the absence of a general covering of snow, extending much further from the poles than appears to be the case, may be explained. There are, however, many difficulties involved in the whole question, which we must not discuss any further. We can only say, that, if the atmosphere upon the surface of the planet be of less than two-fifths of the density of that to which we are accustomed, or possibly of a much rarer density still, any beings who may breathe it must be of a decidedly different constitution from ourselves.

It seems, therefore, upon the whole, most reasonable to suppose that Mars, as a smaller planet than the Earth, may have passed more rapidly through its process of evolution, and be now no longer so suited for habitation as it may once have been. At the same time we must allow, that the difficulties connected with the supposition of the habitability of Mars by beings not altogether dissimilar from ourselves, are, so far as we can judge, less formidable than in the case of any other of the planets.

It may consequently be not altogether out of place to remark, that any such inhabitants, if actually existent, would probably be of a considerably larger build than those who people the Earth, and might, therefore, in their larger bodies maintain a higher temperature than ourselves, and for their larger eyes need less light. The supposition of their larger size is, of

* See the articles, previously referred to, by Mr. Maunder, in the first three numbers of the *Sunday Magazine* for 1882.

course, consistent with the diminished attraction of gravity, to which we have previously referred, which would cause a man weighing ten stone here to weigh less than four stone upon Mars, and enable one fifteen feet high to be as light and agile as a man six feet high would be upon the Earth.

In connection with the above reference to the diminution of the attraction of the force of gravity upon the surface of Mars, —a result which depends upon its size as well as upon its density,—it may be well to mention, that the planet is really smaller than the occasional brilliancy of its light might lead many persons to suppose; its volume being only about $\frac{1}{8}$th of that of the Earth, and its weight between $\frac{1}{9}$th and $\frac{1}{10}$th.

It is, indeed, a somewhat curious fact that, as time has gone on, successive observations have given smaller and smaller values for the diameter of the planet. It may be remembered that we stated that this was also the case with Venus. In both instances, this result is probably related to the irradiation which surrounds the bright images seen in a telescope, from which the measures of their diameters have been made. The better the telescope employed, the more accurately can the boundary of their discs be determined. Consequently the continued improvement in modern instruments has diminished our estimates of their size. The diameter of Mars is now believed to measure about 4,200 miles, or between that and 4,300 miles; instead of 4,800, or 4,900 miles, which was until recently the accepted estimate.

It should also be remembered, that the diminution in the value assigned by recent investigations to the distance of the Sun from the Earth diminishes our estimate of the distances of all the planets, and consequently of all their diameters which have been deduced from those distances combined with their apparent sizes. If, for instance, we alter the Sun's distance from 95,000,000 miles to 93,000,000 miles, we must take $\frac{2}{95}$ths off our previous value for every planet's diameter. We have mentioned these facts lest any of our readers should find it difficult to reconcile the values assigned to the diameter of Mars in text-books of a few years since, with some of the statements made in this lecture.

We will now give a short account of what is undoubtedly one of the greatest, as well as one of the most recent, triumphs of astronomy. We refer to the discovery of two moons, or satellites, of Mars in the year 1877. On the night of August 11th, 1877, Professor Asaph Hall was searching with the magnificent refracting telescope of the Washington Naval Observatory, in the neighbourhood of the planet, with the hope that the application of a more powerful and effective instrument than had ever before been used for its observation, combined with its especial proximity to the Earth at the time, might enable him to discover a satellite, if one should exist which, owing to its small size, had not been previously seen. It is very remarkable that he almost immediately perceived a small star-like object, at an apparent distance from Mars equal to between two and three times the diameter of the planet's disc. Perhaps we can best indicate the skill and ability with which the discovery was completed and confirmed, by the following quotation, which is taken, with a very slight alteration, from Professor Newcomb's "Popular Astronomy."

"On the night of the 16th of August this small object was again seen; and two hours' observation showed that it followed the planet in its apparent orbital motion. This showed conclusively that it was not a fixed star, and must therefore be a satellite, unless by chance one of the group of small planets between Mars and Jupiter happened to occupy this position. Upon examining an ephemeris it was found that the small planet Europa was calculated to be only two or three degrees distant from Mars, and if the ephemeris were erroneous by this amount, the object observed might be this very body. A rough calculation, from the observed positions of the satellite and the known mass of Mars, showed that the period of revolution would probably not be far from twenty-nine hours; and that, if the object were a satellite, it would be hidden during most of the following night, but would reappear near its original position towards morning. On the contrary, if the object were the small planet Europa, it would, on the next evening, be a little south-east of the planet.

"The following night was beautifully clear, and when Mars

rose high enough to be well seen, the telescope was pointed at it. A small star was soon seen quite near the computed position of the hypothetical small planet, while no satellite was visible. But a few minutes of observation with the micrometer showed that Mars was passing by this object, and that the latter was therefore a fixed star, and not the moving object seen on the preceding night. The appearance of the satellite was therefore looked for with much confidence, and at four o'clock on the following morning it emerged from the rays of the planet as predicted; so that no reasonable doubt of its character could remain.

"But this was not all. The reappearance of the satellite was followed by the appearance of another object much closer to the planet, which proved to be a second and inner satellite. The reality of both objects was abundantly confirmed by observations on the following nights, not only at Washington, but at the Cambridge Observatory by Professor Pickering and his assistants, and at Cambridgeport by Messrs. Alvan Clark and Sons."

The movements of these two satellites were, of course, most carefully followed, both in America and in Europe, until the increasing distance of Mars rendered them invisible. The observations of the inner satellite were, however, very few, except at Washington, and by Professor Pickering at the Havard Observatory (Cambridge, U.S.), and at Cambridgeport. It was, in fact, very difficult to obtain accurate measures of its position. But the observations made at the above-named places suffice to give valuable elements for the orbits of *both* of these moons; so that, in 1879, when the planet again approached near to the Earth (although not so near by about ten millions of miles as in 1877), they were found very close indeed to their expected places.

In 1877 the great reflectors of Lord Rosse and of the Melbourne Observatory did not show them very successfully. On the other hand, in 1879, they were excellently observed, not only by the large refractors in America, but also by means of the new reflecting telescope belonging to Mr. Common, of Ealing, three feet in aperture, which appears to be of the highest excellence, and to possess remarkably good defining power. Indeed, both in America and in England the two satellites were much better seen in 1879 than had been anticipated; the reason perhaps being, that in practice it is always easier to see a faint body when we know where to look for it, than to discover it for the first time when our gaze is roaming about in a somewhat vagrant way over the field of view. They have again been visible, but only for a few days, during the Opposition of 1881. We can hardly, however, hope that they will be seen any more before the year 1888, possibly not until 1890.

To these two satellites, the smallest by far of all the known members of the Solar System (with the exception, perhaps, of some of the numerous family of minor planets), the names of DEIMOS and PHOBOS have been assigned. These names have been variously rendered *Dread* and *Terror*, or *Fear* and *Flight*, or *Dread* and *Fear*. They are to be found in a passage in the fifteenth book of the Iliad, in which Homer uses them to describe the two companions of Ares, or Mars, from which, at the suggestion of Mr. Madan, they were adopted for the satellites. In Pope's translation we read:—

> "With that he gives command to Fear and Flight
> To join his rapid coursers for the fight.
> Then, grim in arms, with hasty vengeance flies;
> Arms that reflect a radiance through the skies."

These two minute attendants of their primary are so small that it is impossible to measure any appreciable diameters of their discs, or to judge of their sizes, except by estimating as carefully as may be the amount of light which they reflect. This has been done with every possible precaution by Professor Pickering (see vol. xi., Annals of Harvard College Observatory); and it has been concluded by him, and by some other observers, upon the assumption of their light-reflecting power being similar to that of Mars itself, that their diameters most probably lie between six and seven miles. There seems to be little, if any, doubt that Phobos, the inner one, is the brighter of the two; especially when due allowance is made for its nearer proximity to the glare of the planet in the field of view of the telescope. Phobos is therefore, in all probability, rather larger than Deimos.

We cannot but feel that there is altogether something very strange about these little members of the Sun's family. It is most likely that some of our great landowners would hardly find sufficient room for such an estate as they occupy upon the Earth, either upon Deimos or Phobos. They seem to bear somewhat the same relation as regards size to the rest of the satellites, that the minor planets revolving outside the orbit of Mars bear to the other planets. Nor can we but be struck with the fact, that we meet with them upon the confines of the very same region in which the minor planets circulate. And, although they may have long belonged to Mars, and have only been so recently discovered, owing to the previous want of sufficient telescopic power, we may, perhaps, be allowed to suggest, that it is just possible, that these two satellites may be two of the minor planets, to which Mars may have passed so near that it has picked them up and attached them to itself.

We are of course aware, in making this suggestion, how very unlikely it is that a comet, or a meteorite, should at any time pass so close to a planet, and under such conditions as to its velocity and direction of motion, that its orbit should not only be perturbed (or, it may be, *very greatly* changed, as has happened in the case of certain comets which have passed near to Jupiter or to Saturn), but that it should actually become a satellite of the planet, and remain permanently attached to it. But this does not seem to be quite so unlikely in the case of Mars and a minor planet, inasmuch as it will be shown in our next lecture, that we know at the present time of at least one minor planet which, in the nearest portion of its orbit to the Sun, comes about five million miles nearer than the maximum distance of Mars. Such an one, if its orbit were only moderately oval, might therefore, at any given time, be moving nearly in the same path and with the same speed as the planet itself, and thenceforth remain under its permanent control.

The small size of these bodies is not, however, their most remarkable feature. Their nearness to the planet's surface, and the rapidity of their motions round it, are still more curious.

Mars turns upon its axis, as we have already stated, in 24 hours 37 minutes $22\frac{3}{4}$ seconds. The nearer of the two satellites goes round it in 7 hours 39 minutes 14 seconds; the outer in 30 hours 17 minutes 54 seconds, the latter period being almost exactly four times the former;—a fact which may some day prove to be of special interest and importance. It follows from the above-mentioned statements, that the apparent motion of the outer one, as it would be seen by an observer upon Mars, although it would be from east to west across the sky, would be very slow. On the contrary, the inner one would possess the most extraordinary property of appearing to cross the sky day by day from *west to east*; *i.e.*, in the reverse direction to that pursued by all the other heavenly bodies.

Let us endeavour to explain why this would be the case. In order to do so we may recall the statement made in Lecture III., page 56, with regard to the way in which the apparent daily circuit, which our own Moon would otherwise appear to perform from east to west once in every 24 hours, is modified by its orbital motion round the Earth.

We showed that in 24 hours the Moon describes about $\frac{1}{30}$th of its own path around the Earth from west to east; from which it results that its apparent daily motion across the sky from east to west is somewhat slower than it else would be, the velocity which it would otherwise appear to have being diminished by about $\frac{1}{30}$th part. At the end of 24 hours after it has risen on any given day, it does not rise again, but is in general about $\frac{1}{30}$th part of a whole revolution below the horizon. It therefore rises upon an average about $\frac{1}{30}$th of 24 hours, or from 40 to 50 minutes later, on successive days.

But let us make another supposition, and imagine the Moon to go round the Earth in 48 hours. It would in that case go back just half as quickly as it would otherwise seem to go forward. If it rose at 6 p.m., and (apart from its own motion in its orbit) would appear to go across the sky so as to set at 6 a.m., it would, instead of so doing, be seen only half-way across at that hour; in other words, it would be upon the meridian at 6 a.m.

Once more, let us suppose the Moon to describe its orbit round the Earth in a single day, and let us also (in order to avoid certain complications) suppose its path to be in the plane of the equator, a position which very nearly corresponds to that of the satellites of Mars. In this case a very curious result would follow; viz., that the Moon would be seen by one half of the Earth, but never by the other half; and any place that would see it would always see it in one fixed position in the heavens.*

But lastly, let us suppose the Moon to circle once round the Earth in less time than the Earth occupies in one revolution upon its axis. What would then happen? Its real motion from west to east being greater than the apparent motion from east to west imparted to it by the Earth's rotation, it would appear to go round the sky from *west* to *east*, with a velocity equal to the excess of the one motion over the other, instead of apparently travelling from east to west, as all the other heavenly bodies do. Now this is just what happens in the case of the *inner* satellite of Mars, and for an exactly similar reason. The rotation of the planet would make it appear to go once round from *east* to *west* in 24 hours $37\frac{1}{3}$ minutes; while its own motion would take it round from *west* to *east* in 7 hours $39\frac{1}{4}$ minutes. The combined effect is, that in about 11 hours † it seems to go once round the Martial sky from *west* to *east*.

The outer satellite travels round its orbit in about 30 hours

---

* This may perhaps be best understood by a simple example. Let the Moon, for instance, be supposed to be upon the *meridian* of any place at any given time; then it is not difficult to realize that it would (upon the above supposition) just keep up with the movement of the place produced by the rotation of the Earth, so that it would appear to remain fixed upon the *meridian* in question. Similarly for other positions in which it might be continuously seen from other places of observation.

† The former velocity being rather less than $\frac{1}{3}$rd of the latter, inasmuch as $24^h 37^m$ is rather more than 3 times $7^h 39^m$, it follows that the velocity corresponding to a rotation in $7^h 39^m$ will be diminished by rather less than one-third part. We must consequently increase this period by barely one-half (or, in other words, increase it in a ratio rather less than that of 3 to 2) in order to obtain the duration of the apparent circuit of Phobos. This gives, as above stated, about 11 hours.

from west to east. The apparent motion from east to west caused by the rotation of Mars being somewhat greater than this, it does not appear absolutely to stand still, as it would if its period were exactly 24 hours 37 minutes; but it has a comparatively slow movement from east to west—in fact, so slow, that about $131\frac{1}{2}$ terrestrial hours are occupied in its apparent circuit once round the Martial sky.

It may therefore be quite possible for the two, on certain occasions, to be seen approaching each other from opposite directions, until one may pass in front of the other and eclipse it; or, as would more often be the case, the one might pass slightly above or below the other.

We cannot now repeat the calculations with regard to various other points of interest connected with these satellites, which we made in a course of lectures delivered at GRESHAM COLLEGE shortly after their discovery. It may suffice to mention, that we found that Phobos altogether escapes eclipse about once out of every twice that it is in the position of Full Moon, owing to the tilt of its orbit to the plane of that of Mars, while Deimos does so four times out of every five. We showed that the maximum duration of an eclipse of Phobos is about 53 minutes, and of one of Deimos about 84 minutes. We also stated that, owing to its slow apparent motion across the Martial sky, Deimos may, in the interval of about 66 hours between its rising and setting, pass three times through the phase of Full Moon, and thrice suffer eclipse.

An example or two of what might occur in the course of a single night may, perhaps, not be uninteresting. Supposing sunset to take place near to such an hour as upon the Earth we should call 6 p.m., Phobos might have risen a short time previously in the *west*. After the lapse of about $3\frac{2}{3}$ hours, if we suppose the observer to be at the time in question at a place upon the equator of Mars, and the satellites to be apparently moving nearly in the equator of the heavens, the Sun would be about 55° below the western horizon, and Phobos about 55° above its eastern point; and Phobos might now be in the middle of an eclipse, which (having commenced at about a quarter past 9 p.m.) might, under favourable circumstances, continue until soon after 10 p.m. At about half-past 11 p.m. Phobos would set in the east. At about 5 a.m. it would rise again in the west, and for a considerable part of the time between this hour and the rising of the Sun at about 6 a.m. in the east, another eclipse might be going on; *i.e.*, two eclipses might occur during the same night, but only *one* between a rising and setting of the satellite.

In the preceding paragraph, for the sake of simplicity, we have spoken approximately of Martial hours as though they corresponded with those of the Earth, the difference in the length of the diurnal rotations being small. If, however, by the Martial hour we mean $\frac{1}{24}$th of the planet's rotation-period, its value would of course be about 1 hour $1\frac{1}{2}$ minutes of terrestrial time.

Again, we might suppose the Sun to set in the west at about 6 p.m., and Deimos, the outer satellite, to have risen about an hour previously in the east. A total eclipse, which in an extreme case would last for about 84 minutes, might soon afterwards occur. Upon the occurrence of the next sunset Deimos would have increased its distance above the horizon by about 68°, and a little before the following midnight it might be undergoing another total eclipse. The Sun would then rise and set again, during which interval Deimos would have described another 68° in the sky. Before a third sunrise Deimos might be again eclipsed, and some time after that third sunrise it would set in the west.*

A few other interesting calculations relating to these Martial satellites may also be made. For instance, the fact that our own Moon really appears to be larger when seen near to the zenith than when seen near to the horizon, which we explained in Lecture IV. (see Fig. XXIII., page 90), may be still more clearly illustrated in the case of Deimos and Phobos, and especially in that of the latter.

Phobos being about 5,800 miles distant from the centre of

* The above calculations easily follow by taking the apparent motion produced by the rotation of Mars in a terrestrial hour to be 14·62°, while those of Deimos and Phobos in their orbits are respectively 11·88° and 47·03°.

the planet, would, if in the zenith of an observer, be only about 3,700 miles above him. But when seen in the horizon, it would be more than 5,400 miles distant. Its apparent breadth would therefore be increased in the former position in the ratio of about 3 to 2, and the area of its apparent disc, and the amount of light received from it, in the ratio of the squares of these numbers, or about as 9 to 4. When seen in the zenith, the light received from it would consequently be more than twice as great as if the satellite were in the horizon. In the case of the outer one the similar increase of light would be rather less than one-third.

The variations in their light as they pass across the heavens would therefore be very considerable. But, even at the best, that light would be *extremely feeble*, notwithstanding their nearness to the planet.

Let us endeavour roughly to calculate what it might amount to. The outer satellite, when seen in the zenith of an observer, would be about 20 times as near as the Moon is to the Earth. If, then, we suppose it to have a diameter 300 times as small, it would result, since 300 is 15 times 20, that it would have an apparent disc of $\frac{1}{15}$th of the diameter of that which our Moon presents, and the area of this disc (225 being the square of 15) would be $\frac{1}{225}$th of that of our Moon. At the same time it would only be enlightened by a sunlight which varies, according to the position of Mars in its orbit, from about $\frac{1}{2}$ to about $\frac{1}{3}$ of that which our own satellite receives. If we multiply 225 by 2 and 3 respectively, we obtain the numbers 450 and 675. The light given by Deimos would therefore vary from about $\frac{1}{450}$th to $\frac{1}{675}$th of that of the Earth's Moon, and even this faint light would be diminished by $\frac{1}{4}$th part if Deimos were seen near to the horizon. It would be very difficult, if not impossible, for the phases of so minute a disc to be distinguished without telescopic aid, by eyes such as our own. We have, however, already suggested, that those who strongly believe in the habitability of Mars, may very reasonably suppose Martial eyes to be on a larger scale, and altogether superior to terrestrial eyes.

The distance of the inner satellite from the surface being

less than ⅓rd of that of the outer one, the light which it would give, supposing it to be of the same size, would be about ten times as great, and would consequently vary about $\frac{1}{15}$th to about $\frac{1}{87}$th of that of our Moon. This amount would again be diminished by more than one-half when the satellite might be seen close to the horizon.

The outer one would therefore be of practically no use as a light-giver, the inner one of very little. In fact, it is hard to imagine any useful purpose which these minute Moons can serve, except it be, that their motions are so rapid, that they may to some extent supply the place of watches and clocks upon Mars; or afford voyagers upon it an opportunity of calculating lunar distances, by which they may determine their longitude with much greater precision than terrestrial navigators can attain.

When the satellites were first discovered, some exaggerated statements were made as to the tides which they might produce in the Martial waters. It was not long, however, before these statements were corrected. Very little calculation showed that the smallness of their masses, notwithstanding their proximity to the planet, would prevent them from forming tides of any importance; and that they would not, in this respect, have the very beneficial influence which some writers were inclined to claim for them, of removing all tendency to stagnation in the seas and oceans of the planet.

It is not a very difficult matter to make a rough calculation that, in all probability, the outer one would not generate, in the open sea, a tide of an average height of more than about $\frac{1}{5000}$th part of that which is generated by the Moon upon the Earth;*

---

* This statement may be approximately confirmed as follows:—The Moon's distance from the Earth's centre is equal to about 60 times the Earth's radius; while the outer satellite is distant from the centre of Mars about seven times the radius of the planet. Its tidal power, owing to its nearer proximity, would consequently be greater in comparison with the attraction of gravity upon the surface of Mars, than that which our own Moon exerts upon the Earth, as the cube of 60 exceeds the cube of 7; *i.e.* as 216,000 exceeds 343, or about 630 times. (See Sir E. Beckett's "Astronomy without Mathematics," p. 163.) But its power would at the same time be less, in the proportion in which the while that caused by the inner one might be, upon an average, about fifteen times as great ; but still more than 500 times less than one of our lunar tides.

It is hardly worth while further to discuss tides of so small a magnitude. We may, however, remark that we have used the phrase "average height of tide," in the preceding paragraph, because we have calculated, that the fact that the inner satellite is at a distance from the centre of Mars less than three times the radius of the planet, while the distance of the outer one is only about seven times that radius, would involve a remarkable result. It is well known that the tidal power of our own Moon is much the same upon water, either situated immediately beneath it or upon the very opposite side of the Earth, so far as that power may be estimated by the difference of its attraction upon the water in question, and upon the mass of the Earth as a whole,* which latter attraction depends upon the Moon's distance from the Earth's centre. In fact, the difference of the tidal power of the Moon, in the two cases supposed, only amounts to $\frac{1}{26}$th of its value in either case. But the comparative distances of the satellites of Mars from the nearest point of the surface of the planet immediately beneath them at any given time; from the centre of the planet; and from the furthest part of its surface; differ from one another to such an important extent, that we found, when lecturing upon them some time since at GRESHAM COLLEGE,

ratio of the mass of the satellite to that of Mars falls short of the ratio of the Moon's mass to the Earth's. The latter ratio is about $\frac{1}{76}$. The satellite's diameter, if taken as seven miles, is $\frac{1}{603}$ of that of Mars ; and if its average density be supposed the same (although it may probably be much less), its mass would be a $(\frac{1}{603})^3$ part of that of Mars, i.e. a $\frac{1}{218600000}$ part. Instead, therefore, of a tidal effect 630 times as great, we should for this reason only have an effect $\frac{630}{218600000}$ of 630 times as great, which would amount to about a $\frac{1}{1000}$ part. And upon a globe of about one-half of the radius of the Earth, the height of the resulting tide would again be diminished by about one-half, so as to be about $\frac{1}{2000}$ of the height of the Earth's lunar tide.

* This is, of course, according to what is called the statical theory of the tides, which, although altogether unsatisfactory as a real explanation of their formation, is sufficient to illustrate the peculiarity to which we now refer.

that of the two corresponding tidal effects for the inner Moon, the one would be three times as great as the other; while for the outer Moon the one would be fully half as large again as the other.*

An interesting consequence, which may probably result from the solar tides upon Mars, although they are very likely only about $\frac{1}{8}$th† of the height of those upon the Earth, has recently been pointed out by Dr. Ball; viz.—that their friction may be one cause which, by gradually retarding the speed of the planet's rotation upon its axis, has helped to bring about the remarkable fact that it is now more than three times as slow as that of the inner moon in its orbit around it. Dr. Ball also states, in his lecture published under the title of "A Glimpse through the Corridors of Time," that Mr. Darwin's recently developed theory of tidal evolution indicates, that a similar effect may be produced by the solar tides upon the Earth's rate of rotation, in the course of long ages yet to come, after the Moon's month and the Earth's day have each

---

* The Moon is distant from the centre of the Earth about sixty times the Earth's radius. Its distances from water immediately beneath it; from the Earth's centre; and from water upon the opposite side of the Earth, are therefore as the numbers 59, 60, 61. The tidal powers above referred to are therefore in the two cases, respectively, as $\frac{1}{59^2} - \frac{1}{60^2}$ and $\frac{1}{60^2} - \frac{1}{61^2}$. A little arithmetical work will show that the difference of these two values is about $\frac{1}{15}$th part of either. The corresponding distances for the inner satellite of Mars are 3700, 5800, 7900 miles; and for the outer satellite, 12,400, 14,500, and 16,600 miles. From these numbers it is easy to verify the statement in the text.

† The Sun's mass being about 3,100,000 times that of Mars, and only about 330,000 times that of the Earth, its tidal power, in comparison with the attraction of gravity upon the surface of the Earth or Mars, would be about $9\frac{1}{3}$ times greater upon the latter planet (see "Astronomy without Mathematics," p. 163). But the distance of the Sun from Mars is about 72,000 times the radius of Mars, while its distance from the Earth is about 23,000 times the Earth's radius. Its tidal effect will therefore be diminished in the cube of the ratio of 23 to 72, or about 30 times. Owing to the radius of Mars being about one-half of that of the Earth, the height of tide would also be diminished about one-half. Hence we have height of solar tide upon Mars equals about $\frac{28}{3}$ of $\frac{1}{30}$ of $\frac{1}{2}$ of its height upon the Earth. In other words, it would be approximately of about $\frac{1}{8}$th of the height.

been lengthened by the present lunar tides to about 1,400 times twenty-four hours. The lunar tides would then for a time remain stationary; but the solar tides continuing would still further lengthen the Earth's day, so that it would at last exceed its satellite's month, as the day of Mars exceeds more than thrice the month of its inner moon.

If there be any astronomers possessed of good telescopes upon Mars, there is little doubt that they can determine whether its satellites are inhabited. It would, in fact, be just as easy as it would be for us to determine the like question with regard to the Moon, if it were brought, in the two cases respectively, within about $\frac{1}{21}$th or $\frac{1}{60}$th of its present distance. If, on the other hand, there are astronomers upon those little moons, most wonderful must be the view of Mars which they enjoy! From the inner satellite its disc would subtend an angle representing more than $\frac{2}{8}$ths of the width of the whole heavens, i.e., nearly $42\frac{1}{2}°$; while from the outer it would still subtend an angle of about $16\frac{1}{2}°$, i.e., about $\frac{1}{11}$th of the width of the celestial vault. The apparent area of its disc in the two cases would respectively be about 6,400 and 1,000 times that which our Moon presents to us; its light some 400 times as great when viewed from the outer satellite, and 2,500 times as great from the inner one. And yet the sight would not be so wonderful as that of Saturn to an observer placed upon the inner edge of its bright ring, at a distance of only 18,000 miles from its surface, where its huge globe would subtend an angle of about $84\frac{1}{2}°$.

Since the attraction of gravity upon the surface of Mars is only about $\frac{2}{5}$ths of its value upon the surface of the Earth, we have explained that any inhabitants upon it might well be two-and-a-half times as tall as ordinary men; but we can hardly conceive how very huge those would have to be who should be suited to the smallness of the gravity upon these little moons. If their diameters are $\frac{1}{600}$th of that of the planet, the attraction of gravity upon their surfaces would be $\frac{1}{600}$th of that upon the surface of Mars, or about $\frac{1}{1500}$th of that upon the Earth, supposing their densities to be equal to that of Mars. We may, however, remark that their density, and consequently the attraction exercised by them, may possibly be much less. But, taking the above supposition, let us imagine a man of fifteen hundred times the height of one of ourselves, and suppose such a man to be upon a globe of about twenty miles in circumference. He certainly would not be very long in walking round it, nor would he wish to have many friends upon it to keep him company.

Altogether we must confess, that these exceedingly minute attendants of a planet whose own size is in no wise remarkably small are very puzzling and very inexplicable. They may perhaps suggest the thought:—Can it be that, just as we find a large number of small minor planets in the region between Mars and Jupiter in the midst of which we might otherwise expect to find one larger planet, so there may be a number of little satellites revolving around Mars where otherwise there might be a single larger one? Can it be that there are many others so small that our telescopes will never see most of them, but will only, by their gradual increase of power, reveal to us a few that are not quite so tiny as the rest?

Minute as these satellites are, they have, however, already done good service in enabling us to weigh the planet Mars much more accurately than was possible before their discovery. By Kepler's third law (see Lecture V., page 122) it follows, that the product of the sum of the masses of Mars and of either of its satellites, multiplied by the cube of its distance from the planet, must bear exactly the same ratio to the product of the sum of the masses of Mars and of the Sun, multiplied by the cube of their distance apart, as the square of the periodic time of that satellite round Mars bears to the square of the periodic time of Mars round the Sun. And the mass of the satellite is so small in comparison with the other quantities involved in this statement, that our ignorance of its value does not matter. We may neglect it, and nevertheless have an equation, or rule-of-three sum, which will give a very accurate value of the mass of Mars as a fraction of that of the Sun; in fact, the smaller the mass of the satellite, the less error is produced by neglecting it in the course of the calculations. The very minuteness of the moons of Mars consequently makes them of all the more use in helping us to weigh the planet. The methods employed before their discovery are much too complicated for us to explain, and after most laborious investigations still involved considerable uncertainty as to the true value. As soon, however, as the rotation-period of either of the satellites was known, it only needed a very brief calculation in order to obtain a value for the weight of Mars. Professor Asaph Hall estimates it to be equal to a $\frac{1}{3,093,500}$th part, and Professor Pritchett to a $\frac{1}{3,075,440}$th part, of that of the Sun. We may feel sure that either of these results is very nearly accurate.

Our space forbids that we should attempt to discuss the astronomy of the heavens as it would be seen from Mars, or from its satellites. We cannot, however, resist the temptation to quote from *The Observatory* of December 1878 a most interesting description (which it was our good fortune to hear Mr. Marth narrate at the preceding November meeting of the Royal Astronomical Society) of certain phenomena which may from time to time be seen by any possible inhabitants of the planet.

"On November 12th, 1879, shortly before two o'clock, Greenwich mean time, a small black body would make its appearance on the south-following side of the disc of the Sun; in six minutes it would have fully entered upon the disc, and would proceed slowly from left to right. About a quarter past four o'clock, another and bigger black body would encroach upon the disc, and would occupy twenty-one minutes before it had fully entered upon it. The two bodies would be the Moon and the Earth, and they would be visible from all parts of Mars where the Sun was above the horizon. But observers placed upon a certain zone, or track, would have the opportunity of seeing a third and apparently much bigger body cross the Sun's disc. It would come from the right-hand side in a direction at a considerable slant to the south, and Martial observers would have to look sharp to observe all its contacts with the disc, since the time for doing so would be limited to some twenty or thirty seconds. The third body would be Phobos, the inner satellite. Martial astronomers would, however, be much more interested in the transit of the Earth and Moon (across the Sun's disc) than in that of Phobos, since, in the course of a Martial year, there would be no less than about 1,388 transits of Phobos visible from some part or other of the planet, while the number of the transits of Deimos would be about 133. On the other hand, the transits of the Earth would be rare occurrences, the last one having taken place in A.D. 1800, while the next to follow would happen in A.D. 1905. About a quarter to ten o'clock, the Moon, which in the meanwhile would have drawn nearer to the Earth, would quit the Sun's disc, and the last external contact of the Earth with the disc would take place at Greenwich midnight. But before internal contact, about half-past one o'clock, Phobos would again make its appearance upon the disc for certain stations, after having meanwhile performed a whole revolution round the planet."*

We must now bid adieu to Mars and its satellites. We would fain hope that the various calculations which we have placed before our readers, and the attempts which we have made to enable them to realize some of the effects resulting from the remarkable orbital movements of these bodies, may not have been uninteresting. In any case, we feel confident that those who have exerted the mental effort necessary to follow them, will be rewarded by the greater facility with which they will be able to realize the relative movements and the varied phenomena belonging to those members of the Solar System which still remain for our discussion. If so, the somewhat elaborate study of the Martial system, to which we have invited their attention, will not have been altogether in vain.

* Two or three very slight omissions, or alterations, occur in the above quotation.

LECTURE XI.

THE MINOR PLANETS.

"And is creation not a work of skill
In its grand outline, in its parts minute,
That we should mark its movements, trace its laws,
Observe its fine consenting harmonies?"—ADDISON.

A TRAVELLER journeying round the Sun upon the planet Mars, and accustomed to the two little moons whose minuteness concealed them from our gaze until the autumn of the year 1877, which, as explained in our last lecture, will ever be memorable for their discovery, would perhaps not be surprised to find many other small bodies comparable to them in size revolving round the Sun within the orbit of the huge mass of Jupiter. A few of them would at times approach him somewhat closely; others would, at their nearest, be separated from him by an interval of 100 or 150 millions of miles. Amongst the whole number, some of unusually large bulk would be seen, but the volume of the largest would be many hundreds, or it may be several thousands, of times less than that of the planet from which he would watch them. We wish that we could take our readers, by some magic power, across the great gap that divides the orbit of the Earth from that of Mars, and make them comfortable upon that latter planet, while with us they might study the remarkable little members of the Sun's family to which we refer.

But we can only observe them from a far less favourable point of view; the result being that, in some respects, the Minor Planets, which is the name by which we shall call them, are the most uninteresting of the heavenly bodies. We can never hope, even in the most powerful telescope that may be constructed, to detect any physical features of their surfaces, or (except in a few rare instances) even to measure the diameters of their discs. Our knowledge of their individual peculiarities and constitution must inevitably be most limited. Nevertheless, their study is in other respects both important and instructive.

Even if they had no other claim upon our attention, we can hardly refuse it, when we remember the startlingly rapid rate at which they have of late been discovered. So great has been the skill and industry of those who have watched for them, that no year has passed since 1846 without at least one new one being found; while the total number known, which at the beginning of the year 1847 was 5, is now, in August 1882, 229. We therefore ask our readers to tarry with us a little, before we go on to discuss the wonders of the mighty planet Jupiter, or the fascinating rings of Saturn, while we endeavour to show that some statistical results involved in the tables of the elements of the orbits of the Minor Planets, which may at first sight appear to be very uninviting, are really most important and suggestive.

But we ought perhaps to explain, before we make any further remarks with regard to these little bodies, why it is that we adopt for them the appellation *Minor Planets*, in preference to any other. We do so, because the orbits in which they travel round the Sun are not only governed by the same laws, but in many other respects are similar to those of the larger planets. At any rate, we may confidently say, that in no one respect, except in the minuteness of their discs, can they be justly described as star-like. The name of Asteroid, which has this meaning, and which was originally assigned to them, is therefore about as unjustifiable a title as could well be selected.

Another inaccurate way of speaking of them also deserves a little preliminary notice. It is often stated that they are a sort of swarm, or ring, of small bodies lying between the orbits of Mars and Jupiter,* and approximately occupying the place of a

---

* See Fig. XXXIII., p. 118; in which it may also be noticed that the appellation "Planetoid" is used. This, however, is not so appropriate a title as "Minor Planet," although much better than "Asteroid."

Planet which seems to be wanted to fill up a gap in that part of the Solar System. But this is a very loose way of defining the locality of their orbits, although it is no doubt true, that the mean distances of the *four first discovered* are respectively 2·767, 2·771, 2·67, and 2·36 times the Earth's distance from the Sun; while the so-called law of Bode suggests that a planet should revolve at about 2·8 times that distance.

The inaccuracy to which we refer is, however, at once evident, if we notice the immense range of space within which the Minor Planets are found. We cannot too strongly impress upon our readers, that, so far as we at present know, it begins *within the orbit of Mars*, and extends comparatively near to that of Jupiter. The first part of this statement depends upon a fact which we pointed out in a lecture delivered in GRESHAM COLLEGE, early in 1879; and which, so far as we are aware, had not at that time been noticed in astronomical text-books, although it has since been independently mentioned in a most interesting and learned article by M. Niesten published in the Annuaire of the Royal Observatory of Brussels for the year 1881.* It is, that one of the Minor Planets, viz., Æthra, No. 132, when at its nearest distance from the Sun, approaches fully 5,000,000 *miles nearer to it than the maximum distance of the planet Mars*. It also happens that Æthra, when in such a position, occupies just the same longitude as is occupied by Mars when at its furthest from the Sun. At times, therefore, a most remarkably close approach, or even a collision, of the two might occur, were it not that the orbit of Æthra is greatly tilted from the plane of the ecliptic, so that this little planet is thereby depressed far below the orbit of Mars, when it would otherwise pass very near to it, and even within a certain portion of it.

The fact, however, to which we wish at present to draw special attention is that, apart from the many thousands, or it may even be millions, of Minor Planets which are not known to us, we know of one which, in a portion of its orbit, approaches

---

* The above article contains an immense amount of information with regard to the Minor Planets, accompanied by tabular statements most conveniently arranged for reference.

so near to the Sun that its distance is considerably less than that of the planet Mars when in the same longitude.

Here then their domain begins. Let us next see where it ends. Upon consulting the tables, we find that there is one named Hilda, No. 153, which recedes from the Sun to a distance of about 428 millions of miles, while the nearest distance of Jupiter is only about 461 millions of miles. If Hilda and Jupiter, when at these distances respectively, had the same longitude, they might consequently approach one another within about 33 millions of miles. But their longitudes under these circumstances differ by about 87°, from which it results, that the ellipticity of Jupiter's orbit removes it about 22 millions of miles further away (or very nearly to its mean distance of 484 millions of miles from the Sun), when it is in the same longitude in which Hilda's distance is, as above stated, 428 millions of miles.* The preceding statements are illustrated by Fig. LXI., in which the mean distance of Jupiter from the Sun is compared with the greatest distance of Hilda; and the greatest distance of Mars with the nearest distance of Æthra.

Without, therefore, attempting any great accuracy of statement, we may justly describe the home of these little members of the Sun's family as extending *from within the orbit of Mars comparatively near to that of Jupiter;* or, in other words, as embracing a zone of about 280,000,000 miles in width. At the same time, it must in no wise be imagined that the Minor Planets are scattered with any approach to uniformity through this wide zone. On the contrary, as we shall presently more fully explain, the great majority of those with which we are acquainted are located at distances

---

* The orbit of Hilda has so great an ellipticity that the opposite portion of it approaches nearer to the Sun, and recedes further from that of Jupiter, by about 120 millions of miles, the nearest distance of Hilda from the Sun being about 308 millions of miles. It should also, of course, be remembered, that the closest possible approach of the two planets will be to some extent dependent upon the inclination or tilt of their orbits to one another. This tilt, however, is very moderate, and only amounts to about 9°. In an approximate statement it need not be considered.

which lie between $2\frac{1}{4}$ and $3\frac{1}{4}$ times the Earth's mean distance from the Sun, while a still closer aggregation is found at from $2\frac{1}{2}$ to $2\frac{3}{4}$ times that distance.

We must refrain from repeating the often-told story of the first discovery of any of these bodies, and of the excitement which it caused. It may be read in Grant's "History of Physical Astronomy," in Mr. Proctor's "Poetry of Astronomy," and elsewhere. At present we will only state that the first, to which the name of Ceres was assigned, was found on January 1st, 1801,—a very easy date to remember. Three more, Pallas, Juno, and Vesta, were

Fig. LXI.—A comparison of the greatest distance of the Minor Planet Hilda, the nearest distance of the Minor Planet Æthra, the greatest distance of Mars, and the mean distances of Jupiter and of the Earth from the Sun.

respectively discovered in 1802, 1804, and 1807. We may also remark that these not only claim the honour of being the first four detected, but probably owe it to their superiority in size and in the consequent brightness of their appearance. No more were seen until a fifth, Astræa, was added to the list in 1845. But since that date the rate of discovery has been so rapid that at the end of 1852 twenty-three had been found. The number was increased to fifty in the course of the next five years; while in 1878, 1879, and 1880, the respective additions were twelve, twenty, and eight; the total amounting to 219 at the end of the year 1880. In 1881 only one was found. It is rather remarkable that in the first eight months of 1882 the rate of discovery has again increased, nine new ones having been met with.* The number discovered in 1879, viz., 20, exceeds that of any other year; the next greatest being 17 in the year 1875.

The accurate formation of star-charts, embracing all stars down to about the twelfth magnitude, has much assisted those observers who, as the above numbers show, have so successfully searched for these little planets. When in possession of such assistance, there is still room for the exercise of much skill and perseverance; but, apart from the necessary delicacy of the observations, all that is requisite is, accurately to compare with the charts the stars that are seen in any portion of the sky, by means of a powerful telescope, and carefully to notice any one whose place is not recorded. Then, if the star be found upon a subsequent night to have changed its position, it will follow, that it is not one of those to which the name of fixed stars belongs, but that it is really *a planet*, or (to take the literal meaning of the word) a wanderer amongst them. By further observation it can next be determined whether it is a Minor Planet that has already been discovered, or a new one. It may be interesting to mention, that those who have persistently laboured in such a search find, that they meet with several which prove to have been already known, for every one that is new.

It is worthy of remark that no Minor Planet has ever been discovered without the help of a telescope, although it is just possible to see Vesta (the fourth that was found) with the naked eye, if it be at its nearest distance from the Earth. At such a time the light received from it does not differ much from that of a sixth magnitude star. Pallas and Ceres are less bright,

---

* Dr. J. Palisa, now of Vienna, has discovered 8 out of the above-mentioned 9, in the first eight months of 1882; the only one in 1881; 5 out of 8 in 1880; and 10 out of 20 in 1879; which, with 13 previous discoveries, raises his total to 37. This number has, however, been exceeded by Professor C. H. F. Peters, of Clinton (U. S.), who up to the end of 1880 had discovered 41. Other discoverers of 10 or more are as follows:—Borrelly, 11 (*Marseilles*); Goldschmidt, 14 (*Paris* and *Châtillon*);

and fall somewhat below the extreme limits of ordinary vision. The light of Juno is weaker still; while those more recently discovered have, in general, been increasingly faint. This is shown in the following statement, viz.,—that the average light of the first 10 was about the same as that of a star half-way between the 8th and 9th magnitudes; the second 10 came between stars of the 9th and 10th magnitudes; the third 10 between the 10th and 11th magnitudes; while the light of the remainder, nearly 200 in number, approaches that of a 12th magnitude star. The gradual decline in brightness has, however, of late been very slow.

In astronomical parlance, the Minor Planets are generally denoted by numbers, corresponding with the order of their discovery. They have also at present (with the exception of a few of those most recently found) been dignified by names; but, if their number continue to increase, it will, we think, soon be almost impossible to find suitable appellations for many more. A few of the more recently adopted names, such as NUWA, ATHOR, CELUTA, PHTHIA, KOLGA, BYBLIS, LILÆA, may indicate that some such difficulty has already been experienced. Dr. Palisa, to whom the honour of nearly all the most recent discoveries belongs, has suggested that the last two of those found in 1880 be called Bianca and Thusnelda. It may, perhaps, be of some use to mention that, in reading the second volume of Du Chaillu's "Midnight Sun," we have found, in page 445, a remarkable list of names of Norse, Swedish, and Danish damsels, from which we think appropriate selections may be made.*

A few curious coincidences and interesting facts in the

Hind, 10 (*London*); Luther, 20 (*Bilk*); Watson, 22 (*Ann Arbor, U.S., and Pekin*). It would hardly be of any use to give a more complete list of the Minor Planets and their discoverers without appending the elements of their orbits, for which we have not space. Such a list would also doubtless very soon become *incomplete*. For the particulars of 172 we may refer our readers to Professor Newcomb's "Popular Astronomy." The latest information may always be found in the "Berliner Astronomisches Jahrbuch," or in the "Annuaire du Bureau des Longitudes" (Paris).

* We append a few as specimens. Ulla, Helga, Hedvig, Yrsa, Elvira, Engela, Selma, Disa, Karna, Valborg, Signild, Yra, Astrid, Signe, Ebba.

history of these discoveries deserve our notice. For instance, it has happened in several cases that the same Minor Planet has been discovered by two independent observers; while on one occasion, in 1857, the late M. Goldschmidt found two new ones during the same night. And not only (as we have already stated) have some, which at first appeared to be new, after a while, when more careful calculations have been made, proved to have been already known;* but one really new has for several months been thought to be an old one, until watched through a considerable portion of its orbit. Such instances as these may serve to indicate the difficulty which is involved in keeping count of the separate individuals of so numerous a family, and in making the elaborate calculations necessary for the accurate determination of their orbits.

But sometimes this difficulty is still further increased. This is the case if any orbit is so situated that it undergoes special perturbations, owing to the nearness with which the Minor Planet moving in it approaches either to Jupiter or to Mars; but especially if it approaches the former planet, whose huge bulk causes its perturbing influence to be very great. Owing to such effects, as well as to the great decrease in the apparent brightness of most of these little travellers as they recede from the Earth, which makes it useless for us to attempt to observe them except at times when they are not far from their nearest possible proximity, it has more than once happened that a Minor Planet has been lost for several years. In such a case, it may make its nearest approach to the Earth; it may then be looked for, but not detected; it may then again recede, to undergo, before its next near approach, further perturbations of its orbit, which will increase yet more the difficulty of its rediscovery.

We mention this fact because, in one instance, such a loss was especially to be regretted. It was that of the planet Hilda, No. 153; to which we have already referred as the most remarkable of all in its comparatively near approach to the orbit of Jupiter, and in its mean distance from the Sun, which

---

* This was the case with one of two supposed to be new in the year 1881, so that there was after all only one discovered in that year.

is greater than that of any other with which we are acquainted. After its discovery by Dr. Palisa in 1875, its orbit was calculated. But no success attended the many attempts which were made to see it again, and it was almost given up as lost; until Dr. Palisa himself detected it once more in 1879, at a distance of several degrees from its calculated place; a discrepancy which may be chiefly attributed to the perturbing effect of Jupiter's potent attraction.

One important element affecting the form of the actual path of any Minor Planet, and the planet's own position at any given time, is the eccentricity, or the ovalness, by which the curve of the path differs from a circle. A careful investigation of the tables to which we have already referred, shows, that for about 50 of the orbits the value of the eccentricity does not exceed that which may be represented by the fraction $\frac{1}{10}$th,* which is only slightly more than in the case of the orbit of Mars; while that of about 100 others, although greater, does not exceed $\frac{1}{3}$th, which, it may be remembered, is about equal to that of the orbit of Mercury. In about five instances, however, the measure of the eccentricity exceeds $\frac{1}{3}$rd; while in the orbits of Polyhymnia, No. 33, and Æthra, No. 132, we meet with the exceedingly large values of $\frac{17}{50}$ and $\frac{19}{50}$ respectively.

It is important to notice how remarkable are the changes in the distance from the Sun of such Minor Planets as these. If we take the latter of the two, we find that its greatest distance is 334,000,000 miles; its least 150,000,000 miles. The *difference* between them therefore exceeds the least distance by considerably more than 30,000,000 miles. Indeed, the eccentricity of such an orbit actually amounts to more than

* See Lecture V., p. 120, where it is explained, that an eccentricity of $\frac{1}{10}$th causes the distance of the Sun from the centre of the orbit to be $\frac{1}{10}$th of its mean radius. In the case of a Minor Planet, moving at a mean distance of 250,000,000 miles from the Sun, in an orbit with an eccentricity of $\frac{1}{10}$th, the greatest and least distances would therefore respectively fall short of, and exceed, the above value by 25,000,000 miles; the planet would consequently sometimes recede to a distance of 275,000,000 miles from the Sun, and at other times approach within 225,000,000 miles of it; the total change in its distance amounting to 50,000,000 miles.

two-thirds of that of the well-known periodic comet named Faye's; which goes round the Sun once in every $7\frac{1}{2}$ years, and at its Aphelion passes about 45,000,000 miles beyond the maximum distance of Jupiter. On the other hand, the path of such a Minor Planet as Lomia, No. 117, has an eccentricity less than half as large again as that of the Earth's orbit, and only about $\frac{1}{5}$th of that of Mercury; so that its greatest and least distances from the Sun, which are respectively about 285 and 270 millions of miles, differ by only 15,000,000 miles.

It may be gathered from the preceding remarks that the orbits of the Minor Planets are not only in general more eccentric than those of the greater planets, but in a few cases exceptionally so.

It is also found that almost exactly the same statement may be made with regard to another element of these orbits, which has a very important effect upon the actual path in space of each Minor Planet. We refer to the tilt, or inclination, of the plane of the path to the plane of the ecliptic.

It is known that the inclinations of the orbits of comets may be of any magnitude between coincidence with and perpendicularity to that plane; although those of most of the comets of small period lie within comparatively moderate limits. But the inclinations of the orbits of all the Major Planets to the ecliptic are decidedly small. In no instance do they exceed $2\frac{1}{2}°$; except in that of Venus, for which the tilt amounts to about $3\frac{3}{8}°$; and in that of Mercury, for which the value is larger still, viz., almost exactly 7°. But one of the Minor Planets—viz. Pallas, the second in order of discovery—moves in an orbit which has the very exceptional inclination of 37° to the plane of the ecliptic; and about one in every ten out of the whole number of orbits has a tilt of more than 15°. At the same time, it should be noticed, that, for about one-half of the whole group of orbits, the tilt is less than 7°, and for about one-fourth less than 5°; while for eight of the least inclined it varies from rather less than $\frac{1}{2}°$ to 1°.

We shall find further on that all these statistics have a weighty bearing upon any theory which we may formulate, as to the origin, or the past history, or the present usefulness, of this

THE MINOR PLANETS.

remarkable collection of bodies. Indeed, even though it may almost involve the risk of some weariness to our readers, we must request them to notice, before we go on to the consideration of any such hypothesis, a few more facts with which it is essential that they should be familiar.

For instance, it is to be observed that the nearest of the Minor Planets in mean distance from the Sun, appears to be Medusa, No. 149; which revolves, upon an average, at about 2·13 times the Earth's distance from it, or, in other words, at a mean distance of about 198,000,000 miles. It should also be remembered that it is the very oval form of the orbit of Æthra No. 132, that brings it (as we have previously stated) much nearer to the Sun than this in one part of its course, viz., within about 150,000,000 miles, although its mean distance is some 44,000,000 miles greater than that of Medusa.*

On the contrary, it must not be forgotten that the very *nearest* approach of Hilda, No. 153, still leaves it about 308,000,000 miles from the Sun, or considerably more than three times as far away as the Earth. The nearest distance of several others is also very nearly three times that of the Earth, or about 280,000,000 miles; whereas some, such as Harmonia, No. 40, when at their furthest distances, are not more than from 220,000,000 to 230,000,000 miles away.

These two last statements deserve very careful attention. They involve the important fact that the *furthest* distance of certain Minor Planets is from 50 to 70 millions of miles *less* than the *nearest* distance of others.

Again, it may be well to remark that Æthra, which approaches the nearest of all to the Sun, is not that which may

---

* The orbits of these, as also of several others, including that of Hilda, which are especially interesting, are accurately represented in Fig. LXIII., p. 290. In such a diagram it is of course impossible adequately to indicate the tilt of the orbits. The portion of each, which lies below the plane of the ecliptic (which is taken as the plane of the paper), is, however, dotted ; the other half, drawn in unbroken line, is supposed to be above the same plane. It may therefore be seen in the diagram, that, when the distance of Æthra from the Sun is less than that of Mars, the orbits of these two planets are upon opposite sides of the ecliptic, as we mentioned in page 267 of this lecture.

approach most closely to the Earth, owing to the great inclination of its orbit to the ecliptic, which, amounting to 25°, depresses it, when it is at its nearest to the Sun, so far away from the plane of the Earth's orbit, that it never comes nearer to the Earth than within about 75½ millions of miles. Otherwise, the nearest distance of Æthra from the Sun being 150,000,000 miles, while the Earth's mean distance is 93,000,000, it would have been possible for Æthra and the Earth, had the planes of their orbits been the same, to have approached within the difference of these two distances, *i.e.*, within about 57 millions of miles. As it is, another Minor Planet, named Clio, No. 84, whose nearest distance from the Sun is about 17½ millions of miles greater than that of Æthra, is the one that can approach the nearest to the Earth, viz., within a distance of a little more than 74,000,000 miles, or about 1½ millions of miles nearer than Æthra.*

The periods of the Minor Planets in their orbits, or, to speak more familiarly, their years, vary from rather under 1140 to nearly 2900 of our days; or from 3 years 43 days to 7 years 313 days of terrestrial time. Several pairs may be selected in which the difference of period is very slight, only amounting to a few hours, or to half a day; while the above numbers show that the length of the year for some is much more than double what it is for others. Hilda, No. 153, no longer retains without rivalry its pre-eminently long period of nearly eight of our years. Another, named Irene, No. 190, which was discovered by Professor Peters, in September 1878, is found to make a longer journey than was at first supposed, so that its year, according to present calculations, only falls short of that of Hilda by six days.

It seems that Medusa, No. 149, may claim the distinction of having the shortest period, viz., 1139 days, or less than 3¼ terrestrial years.

It is probable that some exceedingly interesting results, depending upon the mathematical theory of mutual perturbations, may arise in the case of those whose periods are very nearly integral multiples of one another. This may likewise

* See an interesting paragraph in *Nature*, June 27th, 1878.

be the case when certain parts of the actual orbits approach very closely, if the corresponding Minor Planets should happen to be in those parts at the same time. M. Niesten points out instances of the latter kind in the orbits of Fortuna, No. 19, and Metis, No. 9; which approach nearer than those of the Earth and the Moon. He also mentions that those of Juno, No. 3, and Clotho, No. 97, come within about 621 miles, but are not quite in the same plane; while in Fides, No. 37, and Maïa, No. 66; as also in Alexandra, No. 54, and Lumen, No. 141, we have pairs of Minor Planets, which revolve respectively very nearly in the same plane, and in orbits of the same size and shape.

As regards the actual size, even of the largest of the Minor Planets, it is very difficult to form any satisfactory estimate. The images formed in some of the telescopes with which several of them were at first studied, seem to have been so far vague or indistinct, under the high magnifying powers used, as occasionally to suggest the existence of nebulous atmospheres surrounding them, or some deviation from a globular shape; and certainly were not sufficiently distinct to enable any reliable measures to be made of their apparent diameters. Vesta, No. 4, is undoubtedly the brightest, and therefore in all probability the largest. If, as Mädler, one of the earlier observers, supposed, its diameter be as great as 300 miles, its surface would be equal to about $\frac{1}{700}$th of that of the Earth, or about $\frac{1}{13}$th of that of Europe.

In 1867, Mr. Stone, by carefully comparing the amount of light received from a considerable number of the Minor Planets,* came to the conclusion that Vesta and Ceres (if their surfaces be supposed to possess an equal capacity for reflecting light) are most probably nearly of the same diameter; while, upon the same supposition, he found the probable value of that of Juno to be only about $\frac{3}{5}$ths as great, and that of many others to be very much smaller; with the exception of that of Pallas, the size of which appeared to him to lie between those of Ceres and Juno, but nearer to the former than to the latter.

* See *Monthly Notices* of Royal Astronomical Society, vol. xxvii., p. 302.

We believe that some attempted measurements of the discs of Vesta and Ceres, by Sir W. Herschel and others, have indicated for them a possible diameter of about 200 miles. If this value be accepted, that of Juno, according to Mr. Stone's calculations, would be about 120 miles. Nevertheless, Mr. Lassell, in the year 1856, with a magnifying of 1018 upon his reflecting telescope of two feet in diameter, at Starfield, Liverpool, was unable to detect any signs of a disc in Juno.

Argelander and some others have also paid attention to this subject; and, more recently, Professor Pickering, of the Harvard College Observatory, Cambridge, U.S., whose investigations in photometry are well known to be of the highest importance. He has assigned a somewhat greater preponderance of size to Vesta; and has estimated that the diameter of Pallas is only about one-half, and that of Juno less than one-third, of that of Vesta. But his observations still leave it probable that the diameter of Vesta cannot much, if at all, exceed 300 miles.*

There are many of these little bodies which are most likely smaller than such terrestrial islands as the Isle of Wight, or the Isle of Man. The title of *Minor, or Lesser*, Planets by which we have elected to call them is therefore undoubtedly a very suitable one. If we take into account the huge scale of celestial measurements, we might almost term them, with some French astronomers, *celestial dust;* were it not that, in shooting stars and meteoric fragments, we have a dust much finer still.

It may be interesting to mention that the light of Vesta and Pallas appears to be of a yellowish tinge, while that of Juno and Ceres is slightly reddish. M. Vogel also states that, although some of the older observations indicating nebulous surroundings belonging to the larger ones have not been subsequently confirmed, the light of Vesta and Flora, when passed through a spectroscope, exhibits bands which suggest the existence of atmospheric vapours.

* Measures made not long since by Professor Millosevich at Palermo give to the diameter of Vesta a value only less by one-sixteenth part than that of Mädler; while Professor Tacchini makes the value nearly one-third greater than Mädler.

The intrinsic brightness of some of the larger Minor Planets appears to vary. This may be a result of their rotation upon their axes, which may periodically turn portions of their surfaces of different reflective power towards us; or it may arise from temporary changes in the vapours of a cloud-laden atmosphere. Decided variability has been suspected in the light of Frigga, No. 77; as well as in some which are amongst the largest in the group.*

It is, however, much more important to consider what may be the total mass or weight of these bodies collectively, than to estimate that of any individual one. And from this point of view we may ask;—Is it possible that they may, by their united attraction, perturb the movements of Mars and Jupiter, although individually their influence may be quite inappreciable? If so, such a result would be of the highest interest.

But, if we assume for them, upon an average, a diameter of 80 miles (which is most probably very considerably in excess of the truth), the volume of any one will clearly fall short of that of the Earth by the cube of the number 100; the Earth's diameter being 100 times 80 miles, and the volumes of spheres varying as the cubes of their diameters. It would consequently take 1,000,000 of such Minor Planets to form a globe as large as the Earth; and the 229 already discovered would only form a planet of about $\frac{1}{4400}$th of its size. Moreover, unless they are composed of meteoric iron, or of some other metallic substances, it is hardly likely that their constituent materials are so densely pressed together, as in the Earth's larger globe. In that case their average density would be much less, and the united weight of the 229 would fall considerably short even of the $\frac{1}{4400}$th part of the weight of the Earth. It is therefore in no wise surprising that no perturbations produced by their action have as yet been detected in the movements of Jupiter and Mars.

We understand that Le Verrier has shown that an amount of matter of a weight equal to one-fourth of that of the Earth may possibly exist in the space between Mars and Jupiter. If so,

* See *Astronomische Nachrichten*, No. 2314; quoted in the *Observatory* for August, 1880, p. 518.

supposing their average diameter to be as we have suggested, and their density the same as that of the Earth, the actual number of the Minor Planets may be 1,100 times as great as 229; *i.e.*, it may exceed 250,000. But if they are for the most part far smaller than those which we have been able to see;—in fact, so small that they will ever remain invisible even in our largest telescopes;—and not only so, but also of lighter density than we have supposed, it is by no means impossible that they may amount to several millions in number.

In a recent communication to the Vienna Academy, Herr Hornstein has stated some reasons for believing that the diameter of most of those discovered of late years lies between 5 and 15 miles; and that, by an increase of telescopic power, we may probably detect many additional ones of this size in the outer parts of the zone in which they are found. At the same time, he thinks that the great majority of those in existence, whose orbits lie nearer to that of Mars than to that of Jupiter, have already been detected; and that there are not many in that region, of still smaller size, which increased instrumental means would show.

In quoting these various statistics, we have endeavoured, in accordance with our previously stated intention, to select, out of the mass of figures that lies before us, such as are both important in themselves, and may also be useful in our discussion of the question to which we now proceed; viz.,—What is the probable origin of this remarkable group of bodies?

It is true that in some respects we know but little about them; but we find that they are to be numbered by hundreds, if not by thousands, or even by millions. One is found to come 5,000,000 miles within the orbit of Mars; another travels to a distance from the Sun which is equal to about $\frac{4}{5}$ths of the mean distance of the mighty planet Jupiter. With few exceptions they are exceedingly small. Their orbits for the most part are of moderate eccentricity, and of moderate inclination to the plane of the ecliptic; but they occasionally exhibit an unusually erratic form and position. We certainly are not justified in speaking of these minor members of the Sun's family as a *ring* of planets. They may, so far as we know at present, be rather likened to the scattering of a few grains of dust between two spherical shells, the inner one of which is about 150, the outer about 430, millions of miles in radius.

If the above facts are duly weighed, we think it can hardly seem to be probable that the Minor Planets, as was once suggested, are the fragments of some globe of considerable size, which originally moved at about the mean of their various distances from the Sun, and by some fearful catastrophe exploded or was smashed into pieces. And yet we cannot *at once* say that such a supposition is impossible. We shall, however, endeavour to show that recent investigations have almost entirely destroyed the temporary favour with which it was at one time received.

It is, we must allow, a very remarkable fact that the mean distances from the Sun of the three Minor Planets first discovered (which are also undoubtedly three of the largest), all agree, as we have previously mentioned, very closely indeed with that at which Bode's so-called law indicated that a planet might probably be found; while that of Vesta, the fourth in order of discovery, differs but little from a similar value. And not only so, but it is almost equally remarkable that the paths of these four are found nearly to pass through two opposite positions in longitude in the heavens, the one in the north-west of the constellation of the Virgin, the other in the west of that of the Whale. This is just what the fragments of an exploded planet would do, and for the following reason:—

If, from a point at any given distance from the Sun, a body be supposed to be started in space in any direction whatever, it may be shown, by mathematical investigations, that the effect of the Sun's attraction will bend the path of the body out of the straight line in which it would otherwise proceed, into one of those curves which, owing to their being produced by the intersection of a cone with a plane cutting it in certain well-known directions, are termed conic sections.

If the initial velocity of the body be *exactly equal* to a certain value, the path will be a parabola; if it be less, it will be an ellipse; the ellipse under certain conditions having its two principal diameters equal, in which case it becomes a circle. If the velocity exceed the above value, the path will be a hyperbola. The form of these three curves is shown in the accompanying diagram, Fig. LXII.; the ellipse being a closed curve; while the other two extend to an infinite distance, on either side, beyond the portion which is drawn.

It follows, from the forms of the above curves, that, if the body of which we are speaking move in a parabola or in a hyperbola, it will, according as its motion begins by being away from, or towards the Sun, either without delay proceed further and further from it; or else it will gradually approach it, and, after once sweeping round it, pass away never to

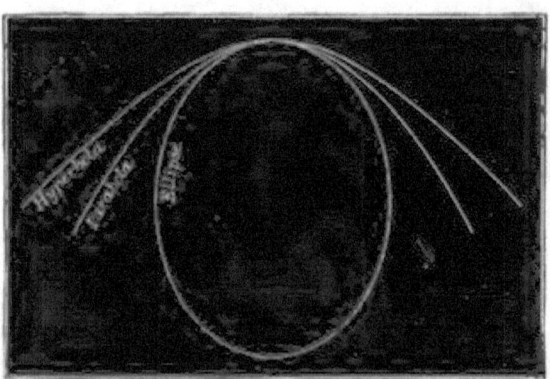

Fig. LXII.—Showing the forms respectively of a Parabola, an Ellipse, and a Hyperbola.

return; which is believed to be actually the case with some comets whose orbits are hyperbolic or parabolic in form.

But, if a planet had exploded, it is more probable that the velocities of the great majority of the fragments, immediately after the explosion, would be such as not to exceed the limit which would permit them thenceforward to move in *ellipses*. These velocities would be compounded of that which the planet possessed at the moment in question, and of those generated by the explosion itself; and it is difficult to conceive of an explosion so violent, that the resulting velocity of any fragment would differ so much from the original velocity of the planet, as to cause its future path to be a hyperbola or a parabola.

We will therefore assume that such fragments would proceed

to move in ellipses round the Sun as a focus. But we must remember that these ellipses might be in many different planes, and of various degrees of ovalness, according as the velocity generated in any fragment by the explosion might be in, or greatly inclined to, the original plane of the planet's motion; and according as it might be in such a direction as to make the resulting velocity greater or less than that possessed by the planet.

For instance, some fragments might be projected exactly in the direction of the planet's previous motion. The whole of the additional velocity communicated to them by the explosion, would in that case be added to the velocity of the planet. Others projected in other directions would have resulting velocities of less magnitude. The velocities which would be the least would be those of fragments projected exactly backwards.

But if, as is probable, the velocity given to every individual fragment by the explosion were much less than that of the planet in its orbit, all would afterwards move in their various paths in the *same direction* round the Sun. The only effect of a different velocity and direction in starting would be seen in the differing size, and ovalness, and inclination of the ellipses described. There would also of course be involved a corresponding difference in each one's period of rotation round the Sun.*

But there are two other facts connected with such a series of orbits which are deserving of special attention. The one is, that, notwithstanding the different durations of their periodic times in their orbits, the various fragments, as each one's orbital period might be completed, would return to, and pass through, the same point from which they all started on the occurrence of the explosion. The other is, that, moving as they all would in planes passing through that point and the Sun, they would also, *once in each revolution*, pass through some point, upon the opposite side of the Sun, *which would lie in the prolongation of the straight line joining the point of explosion with*

* The above statements depend upon Kepler's three laws (see **Lecture V.,** p. 121). and are fully proved in treatises upon dynamics, in which the motion of bodies under the action of a central force is discussed.

the Sun. They would, of course, pass at different distances beyond the Sun, and at different epochs of time; but, to an observer placed at the point of explosion, or directly between it and the Sun, the points of which we speak would all apparently lie in one straight line, and the fragments would consequently appear to pass, once in each revolution, through the point in which the direction of that line would intersect the celestial sphere. This is otherwise expressed by saying that they at such times would have the same celestial longitude.

Now it is exceedingly remarkable, as we have mentioned, that the above two conditions were approximately satisfied by the orbits of the four Minor Planets first discovered; in fact, the third, Juno, and the fourth, Vesta, were actually found, the one in the constellation of the Whale, and the other in that of the Virgin, as the result of a search made in those two opposite parts of the heavens, the advisability of which was suggested, according to the principle which we have just explained, by a comparison of the orbits of the first two.

The hypothesis put forward by Olbers, who at the beginning of the present century was most active and interested in the search for these bodies,—viz., that they might be the fragments of an exploded planet,—was therefore received with much attention, and apparently supported by the above discoveries.

But, as time has gone on, it has been proved that the orbits of very many, instead of continuing to pass near to any one point of the heavens, and through a longitude exactly opposite to it, are so separated, that the nearest distance of some is (as we have stated) from 50 to 70 millions of miles greater than the farthest distance of others; while the ovalness, and the positions of the planes, of the orbits, vary within very wide limits. It has also been found, as we hope our readers by this time remember, that the Minor Planets are spread over a region not much less than 300,000,000 miles in width. The hypothesis suggested by Olbers has consequently been given up by all, or nearly all, astronomers of note. It is no doubt true that, if the orbits had at first been such as we have described, their mutual perturbations would have gradually changed them; but the change would have been in itself so slow, and the amount of change to be produced so great, that, even if it be considered theoretically possible for them to have thus attained their present positions, an interval long beyond all reasonable belief, or almost beyond imagination, would have been necessary for the purpose.

In fact, the differences existing between the various orbits are practically sufficient to forbid our supposing that any conceivable series of perturbations could have sufficed to generate them. At the same time we must allow that it is quite impossible, owing to the number of bodies involved, and the extreme complication of the problem, to work backwards, so as absolutely to determine whether the necessary conditions could ever have existed. An attempt to do so approximately was, however, some time since made by Encke, in the case of a few selected orbits, the result of which, so far as it went, tended to negative the supposition.

We may therefore say, that the orbits of the *first four* Minor Planets undoubtedly supported, to some extent, the hypothesis that their origin might have been due to the explosion of a larger orb. But the orbits of those which have since been discovered in such large numbers, and the great differences found to exist between very many of them, have well nigh given the above-mentioned theory its final death-blow.

We have already stated our reasons for believing that the aggregate mass, or bulk, of the Minor Planets is, comparatively speaking, very small. This is, we think, another argument against the hypothesis of Olbers. So far as we can judge, any planet, by the explosion of which they might have been formed, would have been of so insignificant a size as hardly to be worthy of comparison with the other members of the planetary family; unless we imagine an explosion to have occurred so terrific as to have almost entirely dissipated it into dust, or into millions of fragments far too small for our observation; or that, by some fearful catastrophe, the greater portion of it was converted into gas, so that it disappeared from view, as a nebula, or as a series of nebulæ too faint to be seen. Nothing, however, with which we are acquainted in the physics of the Earth, or of which we see traces in the Moon, in any wise suggests the possibility of the action of any explosive force of so tremendous a character. Our previous arguments (see Lecture IX., p. 219) against the existence of a thin crust covering a seething mass of central fire in the case of the Earth, may also militate to some extent against the theory of the explosion of a hypothetical planet.

We cannot afford space to discuss at any length the possibility of such a supposed planet having been broken up by violent collision with some other heavenly body. If such a body were of suitable size, and the collision sufficiently violent, it is quite conceivable that the heat generated might have almost entirely converted both it and the planet into gas, and have only left a comparatively small number of fragments in a solid form. But, if we assume the existence of such a hypothetical planet, it is very difficult to conceive where the other body should have come from, unless perchance it were a comet with an unprecedentedly large and solid nucleus. Moreover, the same conclusions as before would follow, as to the subsequent passages of all the fragments, from time to time, through the point where the collision occurred, and through an opposite direction in longitude; and a similar difficulty would be involved in the fact that this is not the case.

We therefore consider it to be a much more reasonable supposition (as we shall more fully explain a little further on) that, at some long past epoch, when the matter of the Solar System may have been gradually aggregating from a nebulous, or partly vaporous, condition to form the various planets which now circle round the Sun, a vast quantity of matter was in a long course of ages collected from the neighbouring regions of space into the huge bulk of Jupiter, while a comparatively small amount was left between the orbits of Jupiter and Mars, which only sufficed to generate a number of much smaller bodies.

If so, such bodies may have been at first even more minute than they are at present, and, by collision and mutual attraction, have gradually become fewer and larger. And such collisions, and the consequent fusing together of two or more, may still occasionally occur, although they must be exceedingly rare amongst those with which we are acquainted, owing to the paucity of their numbers compared with the vastness of the space through which they are scattered. But if there really be millions too small for us to detect, such aggregation may still be going on with greater frequency.

It will, however, be noticed, that we have not, thus far, in suggesting such an origin for these bodies, in any wise accounted for the exceptional inclination of a considerable number of their orbits to the plane of the ecliptic. And we may take this opportunity to remark that these inclinations involve a special difficulty in connection with the process by which the planets are supposed to have been formed, in the gradual development of the Solar System according to Laplace's Nebular Hypothesis. He imagined that the various planets were originated by rings thrown off from the equatoreal parts of a nebula, while it was rapidly revolving round a polar axis. But if such a ring, containing less matter than usual, had been thus cast off, and had gradually formed the Minor Planets, we should have expected to find them all revolving in planes inclined at no great angle to that of the ecliptic, as is the case with all the other planets.

The considerable inclination of many of their orbits, therefore, leads us to one of two suppositions. It may be, according to a modification of the Nebular Hypothesis which has been suggested, that the successive portions of the original nebula were separated, not as equatoreal rings, but as spheroidal envelopes, which in most cases, after a while, gradually ran down (if we may be allowed to use the expression), through the descent of their polar regions towards their equators, into the form of rings, by the breaking up and aggregation of which the planets were formed. Upon this supposition, it may perhaps have been possible for a considerable number of the Minor Planets to have been generated in various latitudes of such an envelope, before it degenerated into the form of a ring, the result of which, in some way not very easy to conceive, might be that they would finally move in widely differing planes.

Or, on the other hand, we may imagine, that, during the

whole period of the contraction of the Nebula, from which, according to the hypothesis of Laplace, our system may have originated, a very large amount of meteoric matter existed in its midst, or in the regions through which it was travelling, and that the aggregation of such matter had much to do with the formation of the various planets. It is well known that a considerable quantity of meteoric matter is still scattered throughout the Solar System; but in ages long past it was probably much more abundant.

In those times long ago, may it not be, that comparatively little of such matter as lay within the orbit of Mars escaped the powerful attraction of the Sun, and that what did so escape remained in an especially fine state of division? If so, we can understand why it is that such meteorites as are constantly passing through the Earth's atmosphere, and falling upon its surface in the form of shooting stars, are exceedingly minute in size.

But to proceed with our explanation of this hypothesis. It may also be supposed, that the neighbourhood of the larger planets, and especially that of Jupiter and Saturn, at a distance where the attraction of the central mass of the Sun would have a greatly diminished effect, has supplied a great quantity of such matter towards the formation of their globes. At the same time we may conceive, that, at a distance from the Sun lying between about 150,000,000 and 400,000,000 miles, meteoric masses, somewhat larger in size than those which exist nearer to it, might still remain in their pristine orbits, the *inclinations* of which orbits might in some cases be very considerable. These may be really the bodies to which we give the name of Minor Planets. The Minor Planets, upon this supposition, may perhaps be considered the best examples extant of the original condition of some of the primeval matter of the Solar System.

This same hypothesis, which, however, we must confess is devoid of any certain evidence, also suggests that, in the vast spaces between the orbits of Jupiter and Saturn, of Saturn and Uranus, of Uranus and Neptune,—spaces so vast that the attractions of their great globes would, owing to the effect of distance, lose much of their aggregating power,—there may be myriads upon myriads of Minor Planets having a similar origin, and of a considerably larger bulk than those which we have discussed in this lecture, but all invisible owing to their great distance from us. If so, great as are the wonders which we see in the system of which we form a part, and numerous as are the little planets which we have discovered, there may belong to it wonders still greater, and planets still more numerous, yet unseen.

Let the origin of these minor members of the Sun's family be what it may, there is good evidence that the attraction of Jupiter has, in certain cases, remarkably perturbed their orbits in the course of long, long ages, and in an especially interesting manner. It can be shown, by mathematical investigations hardly suitable for explanation here, that a Minor Planet moving at such a distance from the Sun, that a certain whole number of its periods would be equal, or almost exactly equal, to an integral number of those of Jupiter, would periodically come into such a position, that Jupiter's perturbing effect would be repeated upon it, again and again, in a very similar manner. Its orbit would consequently be so disturbed as to be quite changed in its period, eccentricity, and distance from the Sun. It would no longer be found where it originally was; or, possibly, the planet might be brought into collision and union with some other which it would not otherwise have approached. In either case, gaps would be formed in the series at such distances as would correspond to such periodic times. Such a process would doubtless have been most effective, if it be supposed to have occurred while the Minor Planets were being gradually formed out of nebulous or meteoric matter; in fact, while they were much more numerous and smaller than they probably are at present.

Professor Daniel Kirkwood was, we believe, the first to notice (when not much more than one-third of the present number were known) that such gaps in the series actually existed; and M. Niesten, in his recent article in the Annuaire of the Observatory of Brussels for 1881, has shown that the same fact still holds good, especially at distances equal to $2\frac{1}{2}$ and $3\frac{1}{4}$

times the Earth's distance from the Sun; distances such that they would respectively involve periods which, in the one case, would be equal to one-third of that of Jupiter, and in the other case to one-half of that of Jupiter.

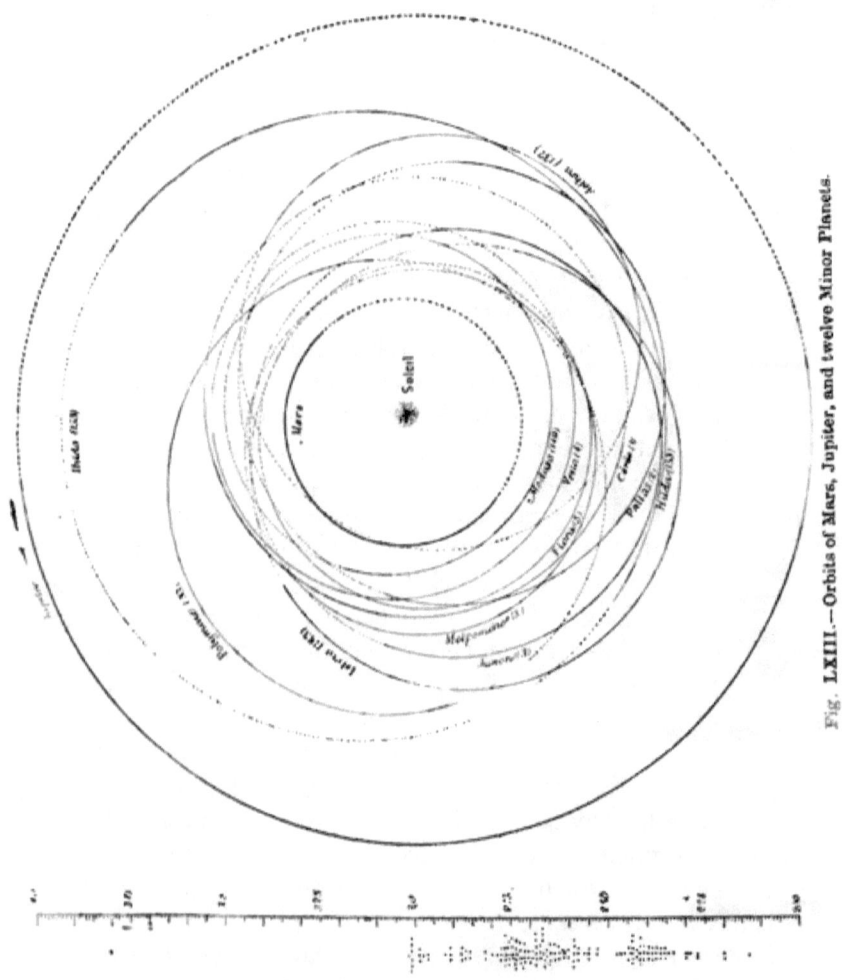

Fig. LXIII.—Orbits of Mars, Jupiter, and twelve Minor Planets.

Scale showing the distribution of 216 Minor Planets according to their mean distances from the Sun. The scale extends from twice to four times the Earth's mean distance.

This is indicated by the scale and the dots which accompany it in Fig. LXIII., which is, by M. Niesten's very kind permission, copied (and at the same time reduced) from one published in his article. The scale is intended to represent mean distances varying from two to four times that of the Earth. The number of Minor Planets known in August 1880, whose orbits correspond to the various distances lying between these limits, is indicated by the small dots which are seen on the left of the scale. The gaps at 2·50 (or $2\frac{1}{2}$), and 3·25 (or $3\frac{1}{4}$), as well as another at about $2\frac{3}{4}$ times the Earth's distance, are especially evident; while some others not quite so decided may be noticed, a little short of, and a little beyond, three times the same distance.

It is a very interesting confirmation of the above hypothesis as to the influence of Jupiter in the production of these gaps in the series of Minor Planets, that the principal divisions in the rings of Saturn, as we shall show in Lecture XIV., correspond to those distances from the planet at which the minute satellites (by myriads of which the rings are probably constituted) would be in like manner affected by the commensurability of their periods with those of some of the nearer of Saturn's moons; while other smaller gaps caused by these, or by some of the other moons, seem on rare occasions to be faintly visible in the indications which have been noticed of additional concentric divisions in the rings.

After all we may ask:—Are the Minor Planets to be looked upon as useless, or nearly useless, members of our system? Certainly not! If our space permitted, we might discuss several important benefits already derived from them.

For instance, we showed in Lecture I., page 18, with reference to the determination of the Sun's distance, that the accuracy with which the minuteness of their star-like discs enables us to measure their apparent distances from neighbouring stars, may be of great advantage if we use such of their number as approach the nearest to the Earth, for observations of the same kind as those which Mr. Gill made of the planet Mars in 1877, in the Isle of Ascension. The minor planet Æthra might have been so used in February of last year (see *Nature*, Feb. 24th, 1881), when it was only about 78,000,000 miles from the Earth. In the present year, 1882, on August 24th, Victoria, No. 12, will be only 83,000,000 miles; and on September 24th, Sappho, No. 80, will be only 79,000,000 miles, from the Earth. The method to be used for the above purpose in the present year, will, however, be that explained upon page 15, which, in its practical application, involves the comparison of observations made at observatories as far apart as possible to the north and south of the Equator. It is not in itself so good a method as that employed in the Isle of Ascension in the case of Mars (which is called the method of *diurnal parallax*, see Lecture I., pp. 16, 17), but it will be the most suitable upon the present occasion, unless a special expedition were undertaken to such a locality as the other method would require. We believe that a considerable number of observatories in both hemispheres of the Earth will co-operate in the endeavour to obtain the best result possible from these favourable positions of Victoria and Sappho.

In the year 1874, the method of diurnal parallax, involving morning and evening observations from one and the same observatory, was applied in a short series, made at Mauritius, to the Minor Planet, Juno, which we have frequently mentioned as the third in order of discovery, and probably the fourth in order of magnitude. The result was a very fairly satisfactory estimate of about 93,000,000 miles for the Earth's distance from the Sun. This value is especially interesting from its close accordance with that obtained by the use of the same method in Mr. Gill's observations of Mars.

Once more, the Minor Planets which pass comparatively near to Jupiter enable us, by the perturbations which their movements undergo, to measure the weight of that great planet with considerable accuracy, and confirm the value calculated by various other processes.

We also hope that this lecture has shown that the continued observation and study of the Minor Planets may in time be found to be fraught with much instruction with respect to the past history of the Solar System. In any case, we may say that, the more exceptional their character, the more unlike they are to the other planets in size, in orbit, in number, the more do they deserve our most careful attention.

We venture, in conclusion, to mention one other point which we originally discussed in a lecture delivered in the year 1878 at GRESHAM COLLEGE, and to which we have already made a slight allusion in the tenth of this series of lectures, viz., the possibility that, as at least one of these bodies is known to come some 5,000,000 miles nearer to the Sun than the maximum distance of Mars, the satellites of Mars recently discovered may really be two Minor Planets picked up by it, and permanently mastered by its attraction.

We suggested that, if such a Minor Planet were at any given moment in a certain position in its orbit, and if Mars at the same time passed very nearly through the same spot, and with almost exactly the same velocity (which is quite possible), it is conceivable that the Minor Planet might thenceforth become its satellite. More recently, however, it has occurred to us that the remarkable ratio which exists between the periods of the two satellites (the one being almost exactly four times that of the other), when compared with the similar almost exact integral multiplication of the periods of the innermost satellite, which we find in the case of those of Jupiter, and also to some extent in those of Saturn, seems to point to some common process in the formation of all these satellites, and to their having been in each instance in some way cast off from the primary to which they belong. We must, however, refrain from any further indulgence in such speculations as these. They may be fascinating, but they may for that very reason be conducive to error.

Here then we must leave the Minor Planets, with the hope, that our arguments against the hypothesis of their origin in an explosion of a larger planet may, at any rate, have sufficed to afford some consolation to those who might otherwise, with a sort of analogical fear, be nervously looking for a similar occurrence in the case of the Earth. Whatever the ultimate destiny of this world may be, we do not expect that it will be broken up into a swarm of Minor Planets.

LECTURE XII.

THE PLANET JUPITER.

> "O telescope, instrument of much knowledge, more precious than any sceptre! is not he who holds thee in his hand made king and lord of the works of God?
> "All that is overhead, the mighty orbs,
> With all their motions, thou dost subjugate
> To man's intelligence."—KEPLER (*translated by E. S. Carlos*).

ABOUT sixty millions of miles beyond the outer boundary of the vast region in which, as we have seen, the tiny Minor Planets are sparsely scattered, we come to the orbit of the huge planet JUPITER. Well may it be called huge, for its apparent volume, although about a thousand times less than that of the Sun, is nearly 1,300 times greater than that of the Earth.

The *real* volume of the planet Jupiter is also doubtless very large. But whether its real volume is nearly the same as its apparent volume, or very much smaller, we cannot say. We have many reasons for believing that the boundary of the globe which we see may be no true boundary of the body of the planet; that what we behold may very likely be only some vast and cloudy envelope, lying like a fog upon a central solid ball, or separated from it by a considerable interval in the same way that the clouds of the Earth's atmosphere are in general raised above its surface. Or it may be that the constitution of Jupiter more or less closely resembles that which many astronomers think to be the most probable condition of the Sun; that the planet possesses no solid nucleus at all, but is wholly fluid or viscous, and perpetually permeated by tremendous currents of gas.

To this last supposition, however, one objection may be urged which does not apply to the case of the Sun. If we consider the vast pressures which must exist within Jupiter as a consequence of its mighty bulk, it would seem that, in order to prevent it from solidifying, it would need to be pervaded by such a high degree of temperature that it would shine with an intrinsic brightness comparable with that of the Sun. But careful estimations of the amount of light received from it do not indicate the existence of any exceedingly high degree of temperature, although it has been suspected that some light is really given out by it in addition to the Solar light which it reflects.

Owing to the absence of any indications of such a temperature as would probably render the whole planet fluid, many astronomers therefore believe, that there is a central globe in Jupiter in a condition more or less like that of the globe of the Earth. If so, such a globe may lie far within the cloudy outer envelope visible to us, and at a distance greater or less according to the temperature which it at present possesses; *i.e.* if the central globe be colder and denser, it will be so much the smaller; if somewhat hotter, and less dense, it will be so much the larger.

Quite apart, however, from the uncertain question of its internal constitution, or of the size of its central globe, we are able, by observations of the periods and movements of Jupiter's satellites, as well as by some other more complicated methods, to determine one undoubted fact connected with it, which is by itself quite sufficient to justify us in speaking of it as very huge; viz., the weight of the planet as a whole. The most careful calculations prove that it weighs rather more than 300 times as much as the Earth.

The above statement, however, necessitates that, if its globe be really bounded by the envelope which we see, it must be formed of matter upon an average more than four times lighter than that of the Earth. This is found to be the case by dividing the number 1,300, which is approximately the number of times that the volume of such a globe would exceed that of the Earth, by the last mentioned number, 300.

But, if the planet be not in a considerably heated condition,

19*

so small an average density seems once more to be hardly consistent with the immensity of its size, and the consequent intensity of pressure which, as we have already hinted, must exist in the vast depths within it. Moreover, the fact, that the calculated value of the attraction of the force of gravity, upon its surface, is found to be nearly $2\frac{1}{2}$ times * as great as upon the surface of the Earth, is another reason for the existence of a much greater, rather than a smaller, average density in the planet as a whole. We may, therefore, conclude that the *joint action* of two causes affords the most probable explanation of the apparently small density of the planet Jupiter: the one, that its real and more solid part lies far beneath the cloudy envelope which bounds our view; the other, that even that central portion may be in a state of very considerable heat (although not sufficiently hot to be decidedly luminous), and be consequently expanded to a greater size and a lighter density than it would possess if it were cooler.

We have been led to make the preceding preliminary remarks touching upon some deeply interesting physical speculations which we shall presently treat at greater length, by our use, in the first paragraph of this lecture, of the expression "the

---

* This may be approximately calculated by the following method, which, with the necessary alteration of figures, will equally well apply to other planets:—

A sphere attracts any external object as if all the matter in the sphere were condensed at its centre. The weight of Jupiter being found, from the rotation-period of any one of its satellites, or by some other method, to be about 300 times that of the Earth, gravity upon its surface would be 300 times as great as upon the Earth's surface, if the respective distances from the centre to the surface were the same in each planet. But the distance in question being about 11 times as great in the case of Jupiter, and gravity acting inversely as the square of this distance, the attraction will be diminished about 121 times. Since 121 will divide into 300 rather less than $2\frac{1}{2}$ times, it results that gravity upon Jupiter's surface is somewhat less than $2\frac{1}{2}$ times as great as it is upon the Earth's. A man weighing 15 stones upon the Earth would therefore weigh about 37 stones upon Jupiter, and a pound avoirdupois would weigh about 39 ounces. A body dropped from rest would fall about 39 feet instead of 16 feet in the first second of its fall, and would at the end of that second possess a velocity $2\frac{1}{2}$ times as great as it would upon the Earth.

apparent size of Jupiter," which naturally suggested the question,—Is what we see really the boundary of the planet or not? It may, however, be well, before we indulge in any further discussions of a similar character, to state a few facts which have been certainly ascertained with regard to the dimensions of the planet and of its orbit.

The surface of the globe, seen by us, which (whether it more or less exceed in size the real body of the planet) is all that we can measure, is about 120 times that of the Earth; more than 6,000 times that of Europe; and more than 400,000 times that of England.

The equatoreal diameter of Jupiter measures about 88,000

Fig. LXIV.—The polar flattening of Jupiter is such that one-half of the planet Mars placed, as in this figure, at one of its poles would lie 500 miles within a sphere of a radius equal to that of Jupiter's equator.

miles.* But the polar diameter is more than 5,000 miles shorter, the proportionate difference being greater than in the case of any other planet, except Saturn.

So vast is the size of the apparent globe of Jupiter, and so considerable the amount of its polar flattening, that, if a diagram be carefully drawn to scale as in Fig. LXIV., and the shape, which one-half of a circular disc of the same diameter as its equator would have, be indicated by a dotted line, the distance between one of the poles of the real disc and the dotted semi-circle would more than suffice to receive one-half of the disc of the planet Mars. In fact, between the centre

* The above value assumes the Sun's distance from the Earth to be about 93,000,000 miles. If that distance be greater, the apparent width of Jupiter's disc will represent a proportionally greater diameter, and *vice versâ*.

of Mars, as we have drawn it, and the semi-circle in question there would remain a vacant space, which, although only just distinguishable in the figure, is really about 500 miles across.

Jupiter pursues his journey round the Sun* at a mean distance very nearly equal to $5\frac{1}{5}$ times that of the Earth. If the Earth's mean distance be estimated at 93,000,000 miles, Jupiter's will therefore be about 484,000,000 miles.

But this mean distance is subject to considerable variation owing to the ovalness of the planet's orbit, which, although only about one-half of that of the orbit of Mars, and less than one-fourth of that of Mercury, is three times that of the orbit of the Earth. Such an eccentricity, or ovalness, is in fact measured

---

* We should more accurately say (see Lecture V., p. 121), round the common centre of gravity of the Sun and Jupiter. For a similar law to that which we explained in the case of the revolution of the Earth and the Moon around their common centre of gravity (see Lecture III., p. 53) holds for any two mutually attracting bodies; and it is worthy of notice, that, so far as the Sun and Jupiter are concerned, the common centre of gravity about which they would revolve is *not within the Sun*, but even when Jupiter is at its nearest to the Sun, and their centre of gravity consequently at its closest approach to the centre of the latter, it is some thousands of miles outside the Sun's surface.

This may be roughly proved as follows. Let s in Fig. LXV. be the

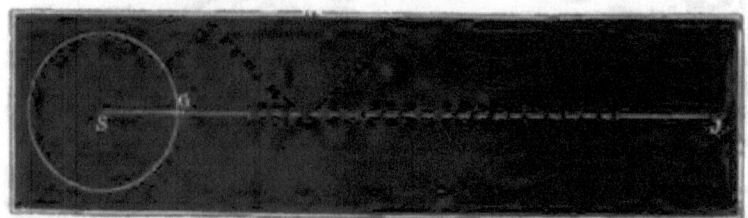

Fig. LXV.—The common centre of gravity of the Sun and of Jupiter lies outside the Sun.

centre of the Sun; J of Jupiter; G their common centre of gravity. Then the weight of Jupiter being about a $\frac{1}{1048}$th part of that of the Sun, GJ must be about 1048 times GS, or GS a $\frac{1}{1049}$th part of SJ. The smallest value of SJ being about 461,000,000 miles, we have for the minimum value of SG, the quotient of this by 1049, or about 438,500 miles; which is about 6000 miles further from s than the Sun's surface, if we take 432,500 miles to be the most probable value of the Solar semi-diameter. It is therefore hardly accurate to state, as is sometimes done, that the weight of the Sun is so enormously preponderant, that the centre of gravity of it and of all the planets lies very close to the Sun's centre.

THE PLANET JUPITER.

by a fraction which is rather less than $\frac{1}{20}$th, so that the value of its greatest distance from the Sun is rather less than a $\frac{1}{20}$th part (or about 23,000,000 miles) more, than the mean value, which we have already stated to be about 484,000,000 miles. Jupiter's greatest distance from the Sun consequently amounts to about 507,000,000 miles. In like manner its least distance is 23,000,000 miles less than its mean distance, or very nearly 461,000,000 miles.

If only approximate numbers be desired, such as may easily be carried in the memory, it may be said that the planet's distance from the Sun varies from about 460,000,000 to about 510,000,000 miles; its mean distance being somewhat over five times that of the Earth. From this it would also follow by Kepler's third law (see p. 122) that, if we cube the number 5 (which gives 125), and then take the square root of this latter number (which is somewhat greater than 11), Jupiter's year, or the period of its orbital rotation, must be somewhat more than 11 times as long as the Earth's year.

Its more accurate value is found to be very nearly $4,332\frac{2}{3}$ days, *i.e.*, about $11\frac{6}{7}$ years, or about 50 days less than 12 years.

Jupiter travels in an orbit, the plane of which is inclined at less than $1\frac{1}{3}°$ to that of the ecliptic, or to the plane of the Earth's orbit; an inclination smaller than for any other planet, except Uranus. While travelling in such a plane it keeps its polar axis very nearly perpendicular to it, the deviation from an upright position being only about 3°, instead of nearly $23\frac{1}{2}°$, as in the case of the Earth.

Jupiter's velocity in its orbit, owing to its greater distance from the Sun, is, of course, much less than that of the Earth. While the latter moves at the rate of more than 18 miles per second, the speed of the former is only about 8 miles per second. This, however, amounts to about 29,000 miles per hour. On the other hand, the huge bulk of the planet appears to turn on its polar axis nearly $2\frac{1}{2}$ times as rapidly as the Earth.

With regard to the exact period of its axial rotation, as we shall presently see, considerable doubt exists, owing to the fact that somewhat different results are given by the observation of the times in which spots upon different parts of its surface are seen to go round it. Its most probable value, however, is about 9 hours 55½ minutes.

It may be interesting to contrast the corresponding speed for a point upon the Earth's equator with the vast speed with which a point on the equator of Jupiter is thus whirled round and round, as the result of so rapid a rotation and the large diameter of the planet. Such a point upon the Earth is carried through a distance equal to about 3¼ times the Earth's diameter, *i.e.*, through about 25,000 miles in 24 hours; or with a velocity somewhat exceeding 1,000 miles per hour. But the equatoreal circumference of Jupiter is 3¼ times a diameter of nearly 88,000 miles, or about 275,000 miles, and a point upon it is carried round this distance in less than 10 hours, or with a speed of nearly 28,000 miles per hour.

It also deserves notice, that the Earth rushes round the Sun with a speed of between 65,000 and 66,000 miles per hour, so that the velocity of any place produced by its axial rotation, although very great, is nevertheless more than 60 times less than that produced by its orbital motion; while, in the case of Jupiter, a curiously near approach to equality exists between its orbital velocity and the velocity of rotation of a point upon its equator. The average value of its orbital velocity, in fact, only exceeds the value calculated above for the result of the axial rotation, viz., 28,000 miles per hour, by about 750 miles.

A very remarkable result in consequence arises, which we will endeavour to explain by means of the diagram in Fig. LXVI., in which the direction of the rotation of Jupiter round the Sun from west to east is represented by the arrows round the large circle; and that of the equator of the planet round its axis by the smaller arrows at P and Q.

When any place is in the position Q, its meridian is turned directly away from the Sun, and it is passing through the hour of midnight. The arrows which point in the same direction show that, at such a time, the speed of movement through space of the place in question will equal the sum of that caused by the onward motion of Jupiter in its orbit, and of that due to the rotation of Jupiter upon its axis. It will in fact be about twice 28,000 miles, or nearly 57,000 miles per hour.

But for a place in the position P, which is passing through the instant of noon, the two velocities are in exactly opposite directions, as the arrows again indicate. An observer in such a position (so far as these velocities are concerned) would therefore at noon, or within a brief period before or after it, be almost at rest in space. And, inasmuch as Jupiter's velocity in its orbit varies by as much as 1,500 miles per hour on either side of its mean value, and that mean value only differs, as we have stated, by about 750 miles per hour from the velocity caused by the axial rotation of a point upon its equator, it follows

Fig. LXVI.—Comparison of Jupiter's orbital velocity with the velocity of rotation round its axis of a point upon its equator.

that there must be times in each period of the planet's circuit round the Sun, when the one velocity will at noon precisely neutralize the other.

We have therefore shown that, so far as the above velocities are concerned, an observer upon the equator* of Jupiter would

* That is to say, upon the equator of the surface of the globe, which is visible to us. The compensation of the one velocity, to which we have referred, by the other, would, of course, more or less fall short, if the real equator of the body of the planet were that of a globe situated at a considerable distance within the limits of the sphere which we behold, according to the theory mentioned in the earlier part of this lecture. The nearer such an equator might be to the planet's centre, the smaller would be the velocity communicated by the planet's rotation to any point upon it.

at such a time be at rest in space. Indeed, he would be altogether at rest, were it not that we believe that the Sun and all the planets are together travelling onwards with a common motion round some very distant centre. Those who are accustomed to the use of the spectroscope may, however, understand, that the conditions involved would facilitate some specially interesting observations upon the approach or recession of other heavenly bodies, or upon the motion of the Solar System itself.

We must, however, hasten on to the consideration of some of the many and very interesting *physical features* of this wonderful planet; amongst which we shall find that its Spots and Belts are most important.

These the astronomers of olden time of course never saw, although they so carefully watched the movements of Jupiter that one record has been handed down, dated more than twenty-one centuries ago, in which Ptolemy tells how, upon a certain day in the autumn of the year B.C. 240, the planet was seen to occult the star δ Cancri, the same star which was distinctly seen near to the Sun during the total Solar eclipse of July 1878.

But now that the telescope has come to our help, it is hardly possible to look at the disc of Jupiter with an object-glass of two inches, or even less, in aperture, without being at once struck by certain dark streaks running across its surface, in a direction approximately parallel to that in which its Satellites are also seen to lie.

It is doubtless strange that these streaks, or belts, do not seem to have been noticed by Galileo,* and that the first observation of them was not made until about twenty years after he detected the Satellites of the planet. But from that time onwards, they have been constantly watched. Nor was it long before they were found to exhibit frequent changes; so that

---

* In England we are accustomed to call Galileo Galilei by his Christian name Galileo, instead of using his family name Galilei, as is customary among other nations. We hope it may soon be practicable, possibly by using both names for a while, to correct this peculiarity. (See R. A. S. *Monthly Notices*, vol. xxxviii., p. 258.)

occasionally only one was visible; while, in general, two (one on each side of the equator), or three; and sometimes four, five, six, or even seven, were seen; or, as has happened more than once, the whole surface nearly as far as the poles has been traversed by a succession of numerous and rather faint parallel markings, alternately darker and lighter.

A continuous and careful course of observation has shown that some of the principal features of these belts frequently remain comparatively constant for weeks or months together. At other times they are disturbed by changes so rapid, that the only phenomenon with which they seem to be in any wise comparable (and even that is on a vastly smaller scale), is the extraordinarily rapid formation of a stratum of cloud, by which we occasionally see a clear blue sky entirely overcast from one horizon to the opposite in the course of a few brief moments.

In the various belts, unusually dark spots, or unusually bright spots, of various degrees of magnitude, may also often be watched, while they are carried rapidly round by the planet's rotation. Of these, small bright spots are upon the whole the most rare; but they are, if carefully followed and studied, as some have been which have recently appeared in the midst of the equatoreal regions of the planet, perhaps more full of interest and instruction than any others.

We much regret that the space at our disposal does not allow us to enter upon anything like an exhaustive description, or discussion, of the never-ending variety in the appearance of the belts and spots of Jupiter. We must perforce be content with a brief mention of some of the more interesting observations of recent date, and refer our readers for further information to the Records of the Royal Astronomical Society, and to many interesting articles which have appeared in the *Observatory*, the *Astronomical Register*, and other scientific publications.

In Fig. LXVII., we have an attempt to reproduce a very beautiful drawing of the planet made by Mr. De La Rue, on October 25th, 1856. Although the woodcut is certainly too harsh in its outlines, and does not very well represent the cloudy and ill-defined appearance of the markings usually seen, it may give our readers some idea of the amount of detail which is visible in a powerful telescope. In Plate VIII. are three very interesting views, selected from a large number recently presented to the Royal Astronomical Society by Mr. G. D. Hirst, of Sydney, N. S. Wales, who has paid special attention during the last few years to the observation and delineation of the planet. The view marked No. 1 was obtained

Fig. LXVII.—Jupiter as observed by Mr. De La Rue, October 25th, 1856.

with a $10\frac{1}{4}$ in. silvered-glass reflecting telescope, and a magnifying power of 214; the 2nd and 3rd views with a $7\frac{1}{4}$ in. Merz refractor and a magnifying power of 200. It may be well to mention that the copies of the original drawings have been reversed in the Plate, in order that they may represent the view that would be seen in a telescope by an observer in our northern latitudes.

They all illustrate the great variety and constant change in the belts and markings to which we have already referred;

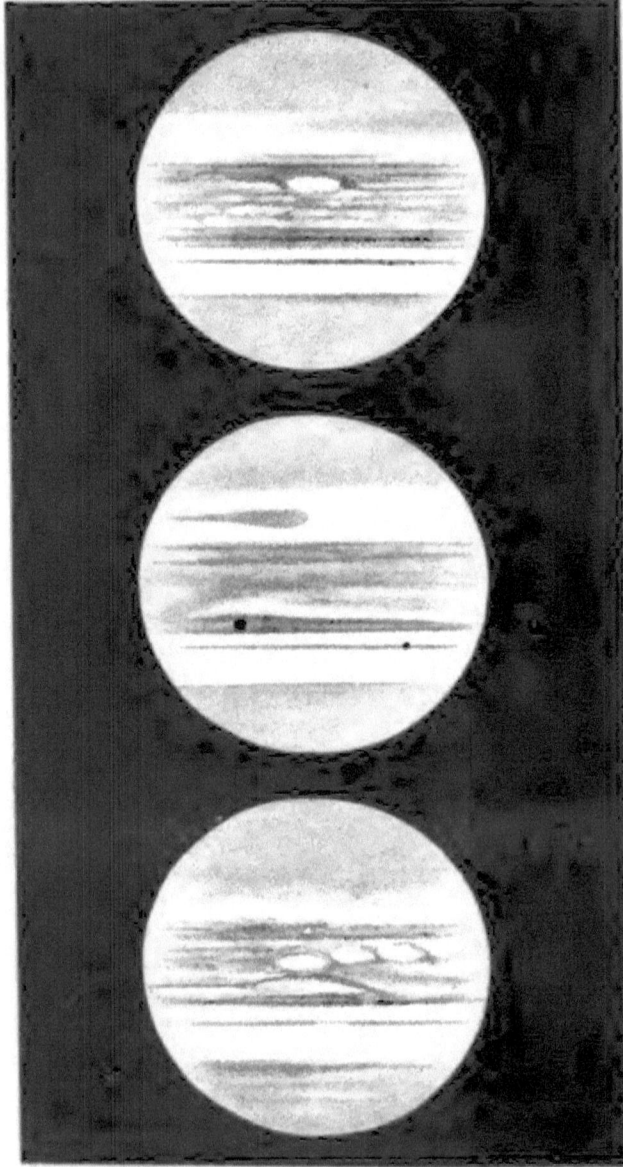

while the third, and the first (in a less degree), also afford indications of the appearance of certain large and moderately bright oval forms, which appear from time to time (and possibly with some approach to a regular periodicity) in the principal central belts. Of such larger egg-shaped markings a regular succession, ranged across the disc at almost equal intervals, and separated by smaller and darker spaces, is occasionally seen; when the effect produced, as is shown in a considerable number of published drawings of the planet, closely resembles that of a series of arches in a bridge. Many observers so described what they saw in the years 1869-1871.* The writer of this lecture also observed the same appearance to be remarkably distinct in the autumn of 1879.

We think that our readers will find it well worth while to study all the drawings of Jupiter to which they may be able to obtain access. A large number may be found in vols. xxxi. and xxxiv. of the Royal Astronomical Society's Monthly Notices; the latter volume containing a very fine series observed at Parsonstown with the great reflector belonging to Lord Rosse, as well as some drawn by Mr. Knobel. There are also many excellent pictures of the planet published in the years 1871 and 1872, in vols. ix. and x. of the *Astronomical Register*. Others are to be found in the volumes of the *Observatory;* while two especially beautiful woodcuts illustrate an important article in *Nature* of January 5th, 1882.

In several of these numerous delineations of the planet's appearance, some of the smaller bright or dark spots, to which we have previously referred, may be noticed; but in the second of the three drawings in Plate VIII. it will be observed that the most remarkable feature is a large oblong spot in the upper half of the view, which has not only caused a great amount of excitement and interest amongst astronomers during the last four years, but has so far exercised the public mind that it has frequently been discussed in the columns of our daily newspapers.

* See drawings in *Astronomical Register* for April and November 1871. Also *Popular Science Review*, vol. ix., pp. 129 and 131. Slight indications of these markings are also shown in Fig. LXX., p. 309, of this lecture.

In the view in question it is seen when situated not quite centrally upon the disc. Indeed, whenever it is upon that side of the planet which is turned towards the observer, it is so prominent a feature of the surface, and so distinct from any other, that it at once rivets his attention. And yet, huge and most extraordinary as it appears to be, it first began to attract special notice so recently as in the summer of the year 1878. Since that date it has remained almost unaltered in size, and in its latitude upon the disc, although it has recently lost some of

Fig. LXVIII.—Jupiter, showing the great red spot on July 9th, 1878, as seen at Morrison Observatory, U.S.*

the very decided reddish colour by which the astonishment which it at first excited was greatly increased.

It is now generally known as the *Great Red Spot*. We believe that it was first publicly mentioned by Mr. F. C. Dennett, in the *English Mechanic* of November 22nd, 1878, who then stated that he had repeatedly observed it since July 27th of that year. It afterwards proved that Professor Pritchett had seen it in

* Figs. LXVIII.—LXXIII. are taken from the *Observatory*, vol. iv., by the kind permission of the Astronomer Royal, who has most liberally lent the author the original wood-blocks. This is also the case with another wood-cut in Lecture XIV.

America on July 9th; and that it had been noticed as early as June 26th at Lord Lindsay's observatory at Dun Echt. We have much pleasure in showing our readers two drawings of its appearance at that time, which were originally published in the *Observatory* of January and April 1879. The first is by Professor Pritchett, of Morrison Observatory, Glasgow, Missouri, dated July 9th, 1878; the second, in which the spot in question is marked *a*, is by Mr. L. Trouvelot, of Cambridge, Mass., U.S., dated Sept. 25th, 1878.

Mr. Trouvelot mentions that its "intense rose colour

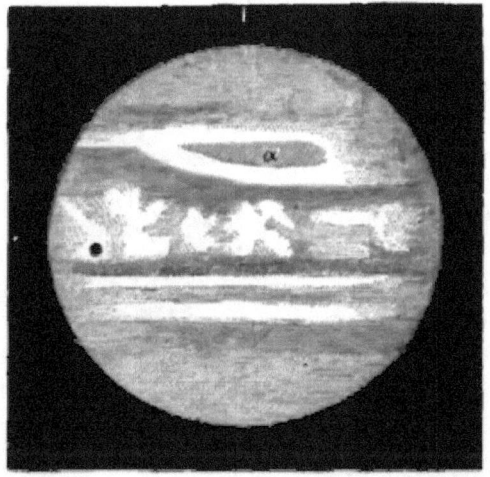

Fig. LXIX.—Jupiter, showing the Great Red Spot (*a*) on September 25th, 1878, as seen at Cambridge Observatory, U.S.

appeared in strong contrast with the white luminous background upon which it was projected." Some other observers, who went so far in the exuberance of their feelings as to compare it to a blood-red flame, must, we think, have been more or less carried away by the unexpected excitement of watching so remarkable an apparition. Whenever we have gazed at it with a refracting telescope, it has appeared to be of a dull brick-red colour. In a reflecting telescope it is, however, likely that its tint may be more brilliant. One well-known observer of great artistic taste and skill (Mr. Brett) has suggested that the best possible idea of the appearance which it presented in 1879 may be obtained by fastening a patch of maroon-coloured velvet on a pale orange, and looking at it from a considerable distance.

To the spot itself there have occasionally been attached certain streaky appearances, either preceding it, or following it, as in the figure before referred to (No. 2), Plate VIII.; and its extremities have sometimes been more pointed than at other times. It has also been thought to be surrounded by a border somewhat brighter than the bright parts of the surface in its immediate neighbourhood.

But, apart from any minor details of this kind, and even from the fact of its remarkable colour,[*] it deserves our most careful attention because of its enormous size, and the persistence with which it has remained not only visible, but almost unaltered in shape. It is hard for one who beholds it for the first time with a telescope, or sees it depicted in a drawing, to realize how large its area really is. Its form is approximately elliptical; but its surface is upon the whole larger than a perfectly elliptic form would involve. The measurements of Professor Hough, of the Dearborn Observatory, Chicago, made with a refractor of $18\frac{1}{2}$ inches aperture, assign to it a length of 29,600 miles, and a maximum breadth of 8,300 miles. It is therefore longer than the equatoreal circumference of the Earth by between 4,000 and 5,000 miles. Its area is about 200,000,000 square miles; which does not differ much from that of the whole surface of the Earth, while it is about 50 times as large as that of Europe.

This Great Red Spot is situated in about 30°, or somewhat less, of South Jovian latitude; and, approximately, it extends about one-third of the way across the disc of the planet in that latitude, when seen in mid-transit. It is therefore so very large a feature of the disc, that we cannot but think that

---

[*] This colour is all the more remarkable when its latitude is taken into account. A reddish or orange-coloured tint is not uncommon in the belts nearer to the equator, but is not so usual in higher latitudes. In 1871 Mr. Lassell (see Chambers' "Handbook of Descriptive Astronomy," p. 114) described the colours of the upper and lower belts as purple and light olive-green, while the equatoreal belt was of a brown orange-tint.

many an observer would most certainly have called special attention to it, if it had been as noticeable, or its colour as remarkable, before the year 1878, as since the summer of that year.

It is, however, very interesting to find, that various traces of its previous existence have been recorded. For instance, the writer of this lecture has noticed, as he has mentioned in the *Observatory*, vol. iii., p. 449, that some of the drawings of the planet made in 1873, and published by Lord Rosse, afford indications of its presence.* It is also a remarkable fact that, near to the same part of the disc, an elliptic form of similar shape was frequently seen by Mr. Gledhill and others in 1869 and 1870.† But this was simply an *outline* ellipse formed by a narrow black line, the interior of which, with

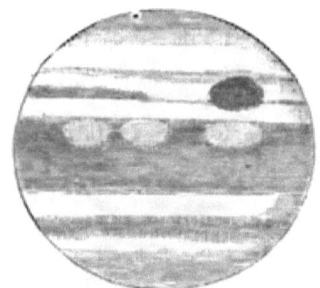

Fig. LXX.—Jupiter, March 13th, 1872.   Fig. LXXI.—Jupiter, April 19th, 1872.

the exception of an occasional streak across it, was as bright as the surrounding portions of the disc.

More recently, however, Mr. H. C. Russell, F.R.A.S., of the Observatory, Sydney, N.S.W., has stated ‡ that he finds, on going through his drawings made during the last few years, that he had frequently observed it in 1876, when it was involved in the equatoreal colour-band of the planet and somewhat different in shape, but not in colour, from its subsequent

* See R. A. S. Monthly Notices, vol. xxxiv.
† See *Astronomical Register*, vol. ix., p. 249; and vol. x., frontispiece and p. 68.
‡ See "Recent Changes in the Surface of Jupiter," a paper read before the Royal Society of N. S. Wales, December 1880, by Mr. Russell.

appearance. He finds its then locality to have exactly corresponded with that which the best recent measurements give. He also mentions a drawing by Dawes, made in 1857, in which he thinks that it may be seen. In the *Observatory* for January 1882, Mr. Corder, of Great Baddow, Essex, has also published two drawings, which he has found in his possession, made in the year 1872, in which he considers that it is depicted. These are shown in Figs. LXX. and LXXI.

Again, in the *Observatory* of February 1882 the two drawings in Figs. LXXII. and LXXIII., one previously unpublished, and the other taken from the "Bulletins de l'Academie Royale de Belgique" of 1872, are shown by M. Terby, who is so well known for his very elaborate study of the planet Mars, as well as for much other important astronomical work. The remarkable spots to which the letter a is attached are undoubtedly of

Fig. LXXII.—Jupiter, January 28th, 1872.   Fig. LXXIII.—Jupiter, January 30th, 1872.

the highest interest, as they strongly confirm the drawings of Mr. Corder, and especially that which is dated March 13th, 1872.

It will be exceedingly interesting if any much older drawings be found in which the spot appears. Nor would such time as might be devoted to a very careful search to find them be ill spent. For such drawings might be of much use in connection with the discussion of various matters of the highest importance relating to the physical condition of Jupiter, and to the determination of the exact period of its rotation upon its axis, which the observation of the Great Red Spot has recently brought prominently forward.

As regards this last-mentioned question, we have already stated (p. 300) that the probable period of the planet's axial rotation lies between $9^h 55^m$ and $9^h 56^m$; one of the best determinations yet made being that by Sir G. B. Airy, in the year 1835, when Plumian Professor of Astronomy at Cambridge, which gave for its result $9^h\ 55^m\ 21\cdot3^s$.

But it seems that any attempt to determine the rotation-period with extreme accuracy must now be altogether given up. We can adopt no other method for the purpose than that which was used in the year 1664, by the very first observer who detected the rotation of Jupiter, the celebrated Dr. Hooke, then Professor of Geometry in GRESHAM COLLEGE. He watched the movement of a spot upon its surface, as the rotation of the planet carried it round; and we must do the same. Our difficulty arises from the fact, that the more carefully we observe various spots, the less are we able to obtain a decisive result, owing to certain irregularities in their movements, and to a very appreciable, although moderate, difference in their speed of rotation, which seems to depend upon their distance from the planet's equator.

It was in pursuance of the above method that Cassini, in 1665, observed a spot which he followed during 29 consecutive revolutions, from which he obtained for the duration of one revolution a period of $9^h\ 56^m$. It was by the same method that Sir George (then Professor) Airy deduced his more accurate result of $9^h\ 55^m\ 21\cdot3^s$, from a dark spot which performed 225 consecutive revolutions, and by the long interval during which it was thus visible diminished any probable errors of observation. We must, however, consider all such determinations, whether of earlier or of more recent date, whether more or less refined and precise, simply as the values which correspond to the individual spots to which they belong, and not as giving us any certain and undoubted information as to the exact duration of the planet's own rotation.

And here we may remark, that it is not only found that, as a rule, spots upon Jupiter go round more rapidly the nearer they are situated to its equator, in which they follow the same law that Carrington noticed in the case of the spots upon the Sun, but individual spots have very decided individual peculiarities of speed, and the speed of the same spot is often from time to time somewhat irregular. So long ago as 1779 Sir

W. Herschel found that some spots gave a period differing from that of others by as much as five minutes. And it is certainly in very remarkable agreement with this statement that we read in the recent Report of the Dearborn Observatory, Chicago, by Professor Hough, dated May 1881;—that the time of rotation deduced from the Great Red Spot was $9^h\ 55^m\ 35^s$, while that obtained from certain white spots seen near to the equator varied from $9^h\ 50^m\ 0\cdot6^s$ to $9^h\ 50^m\ 9\cdot8^s$, a value more than 5 minutes less than the above. On the other hand, from some spots nearer to the poles than the Great Red Spot the period obtained was as large as $9^h\ 55^m\ 39^s$, or even $9^h\ 55^m\ 40\tfrac{1}{2}^s$.

Again, Dr. Lohse of Bothkamp, who has given great attention to the determination of the comparative speed of the spots seen in various latitudes upon Jupiter, mentions one which, in the year 1871, appeared to go round in about $3\tfrac{1}{2}$ minutes less than the $9^h\ 55^m\ 20\cdot3^s$ assigned as the planet's rotation-period by Airy.

Professor Hough's recent observations have been confirmed by those of Mr. Denning, M. Terby, and others, who have watched with especial care one remarkable *white spot*, which persists in appearing from time to time in a certain latitude a little to the south of the equator. From Mr. Denning's frequent observations of it during the past two years Mr. Marth has obtained a rotation-period of only $9^h\ 50^m\ 6\cdot6^s$, while some earlier observations, if taken into account, would still further reduce the period by about $1\tfrac{1}{2}$ seconds.

But perhaps, after all, the greatest interest at present belongs to the determination of the rotation-period of the Great Red Spot, for which the most accurate value yet deduced from the very numerous observations of many excellent astronomers appears to be very nearly $9^h\ 55^m\ 34\tfrac{1}{2}^s$. This value, it will be noticed, agrees very closely indeed with Professor Hough's result.

So considerable a difference of speed in spots seen in different latitudes is certainly very remarkable. And Mr. Denning, to whom the credit is due of drawing special attention to the above-mentioned equatoreal white spot by his long and accurate observations of it, informs us that it gains $8\cdot1°$ of Jovian longitude on the Red Spot in 24 terrestrial hours; so that in rather less than $44^d\ 10\frac{3}{4}^h$, it performs one more revolution round the planet than is achieved by the Great Red Spot. Between 29^m past 9^h on November 19th, 1880, and 43^m past 9^h on December 24th, 1881, he finds that the latter went round 967, and the former 976 times.

But even this rate of rotation has been exceeded in the case of a series of small black spots, which appeared some distance to the north of the planet's equator in October 1880. He found that these travelled once round in so short a time as $9^h\ 48^m$; so that in 117 of our days they completed one more revolution than the previously mentioned white spot performed.

This speed was so exceptional that we may probably attribute it to some extraordinarily violent current in which they were involved; although it is, of course, quite possible, that the increased rapidity of rotation in question may simply have been due to the fact that these spots were at a lower level than those which we most often behold. If so, it may be, that the globe or central portion of Jupiter is rotating, at a still lower level, faster than any such spots.

We believe, however, that Mr. Denning, whose opinion with regard to the equatoreal white spot which he has observed with so much perseverance and skill has a special claim upon our attention, is, upon the whole, rather disposed to regard it, either as a permanent feature of the planet, or as in some way connected with some such feature which may be situated beneath it, notwithstanding that its rate of rotation is slower than that of the last mentioned small black spots.

For our own part, we feel that the greatest possible difficulty is involved in any attempt to formulate a theory of what may be the physical condition of Jupiter, or of the relation of its spots and belts to that condition, and to the regions upon which they are superimposed. So much is doubtful, so much apparently contradictory, so much inexplicable, that we fear that the whole problem must be considered to be one for future solution, rather than one which can at present be solved.

We must confess that we were at one time inclined to think

that the permanency of the Great Red Spot seemed to indicate, that it might be something which, while coagulating or solidifying, in some way caused a gap or break in the cloudy regions above it, or by its cooling condensed the vapours incumbent upon it, and thus increased its own visibility; in fact, that we might be watching in it the gradual formation of a huge continent upon Jupiter. But the recent observations of the equatoreal white spot and of some other similar spots tell us, that, if we see in the Red Spot a part of the body of Jupiter itself, we must find some explanation of the more rapid rotation of these other spots. It might not be so difficult to explain a slower rate of rotation in their case, if they are simply atmospheric phenomena, but it seems hard to imagine by what means they can be carried steadily round, with a speed so much more rapid than that with which the globe below them must be rotating, if the Red Spot is actually a part of it and by its movement indicates the true rotation-period.

If it be the case that the motion of the equatoreal white spot, to which we have so frequently referred, is uniform, and that any slight apparent irregularities in its movement are really due to small errors of observation, it is doubtless possible that such a spot may be the summit of some snow-clad mountain, which from time to time is more or less clearly seen, or is occasionally altogether hidden by masses of overhanging clouds. Or possibly some volcano there placed may periodically eject vast masses of vapour which may hang over it, and alternately be of such a character as to shine with special brightness by reflected sunlight, or to obscure and darken our view. But if so,—if the white spot be either a permanent feature of the surface, or originate from such a feature,— may we not well ask whether the Great Red Spot, upon the ground of its long-continued regularity of rotation, may not claim to have a similar location? And yet this would seem to be impossible, because of its decidedly different period of rotation. In fact, every such supposition seems to involve us in some insurmountable difficulty in one direction or in another.

We may, however, remark that the generality of observers

have been disposed to believe that the Red Spot is probably situated at no very great depth in Jupiter's cloudy envelope. It is also thought that a very slight slackening of its rotation-speed has been recently noticed. If so, this may arise from the opening, or depression, by which it may be formed, having become somewhat more shallow, so that its motion corresponds more nearly with that of the uppermost layers of the clouds and vapours of Jupiter's atmosphere, which would naturally move the most slowly. But, even if so, we can in no wise explain why that Great Red Spot should be so permanent, or in what way it has been generated. We are quite unable to decide whether it has originated where it is at present found, or in some source situated at a vast distance beneath.

We are indeed not only disposed, as we have already stated, to leave the time of Jupiter's rotation upon its axis an unsolved problem, but we must allow that it is even possible that it ought to be described as different in different latitudes. If so, the differing rate of movement of the spots in such localities would be explained; but at the same time it would be necessary to suppose, that the planet is not only surrounded by a vaporous envelope many thousands of miles in depth (as various classes of observations indicate),* but that it may be to a very great depth indeed, and perhaps to its actual centre, vaporous, molten, or in a semi-fluid and viscous condition. In fact, there may be no solid globe in the planet at all; or, in other words, it may be, as suggested at the commencement of this lecture, in a very similar condition to that in which the Sun is believed to be by many of the best authorities.

The only other tenable supposition probably is, that, at some very considerable depth, there is a globe as to whose period of rotation we know nothing; while the outer regions, which are all that we can observe, are in a fluid, gaseous, or cloudy condition; a state of things which we also mentioned as possible, although not probable, in the case of the Sun. But if

---

* We shall presently draw special attention to some of these in connection with the occultations of the planet's Satellites.

so, the *permanence* of some of the features of the planet, contrasted with the incessant, and at times exceedingly rapid *changes* visible in others, is extremely puzzling.

When, however, we know so little, we must not let our ignorance suggest unnecessary difficulties. Rather let it teach us to wait, and watch, and learn. We can at any rate say that, in the increasing speed of the rotation of spots as we approach the equatoreal parts, there is a close resemblance to what takes place upon the Sun. In the comparative permanence of the positions of the principal belts, there is a resemblance to the permanence of the zones upon each side of the Sun's equator in which Sun-spots are found. In the size of the planet there is a nearer approach (although the interval is still immensely great) to the size of the Sun, than in that of any other body in the Solar System. Nor can we tell how much more, in bygone ages, before it had cooled down to its present condition, Jupiter may have resembled the existing state of the Sun; nor how far, in ages to come, the Sun's condition may resemble that which we now see in Jupiter.

We know but little of either; but we can see many points of resemblance between them, many suggestive analogies. And yet we would not for a moment infer that Jupiter, although it may be in a state of considerable heat, and give out some intrinsic light,* is in anything like so hot or luminous a condition as the Sun.

Here we must draw to a close our discussion of the possible condition of this great planet, only detaining our readers for a few additional moments in order to mention two or three especially interesting observations of the transits of Satellites, or of their shadows across the Great Red Spot, which have been recently made. They deserve special attention because they are such that, if they be carefully studied, and contrasted with future occurrences of a similar kind, they may enable us to decide how far the spot is in any degree self-luminous, or the contrary.

---

* See the commencement of this lecture, p. 295. Also Lecture VI., p. 149, with reference to the comparative light of the planets Jupiter and Mercury.

In the Report of Professor Hough previously referred to, and quoted in the *Observatory* of October 1881, p. 303, we read:—

"On July 3rd, 1880, the second Satellite during transit passed almost directly over the centre of the Great Red Spot, when it appeared sensibly as bright as when off the disc."

On November 1st, 1880, a transit of the shadow of the second Satellite over the centre of the Red Spot was seen from the Dearborn Observatory; and at the same time the transit of the shadow of the first Satellite over the disc of the planet. It is stated in the Report that "the shadow of the Satellite, when fully projected on the Red Spot, was distinctly visible, but not quite as black as the shadow on the disc; proving that the Red Spot, although much less luminous than the disc, was yet much more luminous than the shadow of the Satellite."

On the other hand, Mr. Gwilliam mentioned in the *Observatory* of January 1882, that he had seen the shadow of Satellite No. 1 upon the Red Spot, and that he had forwarded a drawing of it, when so seen, to the editor. He has since very kindly informed the writer of this lecture that, when the shadow was half upon the spot and half off it, he could detect no difference in the colour of either part. He has also stated that on November 26th, 1880, he saw Satellite No. 2 upon the Red Spot, when the effect was very curious, the Satellite appearing as if surrounded by a *bright halo*, which partially obscured the dark-red tint of the spot. His observations were made with a $6\frac{1}{2}$ inch Calver speculum, and a magnifying power of 200.

It will be noticed that the above-mentioned observations, being somewhat contradictory, leave the important question of the self-luminosity of the Great Red Spot still undecided.

Upon the whole, we decidedly incline to the opinion that the planet Jupiter is most probably in too highly heated a condition to be fit for habitation, apart from the fact that its great distance from the Sun, and the small amount of Solar light and heat received by it, would otherwise make it very difficult to believe that it can be a suitable abode for beings in any way similar to the human race. It is, nevertheless, quite possible that, at some still far-distant date, it will only retain sufficient heat to make it an agreeable place of residence during a long period of further gradual cooling, until at last, instead of being too hot, it will become too cold for habitation.

As its habitability at the present time seems to be so improbable, we need hardly devote much space to discuss the seasons of Jupiter's year, or the climatic conditions existing upon it. We will only remark in passing, that, its axis being nearly perpendicular to the plane of its orbit, its polar regions only extend for about 3° around each pole, instead of for $23\frac{1}{2}°$, as upon the Earth. Upon nearly the whole surface a resident would therefore find Spring, Summer, Autumn, and Winter to be almost alike; in fact, a sort of never-ending Spring would reign in every latitude, the Sun rising to nearly the same meridian altitude each day in the year; that altitude, however, being less the farther any locality might be from the equator.

The actual intensity of the Sun's heat and light upon Jupiter would be about $\frac{1}{27}$th of that which is enjoyed by the Earth. The Sun's disc would appear to be of a diameter less than $\frac{1}{5}$th, and of an area equal to about $\frac{1}{27}$th, of that which it exhibits to us; and it would seem (owing to the rapid axial rotation of the planet) to move across the sky $2\frac{1}{2}$ times as quickly as it does to an observer upon the Earth, so that it would pass over the width of its own diameter in about every successive 9 seconds of our time.

We may also remark that, if there should be any inhabitants upon such a planet, they would, of course, have to be much smaller than upon the Earth, the attraction of gravity upon its surface being, as we have shown in page 296, about $2\frac{1}{2}$ times as great. This is in accordance, although in a reversed direction, with the reasoning by which we showed, in our previous statements with regard to the satellites of Mars and the Minor Planets, that, the smaller the attraction of gravity upon any surface, the larger should be the scale of any beings upon it. In like manner, the greater the attraction of gravity the smaller must they be, in order to prevent their own weight from destroying them. There is, therefore, an important and fundamental error of principle in a somewhat amusing quotation in the first volume of Admiral Smyth's "Celestial Cycle," which might at first sight seem to contain a reasonable, although somewhat facetiously expressed, suggestion, that so large a planet ought to have inhabitants correspondingly tall. We there read that Wolfius maintained, that the inhabitants of Jupiter must be 14 feet in height, or rather $13\frac{819}{1440}$ feet; and that he *proved* it by the following argument. "Wherefore, since in Jupiter the Sun's meridian light is much weaker than on the Earth, the pupil (of the eye) will need to be much more dilateable in the Jovian creature than in the terrestrial one. But the pupil is observed to have a constant proportion to the ball of the eye, and the ball of the eye to the rest of the body: so that, in animals, the larger the pupil the larger the eye, and, consequently, the larger the body. Assuming that these conditions are unquestionable, he shows that Jupiter's distance from the Sun, compared with the Earth's, is as 26 to 5; the intensity of the Sun's light in Jupiter is, to its intensity on the Earth, in the duplicate ratio of 5 to 26; and that it therefore follows, that even Goliath himself would have cut but a sorry figure among the natives of Jupiter. That is, supposing the Philistine's altitude to be somewhere between 8 and 11 feet."* Wolfius finally concluded the height of the inhabitants of Jupiter to be just $\frac{477}{1440}$ths of a foot shorter than that of Og the king of Bashan, according to the measure of his bedstead given in the Book of Deuteronomy.

The considerations which we have put forward would, on the contrary, lead to the conclusion, that about 28 inches would be a suitable height for an inhabitant of so huge a planet.

It may be remembered that we showed in Lecture IX., p. 205, that a weight of 193 lbs., at the equator of the Earth, would weigh 194 at either of its poles, owing partly to the centrifugal effect of the Earth's rotation, and partly to the fact that the surface of the Earth at the equator is farther from its centre, and therefore the attraction of gravity less, than at the poles. On Jupiter both of these effects would be intensified, owing to its

* See Smyth's "Celestial Cycle," vol. i., p. 173, in quoting from which we have altered the adjective *Jovial* to Jovian, which is the form now generally used.

more rapid rotation and to its greater polar compression; and to such an extent, that we have calculated, that 6 lbs. at either pole would only weigh about 5 lbs. at its equator. If, therefore, the planet should at any future time be inhabited, although there would during the year be very little change of season in any given latitude, it would not only be possible to obtain a change of climate by travelling north or south from any particular locality; but, if any unfortunate individual should suffer from a depression of spirits, he would not only get warmer, but very appreciably lighter by approaching the equator. If, on the contrary, a physician should find his patient to be too elated, he might bid him travel towards one of the poles, with the reasonable hope that his increased weight might steady his feelings. And if temperance principles should not obtain upon the planet, it might be a good rule only to dine out with friends living nearer to the equator than one's own residence, so as to have the advantage of a pole-ward movement, and a consequent augmentation of weight and stability, in returning home after dinner.

We may, however, conclude this portion of our subject, and put on one side all speculations as to Jupiter's habitability, with the remark, that we consider that a much more probable supposition may be made; viz., that Jupiter may afford some subsidiary heat and light to inhabitants who may dwell upon the four moons which circle round it. Their light is, as we shall presently explain, too feeble to be of any great use to Jupiter; it is much more probable that his may be of use to them. In our next lecture we shall invite the attention of our readers to these four important and interesting, although not very large, members of the Solar System.

LECTURE XIII.

THE SATELLITES OF JUPITER.

"Four friendly moons, with borrowed lustre, rise,
Bestow their beams benign, and light his skies."
HENRY BAKER, F.R.S., A.D. 1726.

THE history of the discovery of the four moons, or Satellites, of Jupiter, is inseparably connected with the great name of Galileo Galilei, as the first triumph that rewarded the invention of the telescope. We are very glad to think that it has been recently made possible for those who are not acquainted with the Latin language to read, in the excellent translation of the "Sidereal Messenger," by the Rev. E. S. Carlos (late head Mathematical Master in Christ's Hospital), which has been published by the Messrs. Rivingtons, how modestly and yet how accurately Galileo describes his observations. Amongst much that is interesting and instructive in his narrative, nothing is, to our thinking, more noteworthy than the painstaking and unsparing care with which he used his instrument, until he reaped that rich and complete reward which genius separated from industry never deserves, but joined to it never fails to attain.

Night after night Galileo watched the Satellites, and each night, as his diagrams show, he accurately depicted the position in which he saw them. Some of his earliest observations deserve our special attention. They date from January 7th, 1610. On that night he discovered three out of the four moons, and thought they were ordinary stars, although he wondered to see them all nearly in a straight line, two on one side, and one on the opposite side of the planet, as in the uppermost line of Fig. LXXIV. The following night he again saw three, but they were all on one side of Jupiter, and nearer to one another than on the previous night. This is shown in the second line of Fig. LXXIV. The next night he saw only two, both on the opposite side of the planet. The nights of the 13th and 15th were the first upon which he saw *all the four*, when their positions were such as are indicated in the third and fourth lines of Fig. LXXIV., in which the original drawings published by Galileo are reproduced.*

Galileo continued to observe the Satellites, and to record their places, on every clear night with the utmost care; occasionally marking down, in addition, the proximity of small stars, which, however, were proved not to be Satellites, since

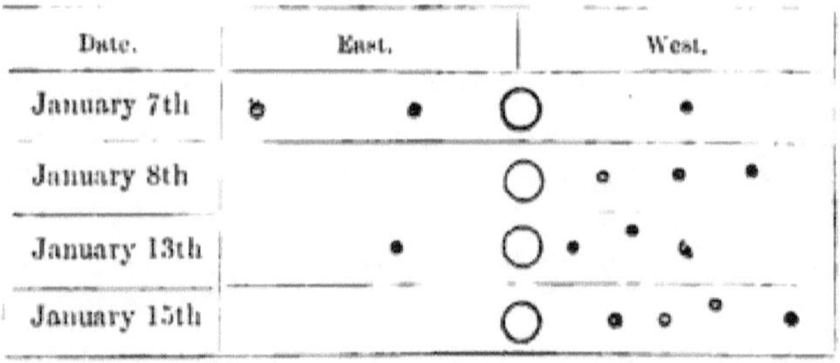

Date.	East.		West.		
January 7th	●	●	○	●	
January 8th			○	●	● ●
January 13th	●		○	●	●
January 15th			○	● ●	●

Fig. LXXIV.—Configurations of the Satellites of Jupiter, drawn by Galileo, in the year 1610.

the Satellites accompanied the planet as it advanced in its orbit, while the stars were left behind.

From the first, he noticed that all the four Moons were not equally brilliant. He was also soon able to conclude that they rotated round Jupiter, alternately passing between it and the Earth, and beyond it to its further side. And he not only came to this conclusion through seeing them sometimes to the east and sometimes to the west of the planet, but he also ascertained that the size of the orbits in which they revolved was much greater for the outermost than for the innermost. This very important fact was discovered as follows:—

Galileo observed that he often found two or three of the

---

* A much more complete reproduction of sixty-four of Galileo's drawings may be seen in the translation of the "Sidereal Messenger" (by Mr. Carlos) previously referred to.

Satellites to be apparently near together, *i.e.*, almost in the same line of view, when they were seen not far from the planet, but that they never appeared close to one another at more than a certain distance from it. This is exactly the effect which the revolution of a series of bodies in concentric orbits, situated nearly in the same plane as the observer's eye, would produce. By painstaking calculation and continued observations he was able to announce, in the year 1612, that the periods * of their apparent rotations round Jupiter, were respectively about 1^d $18\frac{1}{4}^h$; 3^d $13\frac{1}{2}^h$; 7^d 4^h; 16^d 18^h; the most precise values obtained at present being 1^d 18^h 28^m 30^s; 3^d 13^h 17^m 54^s; 7^d 3^h 59^m 36^s; 16^d 18^h 5^m 7^s. It is exceedingly creditable to Galileo that he so soon attained results so accurate. When, however, he proceeded to what was hypothetical,—as in suggesting that the greater or less extent of some dense and far-extended atmosphere of the planet, through which the Satellites were seen at different parts of their orbits, might account for certain variations observed in their brightness,—his conclusions were not so admirable.

Galileo's discovery of these Satellites occurred most opportunely for the support of the all-important theory of Copernicus, which we have explained at length in Lecture V. As we have there stated, men of science were in much doubt, at

---

* These periods are the *synodic* periods (see Lecture V., p. 129), in which the Satellites would respectively appear to an observer situated upon Jupiter to go once round the planet, or to pass from the phase of New Moon to a similar phase again. They are therefore slightly longer than the actual periods of their rotation round it, just as a synodic month in the case of our own Moon is $29\frac{1}{2}$ days long, while the actual period of the Moon's revolution round the Earth, which is called a sidereal month, is only $27\frac{1}{3}$ days; the additional time involved in the former case being that which is required, owing to the onward motion of the Earth in its orbit, to allow the Moon to make up for the different direction in which the Sun is consequently seen from the Earth, and to attain again (relatively to that direction) a position similar to that in which it began the Synodic month. In the case of Jupiter's Satellites, the difference between the two kinds of months is very much smaller, owing to the slower movement of the planet round the Sun, and their own more rapid rates of rotation. It only amounts to about one minute for the innermost, and to about one hour and a half for the outermost.

the beginning of the 17th century, whether to cling to the old ideas of Ptolemy, which for at least fourteen centuries had been almost unchallenged, or to accept the new doctrine that the Sun was the central ruler about which all the planets revolved. So violently was the ancient system maintained, that Copernicus was with difficulty persuaded to publish his great work upon the " Revolutions of the Celestial Orbs " ten years after it was completed, and only received the first printed copy of it when on his death-bed; while Galileo (see Lecture V., p. 106) was forced in his old age to make a humiliating recantation of his belief in the movement of the Earth.

And yet there can be but little doubt that the beauty and simplicity of the Copernican theory was such, that it must have been gradually commending itself to the minds of many, who were, therefore, prepared to realize the important confirmation which it received from the orbital revolution of the Moons of Jupiter. We must no doubt allow, that a far more conclusive argument for its truth was really afforded by the view of the phases of Venus, which Galileo obtained in the December of the same year in the beginning of which he discovered the Satellites of Jupiter. But before the orbits of these Satellites were known, it cannot be denied that there appeared to be a certain symmetry of idea in supposing Sun, Moon, and planets all to revolve around the *Earth*. Until then, the case of the Moon would have seemed a solitary exception to a general rule, if it had been believed that all else circled round the *Sun*, while it alone revolved around the Earth. But even those who had not studied scientific matters deeply, could not but feel that any such difficulty was removed, when Jupiter was found to have a subsidiary system of its own. And it was a still further argument in favour of the theory of Copernicus, that, the more complicated the general view of the Solar System thus became, the more beautifully did that theory explain it.

It is amusing to find that there were amongst Galileo's contemporaries, to whom he announced his discovery, some who denied the existence of the Satellites altogether. Some even refused to look through his telescope, upon the plea that it was an instrument of a Satanic nature. It is of such an unbeliever that Galileo expresses the pious hope that, if he would not believe in them or look at them in this life, he might after death enjoy a good view of them on his way to heaven.

The most powerful telescopes that Galileo constructed magnified in linear dimensions about thirty times, or rather more. With such an one the Satellites were of course very easily seen. They may often be detected with the excellent binocular field-glasses, or opera-glasses, made in the present day. Indeed, if it were not for the imperfection of the image of the planet formed in most eyes,—an image which appears to be surrounded by a certain number of false rays of light,—two at least of the Satellites would be in general visible to the naked eye, when seen sufficiently far from the disc of the planet, inasmuch as their light is quite as great as that of a 6th magnitude star.

In Arago's "Popular Astronomy," in Chambers' "Handbook of Descriptive Astronomy," and in Webb's "Celestial Objects," well-authenticated instances are quoted in which some of them have undoubtedly been seen without telescopic aid; although it should perhaps rather be said that, in several cases, two or three seen very close together have appeared as one, than that they have been separately visible. Their visibility to the naked eye must, however, always be very rare, except under remarkably favourable conditions of climate and locality. It is more a matter of curiosity than of importance, but its possibility cannot be denied. The only point in connection with it which surprises us is, that we have no record of any such observation by the ancient astronomers of Arabia or Chaldæa.

In some books the four Satellites which we are discussing are distinguished by names assigned to them by an astronomer of another nation, who basely sought to deprive Galileo of the honour of their discovery. These names have, however, justly failed to obtain any recognition. Astronomers invariably distinguish the Satellites, in the order of their distance from Jupiter, simply by the numbers 1, 2, 3, 4.

The next question to suggest itself to the reader with regard to these Moons of Jupiter may probably be,—What are their sizes? This is a query which for some time after their discovery was involved in considerable obscurity. Even now we cannot answer it with any pretence to extreme accuracy. From an exact measurement of the angles subtended at our eyes by the apparent diameters of their discs, it is no doubt true, that we can at once calculate the *actual* sizes, which at their known distance from us would correspond to such *apparent* sizes. But the diameters in question lie between 1 and 1½ seconds of angular measure, *i.e.* they only measure from a $\frac{1}{1200}$th to a $\frac{1}{1800}$th part of the apparent width of the disc of the Sun or Moon. They are, in fact, so small, that modern instruments of the very best construction and of the highest power are required, in order that they may be measured with any approach to precision.

The earlier observers were forced to employ other and less direct methods; for instance, Sir William Herschel, by watching how long the second Satellite occupied in entering upon the disc of Jupiter, when about to transit across it, obtained a very accurate value for its diameter. He was also remarkably correct in his judgment as to the probable sizes of the other three. In like manner, Schröter and Harding, by similar observations, effected a very satisfactory estimate of the diameters of all the four Satellites.

Since 1829, however, several careful sets of actual measurements have been made. From some of the best of these, combined with such observations as we have previously mentioned, Professor Engelmann, of Leipzig, deduced in 1871 the following values of the respective diameters, which are most probably approximately correct; although it deserves mention that measurements made in 1881, by Professor Colbert, with the great achromatic telescope of the Chicago Observatory, 18½ inches in aperture, make them to be considerably larger.

For the 1st about 2,500 miles, or approximately 2½ thousands of miles.
,, 2nd ,, 2,100 ,, 2$\frac{1}{10}$,,
,, 3rd ,, 3,550 ,, 3½ ,,
,, 4th ,, 2,960 ,, 3 ,,

It may be interesting to compare the above values with that,

THE SATELLITES OF JUPITER.

viz., 2,160 miles, which belongs to the diameter of our own Moon; and with 3,000 miles, the probable diameter of the planet Mercury. Such a comparison shows that the smallest of the four Moons of Jupiter is almost exactly of the same size as our own, while the other three are decidedly larger. It may be calculated that their united bulk (inasmuch as the volumes of spheres vary as the cubes of their diameters) is between nine and ten times that of our Moon, and between $\frac{1}{5}$th and $\frac{1}{6}$th of that of the Earth. It also follows that the volume of the largest is about two-thirds as large again as that of Mercury; so that this Satellite affords an example of a secondary body in the Solar System which is much larger than another primary body.

It is probable that this is also true of the largest of the Satellites of Saturn, and still more so of that of Neptune. But, so far as we have as yet gone in our outward journey from the Sun, this is the first instance of the kind with which we have met. We certainly cannot consider such a Moon to be in any sense unimportant; especially if we further notice that its diameter is about $\frac{7}{18}$ths, or not much less than one-half of that of the Earth, and about $\frac{3}{5}$ths of that of Mars; while its surface is more than $\frac{2}{3}$rds of the area of Mars. If there be no other reason against the habitability of such a globe, no objection can be raised upon the score of its size.

The diameters of the *orbits* of the Satellites, which are nearly circular in form, next demand our attention. They may easily be deduced by Kepler's 3rd law (see Lecture V., p. 122), from the periodic times in which the orbits are described. By that law they must be such, that the squares of those times are in the same ratios as the cubes of the respective distances of the Satellites from Jupiter. The resulting distances from the centre of the planet are found to be about 262,000 miles; 417,000 miles; 666,000 miles; 1,171,000 miles; respectively: *i.e.* they vary from about a quarter of a million to about $1\frac{1}{6}$ millions of miles.

From this last statement a very interesting illustration may be obtained of the vast power of the telescope, when used to view the heavenly bodies, in neutralizing the effect of distance by enlarging the apparent scale upon which we see them. The average value of the angle subtended at our eyes by the apparent diameter of Jupiter, when it is passing through the position of Opposition, *i.e.*, through one of its near approaches to the Earth, is about 45 seconds of angular measure. The corresponding angles subtended by the respective distances from its centre of the orbits of the four Satellites are about 135, 215, 340, and 600 seconds. Therefore 1,200 seconds (or 20 minutes of angular measure) would in general cover the whole distance across which the outermost, at such a time, is seen to travel from one side to the other side of Jupiter. As it is, we look at the orbits nearly edgewise, so that the Satellites seem to move backwards and forwards nearly in straight lines; but, if their orbits could be tilted up at right angles to our view, while remaining unaltered in size, so as to be seen as circles surrounding Jupiter, it follows, that a circle of 1,200 seconds of angular measure would embrace them all. And since there are 3,600 seconds in a degree, such a circle would subtend at our eyes an angle of only $\frac{1}{3}$rd of a degree, *i.e.* $\frac{1}{3}$rd of the $\frac{1}{360}$th part of that which the whole circumference of the heavens subtends.

Our own Moon's apparent diameter, when estimated in the same manner, is about 1,900 seconds. Consequently within a space in the sky of an apparent width equal to $\frac{2}{3}$rds of that which we see in the Moon,* the movements of these four Satellites, as well as all the detail of the belts and spots of Jupiter, are seen even when the planet is especially near to the Earth. At other times, when it is further away, the space involved would be smaller still. May we not justly say how wonderful is the power of an instrument, which, in so small a field of view, can show so much!

We may also be assisted in realizing the actual scale of that

---

* The ovalness of the orbit of Jupiter occasionally causes it, when seen in Opposition to the Sun, to approach somewhat closer to the Earth than is supposed in the above statement; in which case its diameter, and the apparent size of the orbits of the Satellites, may appear rather larger than the values we have mentioned as their *mean* values at such a time. The conclusion arrived at in the text remains, however, very nearly true.

THE SATELLITES OF JUPITER.

which distance causes to appear so small, if we notice that the nearest of these four Satellites, which appear to be so close to Jupiter, is really more than 20,000 miles further from its centre than the Moon is (upon an average) from that of the Earth; while the 4th Satellite is at a distance nearly equal to five times that of our Moon; and yet, such is the vast radius of the planet itself, the respective distances of the four are only about $5\frac{9}{10}$, $9\frac{3}{5}$, 15, and $26\frac{1}{2}$ times the radius of Jupiter. Our own Moon's distance being about 60 times the Earth's radius, it follows, that the furthest Satellite of Jupiter is more than twice as near, in comparison with the radius of its primary.

The four above-mentioned distances being in the ratios of the numbers 59, 96, 150, 265, attempts have been made to discover some relationship between them similar to that which Bode's series indicates in the case of the planets. It may not be out of place if we here mention two of the most successful of these attempts, although we consider them to be rather fanciful than useful.

If we take the series 5, 5 + 3, 5 + 3 × $2\frac{1}{2}$, 5 + 3 × $2\frac{1}{2}$ × $2\frac{1}{2}$, we obtain the numbers 5, 8, $12\frac{1}{2}$, $23\frac{1}{4}$. These, if multiplied respectively by 12, give 60, 96, 150, 279.

Or we may take	20	20	20	20
Add	20	40	60	80
And again add	20	40	80	160
The sums of which, viz.—	60	100	160	260

as well as the numbers given by the previous series, do not differ much from the true ratios of the distances.

A much more striking relation may, however, be noticed in the periodic times of the Satellites. If we take these to be *
1^d 18^h 28^m; 3^d 13^h 14^m; 7^d 3^h 43^m; 16^d 16^h 32^m; or in days and decimals of a day 1·77, 3·55, 7·15, 16·69 days; we see that the 2nd is almost exactly double the 1st, and that the 3rd is

---

* In this case we use the actual periods in which they revolve around Jupiter, which are termed their sidereal periods, and are, as explained in the note upon p. 323, somewhat shorter than the corresponding synodic periods.

almost exactly double the 2nd, while the 4th is somewhat more than double the 3rd; the apparent law of duplication failing to a certain extent when we get further away from the centre of the system, just as it is found that Bode's so-called law of planetary distances partially fails in the case of Uranus, and altogether in that of Neptune. In connection with the above relation between the periods of Jupiter's Satellites, it may also be well to remind our readers, that we remarked, in Lecture X., how closely the period of the outer satellite of Mars (*Deimos*) approaches a value which is very nearly four times that of the inner satellite (*Phobos*).

While we are engaged in statistical statements, we may also notice that it has been proved, by means of calculations too elaborate to be here explained, that the densities of the 1st and 4th Satellites are nearly the same as that of Jupiter, or about $\frac{1}{5}$th of that of the Earth; while the density of the 2nd is nearly twice as great, and that of the 3rd more than half as great again as that of the 1st and 4th.

We have already stated that their actual sizes are by no means unimportant; but it may, we think, afford one more instructive exemplification of the enormous mass of Jupiter, if we observe how minute the weights, which the above densities involve, are in comparison with the huge bulk of the planet around which these Satellites revolve. And in this connection it may also be interesting to remember, that the weight of our own Moon (as is shown in Lecture III.) is about $\frac{1}{80}$th of that of the Earth; and that Jupiter itself is five times lighter than it would be if its apparent globe were of the same average density as the Earth. In contrast, then, to the case of our Moon, we find that the four Satellites of Jupiter respectively weigh only about $\frac{17}{1000000}$, $\frac{23}{1000000}$, $\frac{88}{1000000}$, and $\frac{43}{1000000}$ of the weight of Jupiter.

It naturally results that the very large amount of protuberant matter in the neighbourhood of Jupiter's equator causes their orbits very nearly to coincide with its plane. The 4th Satellite, being further away, does not obey this law so accurately as the others; but the utmost inclination even of its more distant orbit to the equator of Jupiter can never exceed about $\frac{2}{3}$rds of

THE SATELLITES OF JUPITER.   331

one degree. For the 1st Satellite the inclination is only about $\frac{1}{12}$nd of one degree, or less than one minute.

It may be understood, by means of our previous explanations in the case of the Moon (see Lecture IV., p. 90), and in that of the Satellites of Mars (see Lecture X., p. 257), that Jupiter's Moons, and especially the inner ones, when viewed from the planet, would vary very considerably in apparent size according as they might be observed near to the horizon, or nearly overhead. In the latter case an observer, situated upon the planet's equator, would be not much less than 44,000 miles (*i.e.* the length of Jupiter's radius) nearer to the Satellite than in the former case; which would reduce the distance of the nearest from about 258,000 miles to about 218,000 miles, or by nearly $\frac{1}{6}$th part. The apparent diameters in the two positions would consequently be nearly as the numbers 5: 6; and the areas of the discs as the squares of these numbers, or as 25 : 36; so that the disc of this Satellite, which is the one whose change of size would be the most remarkable, would appear to be larger by about one-half when seen in the zenith than when seen in the horizon.

If we take Jupiter's day to be about ten of our own hours in length, it may be easily calculated that its four Moons go round respectively in rather more than 4, rather more than 8, about 17, and about 40, of the planet's days. In these comparatively short periods they pass through all their varied phases, with the exception, as we shall presently show, that three of them are always eclipsed when they would otherwise be full moons.

Many very remarkable combinations of their phases must from time to time be visible in the Jovian skies; the interest attaching to which is, however, greatly diminished when we remember that they always reflect so small an amount of Solar light, that it must not for a moment be imagined, as the quotation at the head of this lecture may perhaps seem to suggest, that they can be intended to compensate Jupiter by night for the feebleness of the illumination which it enjoys by day in consequence of its great distance from the Sun. The apparent areas of their four discs, if they could all be seen at once in the zenith of a place upon Jupiter's equator in the phase of full moon, would, it is true, be somewhat more than twice the apparent area of our Moon, but the intensity of the sunlight which they reflect being about $\frac{1}{27}$th of that which our Moon receives, their united light would even then be equal only to about $\frac{1}{13}$th of the average amount which the full moon gives to the Earth.

We will next proceed to the discussion of a certain especially interesting class of observations which may be made in connection with the Satellites of Jupiter. We refer to their Eclipses, their Occultations, their Transits, and the Transits of their Shadows across the planet's disc.

In connection with their *Eclipses,* the first thing to be realized is, that Jupiter continually casts a huge conical shadow behind it while travelling in its orbit; and that the axis of this shadow always points directly from the Sun, in the prolongation of the straight line joining the Sun's centre with that of Jupiter. We may very easily calculate, by the solution of a simple sum in proportion,* that this conical shadow will extend to about 54,000,000 miles, before it narrows to a point. This dis-

* Let P and C in Fig. LXXV. be the centres of Jupiter and of the Sun; JJ' and SS' their semi-diameters; then JJ' is approximately equal to $\frac{1}{10}$th of SS'. If, therefore, the point of the shadow extend to O, the distance PO will be about $\frac{1}{10}$th of CO, or $\frac{1}{9}$th of CP. The mean value of CP, *i.e.*, of Jupiter's distance from the Sun, being about 484,000,000 miles, it follows

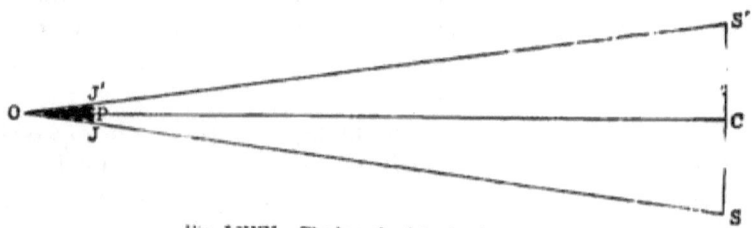

Fig. LXXV.—The length of Jupiter's shadow.

that PO, or the distance to which the shadow of Jupiter will extend behind it, is about $\frac{1}{9}$th of this, or about 54,000,000 miles.

The above style of calculation is comparatively rough, but, as we have often shown in our lectures at GRESHAM COLLEGE, it may be very usefully and instructively applied in many similar cases, including some to which we shall refer a little farther on.

tance is, of course, immensely greater than that of the 4th, or furthest, Satellite, which falls short of 1,200,000 miles. The diameter of a circular section of the shadow, perpendicular to its axis, is, therefore, even at this latter distance, little less than that of Jupiter itself.

The axis of the shadow must evidently lie in the plane of Jupiter's orbit, while the Satellites move nearly in the plane of Jupiter's equator, which is only inclined at an angle of about three degrees to that of the orbit. It is consequently found that the first three Satellites never escape the shadow when they are full moons; and that the 1st and 2nd always pass deeply through it; while the 3rd may, at its greater distance, at times dip but comparatively little into it. But the 4th, for nearly half of its path, is so far removed above, or below, the plane of Jupiter's orbit, that it may, when full, pass

Fig. LXXVI.—Comparative movements and relative positions of the three innermost moons of Jupiter.*

altogether clear of the shadow. When the Satellites travel centrally through the shadow, their eclipses will of course be of the greatest possible duration, and may last in the case of the 1st for $2^h\ 20^m$; 2nd, $2^h\ 56^m$; 3rd, $3^h\ 43^m$; 4th, $4^h\ 56^m$.

The first three can therefore never appear as full moons, but are always eclipsed when they otherwise would be so seen, as would be the case with our own Moon if it moved very nearly in the plane of the ecliptic, instead of in an orbit inclined to it at an angle of about five degrees.

But, in addition to the fact that the three inner Satellites are always thus eclipsed when they would otherwise be seen as full moons, an exceedingly curious relation exists between their motions, which involves the fact that they can never all

* A diagram resembling the above, but more elaborate, may be seen in an article by Mr. Proctor in the *Popular Science Review*, vol. vi., p. 248.

three be in the position of full moon at the same time. Consequently, only two of them can be eclipsed at once. The motion of the 4th is not connected in this way with that of the other three. It may therefore be eclipsed with two of the others; in which case Jupiter would have only one moon visible. And even that might be in the phase of new moon; in fact, if two of the three eclipsed were the 2nd and 3rd moons, the same relation previously named would require that the one not eclipsed (*i.e.*, the 1st) must be a new moon; under which circumstances the Jovian skies would for the time be deprived of all moonlight.

The relation to which we refer was discovered by Laplace. It may be stated somewhat technically as follows:—That the mean angular motion of the 1st Satellite round Jupiter, added to twice that of the 3rd, equals three times that of the 2nd; while it is also the case that, if the 2nd and 3rd are, at any given time, exactly in the same direction in longitude, or (neglecting the inclination of their orbits) exactly in the same straight line as seen from Jupiter, the 1st must be exactly in the opposite direction.

If therefore, as we stated above, the 2nd and 3rd be supposed to be together in the position of full moon, the 1st must necessarily, by the second part of Laplace's theorem, be in that of new moon; while a little consideration of the former part of the theorem in question will show, that the three can never be full moons, or new moons, or in fact be all seen in any given direction, at the same time.

It will be remembered that we stated (see page 329) that the angular movement of the 2nd is not far from being equal to twice that of the 3rd, while that of the 1st is almost exactly double that of the 2nd; their angular velocities being just so much greater as their periodic times are shorter. The numbers 1, 2, and 4 therefore very nearly represent the ratios of their respective velocities. If they did so precisely, it is of course evident, since 4 *plus* twice 1 equals three times 2, that the first part of Laplace's law of their motions would be satisfied.

But that law, as discovered by Laplace, involves the truth

of this relation, when, instead of such approximate numbers, the exact values of the velocities of the three inner Satellites are used. And it results that, if the three be at any given time in the same straight line drawn through Jupiter,—as, for instance, if the 1st be in the position of new moon, and the other two upon the opposite side of the planet (see the first diagram in Fig. LXXVI.),—they will again occupy similar relative positions a very short time *before* the outermost has quite completed one entire revolution (see the fifth diagram in Fig. LXXVI.)

But, for an *approximate* explanation of this peculiarity of their movements, it will suffice to imagine them to start as we have just stated,* and then to suppose that the innermost will go round *exactly* four times, and the 2nd twice, while the 3rd goes round once; which is what would happen if their velocities were precisely as the numbers 1, 2, and 4. Then the four remaining diagrams in Fig. LXXVI. will sufficiently represent in succession the relative positions of these three Satellites when the innermost has gone round once, twice, three times, four times, respectively; and will show, in agreement with Laplace's statement, that all three can never lie in the same direction upon the *same* side of Jupiter at the same time.

We must now explain a few additional technical terms connected with those phenomena of the Satellites, which still remain for our discussion.

When a Satellite is seen by an observer upon the Earth to enter the shadow of Jupiter, and thereby to be eclipsed, the commencement of the eclipse is denominated an immersion, or *ingress;* its end, when the Satellite exits from the shadow, is called an emersion, or *egress*.

But there is another way in which a Satellite may be hidden from our view. It is when the body of the planet is actually between an observer and the Satellite, which is then said to be occulted. The beginning and the end of an occultation are respectively termed a *disappearance* and a *reappearance*.

---

* It may be well to remark that the relative positions thus supposed are such as they really occupy from time to time.

Under such circumstances it is possible that a Satellite may *also* be in the shadow of Jupiter, during a portion of the time which it occupies in passing behind the planet, or *vice versâ;* so that it may be *coincidently occulted and eclipsed*. Or, to explain what we mean a little more fully, a Satellite may become eclipsed, and then enter into the position of occultation before its eclipse is ended; in which case we may begin our observation by seeing its ingress into eclipse, and finish by seeing its reappearance from occultation. Again, it may disappear by occultation, and presently emerge from eclipse, without the end of the occultation or the beginning of the eclipse having been visible. The following diagram, in which it

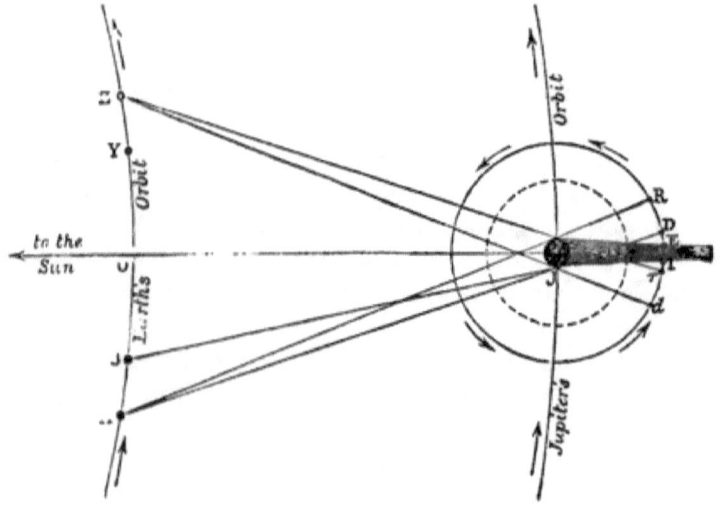

Fig. LXXVII.—Occultations and Eclipses of a Satellite of Jupiter.

must of course be understood that the circles representing the supposed orbits of Satellites round Jupiter are, for the sake of clearness, much exaggerated in size in comparison with the distance of Jupiter from the Earth, may serve to illustrate how such cases may arise.

It is not difficult to understand from Fig. LXXVII. that, if the Earth is approaching the position o, in which it would see Jupiter in Opposition to the Sun;—*i.e.*, if it be in such a position as A,—the eclipses of the Satellites will *precede* their occultations; the arrows showing the direction of the motions involved, and the order of the phenomena that will occur being indicated by the letters I, E, D, R, which respectively stand for Ingress, Egress, Disappearance, Reappearance. After the Earth has passed o, and has attained such a position as z, the order will evidently be reversed, and be *d, r, i, e*.

It may be noticed in our diagram that Jupiter is supposed to be in the same position J, during the whole time that the Earth is moving from A to z; in fact, the onward motion of Jupiter is not shown. But the figure is only so drawn to save complication, and the explanation given will apply just as well as if successive positions of Jupiter were shown. This is the case because the Earth moves round the Sun much more quickly than Jupiter, so that we may neglect the motion of the latter, by supposing the Earth to be moving with the difference of its own angular velocity and that of Jupiter. Or even without so doing, we might first suppose z to be omitted in the diagram, and then, secondly, suppose A to be omitted and z to be inserted: when it would perfectly apply to the two different cases, and correctly represent for each of them the relative positions of the Earth and of Jupiter, and the phenomena that would result.

It may also be perceived, that the nearer a Satellite is to Jupiter, the less likely is it that *all the four phenomena* (indicated by the letters I, E, D, R) will be visible when the Earth is at a given distance from o, without their interfering with one another. For instance, if a Satellite be supposed to move in an orbit represented by the dotted circle, which is drawn at about $\frac{2}{3}$rds of the distance of the circle IEDR from Jupiter in Fig. LXXVII., it is evident that an observer upon the Earth at A would see it pass into the shadow, but he would not see it emerge, since just as it would otherwise do so it would be occulted. Presently, however, he would perceive it reappear from behind the planet, its *emersion* from eclipse and its *disappearance* by occultation having both been invisible. Similarly if the Earth were at z, the *disappearance* of such a Satellite by occultation would be seen, but not its *re-*

appearance, because its eclipse would at the same instant begin. The *egress* from eclipse would, however, be visible.

For a Satellite moving as far from Jupiter as the circle IEDR, the same coincidence of phenomena would occur if the Earth were at B, since the straight line BE joining the Earth and the Satellite at the moment of its egress from eclipse would, in that case, also mark the point where the disappearance by occultation would take place; in fact, for an observer at B the points D and E would come together. Similarly for an observer at Y, the points r and I would coincide, since the line YI, if drawn, would just graze the edge of the planet.

It may be interesting if we state that the comparative distances and dimensions really involved are such, that the 3rd and 4th Satellites are far enough from the planet, for both the beginning and the ending of their occultations and of their eclipses to be in general visible; the 4th, however, about as often as not, as we have previously explained, escaping eclipse altogether when in the position of full moon. If, however, Jupiter be very close indeed to Opposition to the Sun, *i.e.*, if the Earth be very near indeed to such a position as O, the eclipses of none will be visible, as the points D and R will then in every case come slightly beyond I and E on either side of the shadow of Jupiter.

The distance of the 2nd Satellite also on rare occasions permits both the beginning and the end of its eclipses to be seen, although it is so much nearer to Jupiter that the Earth must in such a case be a long way from the line joining Jupiter to the Sun. The 1st Satellite is so close to Jupiter that *both* the beginning and end of its eclipses can never be seen.

And a little further consideration of the figure will show that the general rule for the order and visibility of the phenomena in the case of the 1st and 2nd Satellites may be stated as follows:—If the Earth be moving from a position in which Jupiter has been seen in Conjunction with the Sun, *towards one in which it will be seen in Opposition* (in which case it would be approaching O in the diagram), only the Ingress and Reappearance of the 1st, and (in general) also of the 2nd, will be seen; their Egress and Disappearance being invisible. After the

Earth has passed beyond a position in which Jupiter is seen in Opposition, and is proceeding towards one in which it will see it in Conjunction, the phenomena to be observed in the case of the 1st and (in general) of the 2nd will consist of a Disappearance by occultation and an Egress from eclipse, the end of the occultation and the beginning of the eclipse being invisible.

But, in addition to the eclipses and occultations of the Satellites, we frequently see that they pass between us and the globe of Jupiter. At such times a *transit* of a Satellite is said to occur, as it then appears to travel, or to transit, across the disc of the planet. Nor is this all, for we are able (and with greater facility) to watch the shadow which the Satellite casts upon Jupiter, as it also sweeps across its surface. This is the case in Mr. De la Rue's beautiful drawing in Fig. LXVII., page 304, in which one Satellite is faintly visible in transit, while its shadow is clearly seen at some distance from it against the background of the planet. Again, in the second drawing by Mr. G. D. Hirst, in Plate VIII., p. 304, the 4th Satellite appears as a small, darkish spot on the lower, or northern, half of the view, while somewhat above it and to the left of it the shadow of the 3rd is shown as a darker and much larger spot.

These two kinds of transits correspond respectively to those which might be seen by an observer upon the Sun if the Moon were passing between him and the Earth. He might not only see it apparently crossing the Earth's disc, but he might also see its shadow sweeping over a zone of the Earth's surface, upon which a Solar eclipse would at the time take place. The occurrence of the various phenomena which we have thus described is well illustrated in Fig. LXXVIII., which is almost exactly copied, by the kind permission of Mr. Bentley, from one in the English edition of M. Guillemin's very popular work, "The Heavens." In this figure a Satellite is shown in transit upon Jupiter at t, while its shadow is at the same time seen upon the disc at s. Of the other three Satellites one is undergoing eclipse, and another is just about to do so.

It requires but little consideration to see that, if a straight line joining the Earth and a Satellite meets the planet's disc, it will in general do so, as in the figure, at a different point from one joining the centre of the Sun and the Satellite. The former meets it where the Satellite while in transit across the disc will, if distinguishable, be seen projected upon it. The latter indicates where the shadow of the Satellite falls upon the planet.

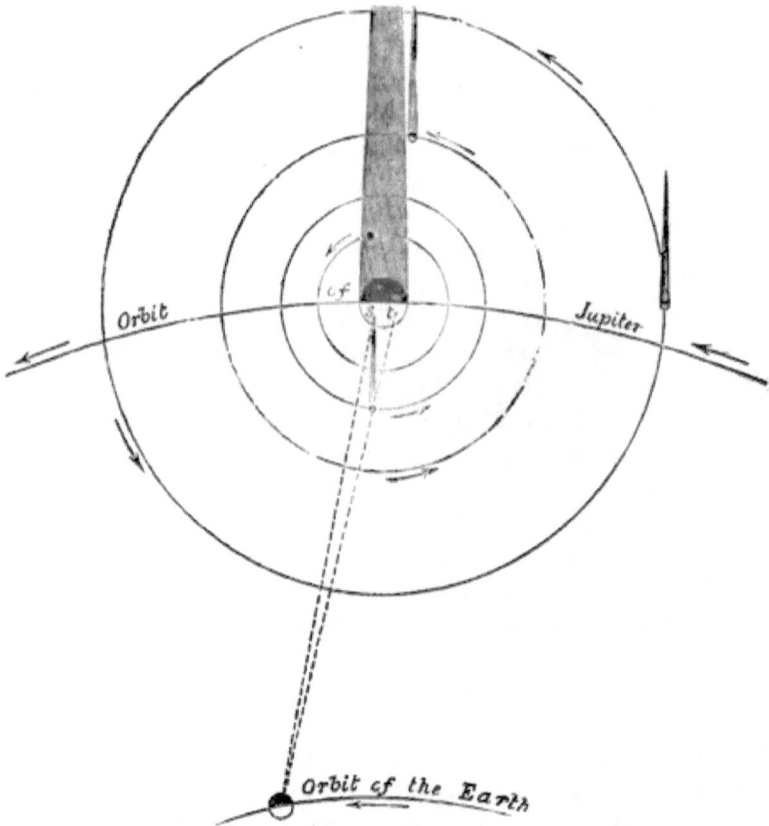

Fig. LXXVIII.—Phenomena of Jupiter's Satellites. The transit of a Satellite, and of a Satellite's shadow.

Nor is it hard to perceive, by noticing the directions of the arrows in the diagram, that, in the position in which the Earth is drawn, the Satellite (seen at t) will *precede* the shadow (seen at s) in its transit. Such a position corresponds to a time, shortly after the Earth has passed such a point as o, in Fig. LXXVII., from which Jupiter is seen in Opposition. And the same order of sequence in the two transits, wherever they are visible, will occur, until Jupiter is so near to Conjunction with the Sun that we are unable to watch it. When it again becomes visible after Conjunction, and until Opposition takes place, the Satellite will be seen to *follow* instead of to precede its shadow as it passes across the disc.

From our previous comparison of the occurrence to that of a Solar eclipse, it will, of course, be understood that, wherever such a shadow falls upon Jupiter, an eclipse of the Sun is in progress. But we have met with no attempt, in any elementary work, to calculate the size of these shadows, or the durations of the total eclipses of the Sun which they produce upon Jupiter; nor can we here devote much space to any such calculation.

We may, however, mention that, in some of our lectures delivered at GRESHAM COLLEGE in which we have considered these Satellites more minutely, we have worked out a few approximate calculations, without any attempt at strict accuracy, which show that the shadow of the 4th (or furthest) Satellite is about 950 miles in diameter when it falls upon the planet. We also found that, at a place upon Jupiter's equator, if this Satellite were in the zenith, a total Solar eclipse would be caused of about 6 minutes' duration. There would, however, be one remarkable difference between the circumstances involved in such a Solar eclipse, and those of one seen upon the Earth; viz., that, instead of the rotation of the place of observation round the axis of Jupiter helping to diminish the speed of the shadow passing across it (as is the case with the rotation of the Earth and the shadow of the Moon upon it), the speed of the shadow (viz., about 18,500 miles per hour) would help to diminish the greater speed of about 28,000 miles per hour, with which the axial rotation of the planet would otherwise carry the observer through it.*

For the 3rd and 2nd Satellites, we found the speed of the respective shadows, under like circumstances, to be about

* The difference of the two speeds being 9,500 miles per hour, the width of the shadow (950 miles) would pass over the supposed point of observation in about $\frac{1}{10}$th of an hour, or in about 6 minutes, as stated above.

24,500 miles and 31,000 miles per hour. But of these two velocities the one only falls short by about 3,500 miles, and the other only exceeds by about 3,000 miles, that due to the planet's rotation. A most remarkable result would therefore ensue; viz., that the shadow in the former case would nearly keep up with the place of observation; while, in the latter case, the place of observation would nearly keep up with the shadow. And the respective widths of the shadows being, for the 3rd Satellite about 2,430 miles, and for the 2nd about 1,450 miles, a total Solar eclipse might, in the one case, last for fully ⅔rds of an hour, or for more than 40 minutes (2,430 miles being rather more than ⅔rds of 3,500 miles); and in the other case for about half an hour, 1450 miles being about one-half of 3,000 miles.

For the 1st Satellite, the speed of the shadow would be nearly 40,000 miles per hour, *i.e.*, quicker than the rotation-speed of an observer by nearly 12,000 miles, over whom the shadow of fully 2,000 miles in width would therefore pass in about ⅙th of an hour, or 10 minutes.

It is a useful and interesting exercise in arithmetic to work out the simple proportions required for the above calculations.*

* We append a rough calculation for the case of the 2nd Satellite, illustrated by Fig. LXXIX., in which, however, the necessities of the

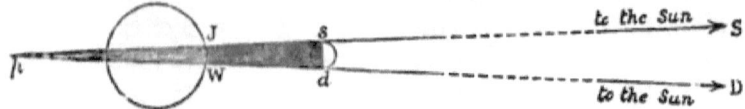

Fig. LXXIX.—Diameter of the shadow of the 2nd Satellite of Jupiter where it meets the Planet.

space at our disposal have forced us very much to distort the relations of size and distance in the bodies involved. The most important point in the calculation is, however, accurately indicated in the figure, viz., that JW is about two-thirds of sd. Let the straight lines terminated in arrows be supposed, at a distance of some 484,000,000 miles from Jupiter, to pass through s and D, the extremities of the Sun's diameter. Let sd be the Satellite's diameter, and spd its shadow. Then sd, being about 2,100 miles, is about a $\frac{1}{400}$th part of sD, the Sun's diameter. Therefore sp is about $\frac{1}{400}$th of sp, or $\frac{1}{400}$th of ss, *i.e.* $\frac{1}{400}$th of about 484,000,000 miles; or about 1,213,000 miles. Let JW be the width of the shadow upon

Of course, in doing so, allowance must be made for the fact that an observer upon the surface of Jupiter would be nearer than the centre of the planet to a Satellite seen, according to our supposition, in his zenith. We cannot now consider what further complications might be involved at places more or less north or south of the planet's equator; or how the eclipses would be affected by the altered speed and shape with which the section of shadow, as it gradually travelled off the disc, would pass over localities from which the Sun and the Satellite would be seen at a less and less elevation above the horizon. Some very remarkable effects might doubtless result, and a considerable lengthening of duration might occur.

It is, however, at all events worth notice, that an observer upon Jupiter would have a great advantage over one upon the Earth, in the much greater frequency with which total Solar eclipses would occur. Every 42 and 85 hours, respectively, the shadows of the 1st and 2nd Satellites would pass across a zone of the surface of about 2,000 and 1,450 miles in width; while that of the 3rd, although less often, would do so once in every 172 hours, and would be of the greater width of some 2,400 miles.

The 4th Satellite is, as we have previously mentioned, the only one which can pass through the position of new moon without casting its shadow upon the planet. In fact, it may travel so far from the plane of Jupiter's orbit, and from the ecliptic, that we may occasionally see the Satellite pass above or below the disc, while the shadow is in transit across it; or

---

Jupiter, then sj equals the distance of the Satellite from Jupiter, which is 417,000 miles, diminished by the radius of Jupiter, which is about 44,000 miles. Therefore sj equals about 373,000 miles. Subtracting this from 1,213,000, the previously obtained value of sp, we have for that of jp about 840,000 miles. And jw will be the same fraction of sd (which latter is 2,100 miles) that jp is of sp. Therefore jw is $\frac{840000}{1213000}$ of sd, or rather more than $\frac{2}{3}$ of 2,100 miles, i.e., about 1,450 miles. As the whole circumference of the Satellite's orbit is about $6\frac{2}{7}$ times 417,000 miles, or about 2,620,000 miles, round which it travels in 85 hours, we also easily obtain (by dividing 2,620,000 by 85) for the speed with which the shadow will sweep across the surface of Jupiter, about 31,000 miles per hour, as is stated above.

it may be that both the shadow and the Satellite escape a transit.

But, notwithstanding the comparatively small number produced by the 4th Satellite, we can hardly feel otherwise than envious when we think of the occurrence of so many and such prolonged Solar eclipses. If we could study the beautiful phenomena which we see so rarely, and in general for only 3 or 4 brief minutes at a time, every two or three days, and at our leisure gaze at them for some 10 or 20, or even 40 consecutive minutes, how much might we discover as to the various envelopes of the Sun, or as to the existence of multitudes of meteorites, or of a limited number of intra-mercurial planets in its immediate vicinity!

To proceed, however, with the actual phenomena of the transits of the Satellites and their shadows which may be observed from the Earth, we may remark that, if Jupiter is seen exactly, or very nearly, in Opposition to the Sun, it is, of course, possible for a Satellite and its shadow to be visible almost upon the same spot, apparently overlying one another, upon the disc. At least one instance [*] has been seen of the 1st Satellite thus so nearly occulting its own shadow as to leave only a crescent-shaped portion of it uncovered. An observation by Mr. F. M. Newton is also recorded in the *English Mechanic* of February 9th, 1872, when the 1st Satellite was seen superposed upon its shadow.

The transits of the shadows are in general much more distinctly visible than the transits of the Satellites themselves. This is only what might be expected, as it may be easily understood that a dark black spot is more noticeable than the small bright body of the Satellite in contrast with that of the planet, which is also for the most part bright. Inasmuch, however, as the planet's disc is not so bright near to its circumference as in the more central parts, the Satellites in general appear to be brighter than the disc when they are entering upon it and leaving it, but when about one-fourth or one-third

---

[*] By Mr. G. D. Hirst, of Sydney, N.S.W., on May 13th, 1876. See *English Mechanic*, August 11th, 1876.

of the way across they are usually lost in the greater brightness of the surrounding surface.

All sorts of variations, however, take place in the phenomena observed during the transits. Sometimes the 3rd and 4th Satellites, and occasionally the 1st, seem to be nearly as dark as their shadows; sometimes they look bright in comparison with the background of the disc throughout their journey across it. Moreover, the same Satellite appears to reflect different amounts of light while in transit on different occasions; and this is the case even when the part of the planet's disc upon which it is seen shows no indications of any decided change in brightness.

When, however, we bear in mind that, at one time, the background against which a Satellite is thus seen may be a bright zone, and at another a dusky, or a very dark, belt of the planet, it may easily be understood that it is well-nigh impossible to compare them when in transit systematically one with another, so as to form a correct judgment as to any periodicity in their changes of brightness, such as might be possible if they passed over an uniformly illuminated portion of the planet. Upon the whole, the 3rd is that which seems most inclined to appear dark, while the 2nd most fully keeps up its reputation for brilliancy.

In connection with this statement we may remark that not only did Galileo notice from the first, as we have mentioned in the beginning of this lecture, variations in the light of the Satellites in different parts of their orbits when not in transit; but Bianchini, so early as A.D. 1714, also mentions the same fact, and states that he sometimes observed the 4th to be so faint as almost to be invisible. Moreover, it is well known that Sir W. Herschel, after a series of careful observations, concluded that the changes of brightness, exhibited by all the Satellites, corresponded to their positions in their orbits round the planet.

If so, the probable conclusion would be, that they rotate upon their axes, as is the case with the Earth's Moon, in times equal to those in which they revolve round Jupiter, the variations in their light being caused by the brighter or darker portions of their surfaces which are periodically turned towards the Earth.

Nevertheless modern observations, except in the case of the 4th Satellite,* have hardly confirmed this conclusion, as may be seen by a reference to the elaborate investigations of Professor Engelmann, of Leipzig (see *Monthly Notices* R. A. S., vol. xxxiii., page 472; and vol. xl., page 169). It may, however, be the case that a general absence of spots or darker portions upon the 2nd, which would agree with its usual brightness when in transit, and a general prevalence of them upon the 3rd, which would account for its general darkness under similar circumstances, may simply make Herschel's supposed law of rotation harder to detect in their case than in that of the other two, so that it may still really hold for all the four.

But, apart from any such periodic effects which may arise from their rotation, their light undoubtedly exhibits other and very irregular changes. These, we think, may most probably be assigned to clouds and vapours in their atmospheres, some other indications of the existence of which have also from time to time been noticed during their transits.

One of the most recent statements which we have met with upon the subject is to be found in "Les Terres du Ciel," by Flammarion, page 511. It is deduced from observations made in 1873, 1874, 1875, and 1876, and is as follows:—In intrinsic brightness for equal surfaces the order of rank is I. (the brightest), II., III., IV.; † although, possibly the 2nd sometimes seems slightly to surpass the 1st. As to variability of the light, the order is IV. (the most variable), I., II., III.; the light of the 4th varying from that of a 6th to that of a 10th magnitude star. On March 25th, 1874, M. Flammarion mentions that a specially remarkable contrast was exhibited. The 2nd Satellite appeared white; the 3rd, dark grey; the shadow of the

---

* Possibly in the case of the 1st Satellite also.

† As the 1st appears, in general, to exhibit the best light-reflecting power, we may observe in passing that, in all probability, it possesses the most cloudy atmosphere of the four. Compare Lecture VI., pp. 148 and 149; and Lecture VII., p. 165, as to the reflecting power of clouds in the atmospheres of Mercury, Venus, and Jupiter.

2nd, grey; and that of the 3rd, black; when in transit at the same time.

And here we may appropriately draw attention to an important class of observations which have from time to time been made; viz., that the Satellites are occasionally seen, so to say, through the edge of Jupiter's disc, for a short interval after they should be, or have been, occulted. Whether this is an optical delusion, or a result of some diffractive effect of the lenses of the telescope, or caused by the refraction of Jupiter's atmosphere, we can hardly say. It may, however, be the case, that movements or disturbances in the vaporous envelope of Jupiter occur, sufficiently vast and rapid to distort the edge of the disc, suddenly, and to such an extent, that we continue to see a Satellite when it would otherwise have passed to a short distance behind it; or that there are occasionally movements in the contour of these vapours so exceedingly violent that they even permit the apparent *reappearance* of a Satellite shortly after its occultation;* so that, after going behind the planet's disc, it seems to have come back again to some distance outside its edge. Some very interesting observations bearing upon this question have of late been made by Mr. Todd, at Melbourne, Australia. Since, however, irregularities in the contour of the planet, when independently and very carefully watched for, have been very rarely, if ever, certainly detected, we must consider this question to be at present undecided.

It may easily be understood from our preceding statements that it is possible that Jupiter may sometimes appear to an observer upon the Earth to be without any Satellites at all, inasmuch as three of them may be occulted or eclipsed, and one invisible while in transit. Or there may be two in transit, and two eclipsed, and so on. Some remarkable and very interesting instances of this kind have been recorded. One so long ago as 1681, when the planet was thus seen by Molyneux; another on May 23rd, 1802, by Sir W. Herschel; another on April 15th, 1826, when this condition lasted for

* See the well-known observations of Admiral Smyth and two other astronomers recorded in his "Celestial Cycle," vol. i., page 184.

two hours; another on Sept. 27th, 1843, when it continued for thirty-five minutes; another on Aug. 21st, 1867, lasting $1\frac{3}{4}$ hours (see vol. vi. *Popular Science Review*, page 248). The next occasion will be on Oct. 15th, 1883. It will, however, only last for nineteen minutes, and will be between the somewhat inconvenient hours of 56 minutes past 3, and a quarter past 4 a.m. (see the *Observatory* of March 1880, page 357).

We need not here discuss the attempts that have been made to determine the distance of the Earth from the Sun by observations of the Eclipses of these Satellites, the principle involved having been sufficiently explained in Lecture I., pp. 13 and 14. We will therefore only recall to the memory of our readers that, when Jupiter is at its nearest to the Earth, the eclipses occur about $8\frac{1}{4}$ minutes sooner than they would if it were at its mean distance, and $8\frac{1}{4}$ minutes later when it is at its furthest distance, the discrepancy depending upon the velocity of light, which takes about 997 seconds, or very nearly $16\frac{1}{2}$ minutes, to traverse the diameter of the Earth's orbit.

It being in these days possible by other means to determine the velocity of light, it follows that the observation and calculation of the eclipses of Jupiter's Satellites, if effected with constantly increasing precision, may, in course of time, give us a value for the Sun's distance from the Earth of great accuracy, and consequently of great importance.

On the other hand, the historical interest of the Satellites is much increased by the fact that, originally, by a reverse process, it was the discrepancy in the observed, from the calculated times, of the occurrence of the eclipses, which led to the discovery, by Römer, in the year 1675, that light possessed a measurable velocity, and that a value might be obtained for it upon the assumption that the diameter of the Earth's orbit was known.

The occultations and eclipses of the Satellites may also be employed for an approximate determination of the longitude at sea, or elsewhere, inasmuch as they are seen to take place at the same instant of time from every spot upon the Earth. They therefore inform an observer of the Greenwich time of their occurrence, by the comparison of which with local time the longitude may be immediately deduced. More exact methods (the eclipses not being of a sufficiently instantaneous character) are, however, necessary when much precision in the longitude is desired.

To conclude our discussion of these four far-distant Moons, we may remind our readers of the proof which we gave, in Lecture III., pp. 58-61, that the Moon's orbit is always concave to the Sun; and we may mention that we have been asked more than once, during the delivery of this course of lectures, whether this is also the case with the orbits of Jupiter's Satellites. It is easy to see that it is otherwise. In the case of our own Moon we showed that the Earth would draw it when new about $\frac{2}{5}$ths of an inch in a second of time out of a straight line, towards itself, and away from the Sun, while we found that the Sun would draw both the Earth and the Moon through nearly $\frac{8}{25}$ths of an inch per second in the opposite direction; the result being that the Moon is at such a time drawn not quite $\frac{1}{15}$th of an inch in a second away from the Earth, while it travels in company with it through a distance of about 94,500 feet around the Sun. The Moon's orbit is therefore concave to the Sun when the Moon is new; and when it is full, since both the Earth and the Sun pull it in the same direction, towards the latter, the concavity is still more marked.

But if we consider the outermost Satellite of Jupiter when it is in the position of new moon, we find, that Jupiter's power over it is sufficient to draw it about $\frac{6}{10}$ths of an inch towards itself in a second, while the Sun would only draw it about $\frac{4}{1000}$ths of an inch in the opposite direction. At that time, therefore, the orbit must be concave to Jupiter, and *convex* to the Sun. And the power of Jupiter over the three inner Satellites being greater still, their orbits will, when they are new moons, be still more *convex* to the Sun than that of the 4th.

Let us investigate this statement a little further; and, to begin with, let us notice that the orbital velocity of Jupiter round the Sun, of which the Satellites partake in addition to their own velocities round the planet, is rather less than 29,000 miles per hour, while those of the Satellites in their respective orbits round Jupiter are,—for the outermost about 18,500 miles per hour;—for the 3rd about 24,500 miles;—for the 2nd about 31,000 miles;—for the 1st about 39,000 miles. The two innermost will therefore, when new, and in their nearest positions to the Sun, be really moving backwards in comparison with the onward motion of Jupiter.* This may be seen by a comparison of the arrows which refer to Jupiter, and to the Satellites respectively, in Fig. LXXVIII.

It will be remembered that the velocity of the Earth's Moon in space varies so little that it only changes from about 64 to 68 thousands of miles per hour, the orbital velocity of the Earth being about 66,000, and that of the Moon around it rather more than 2,000 miles per hour, which latter velocity (see Lecture III., p. 60) is alternately added to and subtracted from the former. But the onward velocity of the outermost of Jupiter's Satellites, when in the position of full moon, will equal the sum of Jupiter's orbital velocity of 29,000 miles per hour, and its own velocity of 18,500 miles per hour round Jupiter; while as a new moon its onward speed will only be equal to the difference of the same two velocities. It will therefore vary from about 47,500 miles to about 10,500 miles per hour. In like manner, that of the 3rd will vary from 53,500 miles to about 4,500 miles per hour; while, as we have just remarked, the 2nd will have an *onward* velocity of 60,000 miles as a full moon, changed to a *retrograde* velocity of about 2,000 miles per hour as a new moon; and the 1st will change an *onward* velocity of 68,000 miles for a *backward* velocity of about 10,000 miles per hour, when similarly situated.

It results that the path in space of the 4th Satellite during one of its months of about 16¾ days is as shown in Fig. LXXX.; and that of the 1st, during its much shorter month of about 42½ hours, as in Fig. LXXXI.; the former path having its convexity towards the Sun when the Satellite is a new moon; while the path of the latter, under similar

---

* The only other Satellites for which this is also the case are the four innermost of Saturn.

circumstances, is not only convex, but is actually looped, owing to the retrograde motion of the Satellite. It may also be noticed, that, in each case, the degree of convexity in the path at the time of new moon is actually greater than the degree of concavity at the time of full moon, although Jupiter and the Sun are pulling against one another in their attractions upon the Satellite in the former case, and conjointly in the latter case. This arises from the fact that the actual velocity of the Satellite through space, when it is a full moon, is so much greater, that even the joint attraction of the Sun and Jupiter is not sufficient to curve the path so rapidly, as it is curved by the difference of their attractions when the Satellite is moving so much more slowly through space as a new moon.

Fig. LXXX.—The actual path of the 4th Satellite of Jupiter in space during one its months.

In Fig. LXXX., J_9J_1 represents a distance of not much less than 12,000,000 miles which Jupiter travels while the 4th Satellite rotates once round it, from the position of full moon at M_9, through that of new moon at M_5, on to that of full moon again at M_1. The distance from Jupiter to the Satellite being about $\frac{1}{16}$th of J_9J_1, we divide this latter into a certain number of parts, as in the figure, in which the points J_9, J_8, J_7, etc., apportion the whole length into eight equal parts. Then from each of these eight points we measure a distance equal to $\frac{1}{16}$th of J_9J_1 to the respective points M_9, M_8, M_7, etc., which will give the corresponding positions of the Satellite as Jupiter travels onwards, provided that the directions of J_9M_9 and J_1M_1 be directly from the Sun; that of J_5M_5 directly towards it; J_7M_7 and J_3M_3 perpendicular to these two last-named directions, or nearly along the path of Jupiter itself; and J_8M_8, J_6M_6, J_4M_4, J_2M_2 at angles of 45 degrees to that path. A curve drawn through the points M_9, M_8, M_7, etc., indicates the actual path of the Satellite in space, which is evidently of the form shown in the diagram.*

Fig. LXXXI. is drawn in a similar manner for the 1st Satellite, only upon a much larger scale, so that J_9J_1, which is the distance travelled by Jupiter in about $42\frac{1}{2}$ hours, only represents about 1,240,000 miles; while the distance from Jupiter to this Satellite is about 262,000 miles; *i.e.*, about $1\frac{3}{5\frac{3}{2}}$, or not quite $\frac{1}{5}$th of J_9J_1. With this exception the method by which the actual path of the Satellite is obtained is the same as in Fig. LXXX. It is evident, as would also follow

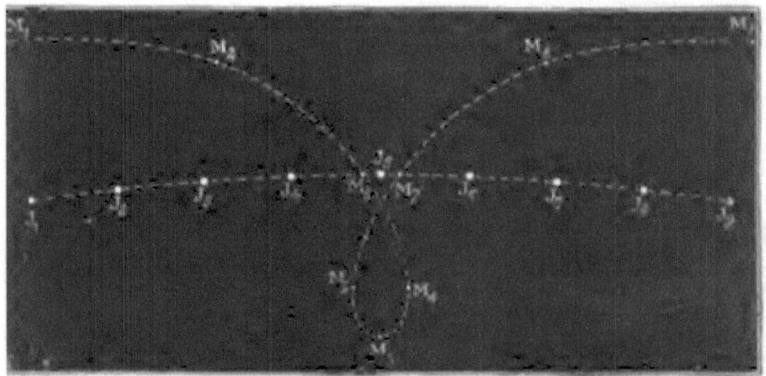

Fig. LXXXI.—The actual path of the 1st Satellite of Jupiter in space during one of its months.

from our previous statement as to the retrograde motion of the Satellite when in the position of new moon (as at M_5), that the orbit in space is looped as is shown in the figure.

We have already exceeded the limits to which we intended to restrict our discussion of these four Moons. Our excuse must be, that in them and the planet round which they circle we have what we might almost be justified in calling a miniature *Solar System*. We have seen in our last lecture, that, the

* It would have been better both in Figs. LXXX. and LXXXI. if the letters had been numbered from right to left, instead of from left to right, as the former direction is that in which in our previous figures we have taken Jupiter to move. The Sun is supposed to be situated beneath each diagram.

more closely we investigate the huge primary to which they belong, the more details we discover to reward our zeal. In this lecture we have endeavoured to show, that the study of the Satellites, although more difficult, and in some respects less encouraging, is hardly less instructive or important. If a new and more accurate set of tables of their places than we at present possess can ere long be constructed, and if instruments of the greatest possible power be applied to their observation, we think it probable that many matters connected both with them and with Jupiter itself, which we as yet regret our inability to explain, may be found to be full of information. The facts which now puzzle us may be the very facts which will teach us what the physical condition of the planet really is, or what mutual good offices Jupiter and its Satellites perform for one another.

We may some day be able to discuss upon much better grounds than at present the possibility, or otherwise, of the hypothesis, that they may be worlds peopled by inhabitants who may look upon Jupiter as a Sun with four attendant planets, rather than as a planet with four attendant Satellites journeying round a far-distant Sun.

LECTURE XIV.

THE PLANET SATURN.

> "But farther yet the tardy Saturn lags,
> And five* attendant luminaries drags;
> Investing with a double ring his pace,
> He travels through immensity of space."
>
> CHATTERTON ("On the Copernican System").

At a mean distance from the Sun almost exactly 400 millions of miles beyond the orbit of Jupiter, we find another planet more fascinating and more marvellous still. Or it may be that we should rather say another planetary system, for Saturn with its eight attendant satellites and appanage of ring outlying ring may well be so termed.

There is perhaps no more beauteous sight in the whole heavens than that of the Saturnian system in a telescope of considerable power. The planet itself surpasses all the rest in a certain calm distinctness which seems to be inherent in its image; while in the peculiar form of the rings, with the mighty globe balanced within them, and in the never-ending variety in the positions of the satellites, there is a charm that no pen can describe. The most experienced observer never wearies of such a sight.

True it is, that we cannot distinguish nearly so much detail of belt, or spot, or physical condition upon Saturn, as upon Jupiter, owing to its smaller size and its much greater distance from the Earth. Several parallel bands, or markings, of lighter or darker colour, are, however, frequently seen on its disc; two at some distance on either side of the equator being generally

---

* The number known when Chatterton wrote the above lines, soon after the middle of the 18th century.

of a brownish hue, while the polar regions, as a rule, are bluish. Spots are rarely visible upon it. Nevertheless some, both dark and light, have at times been sufficiently distinct and permanent to afford a determination of the period of the planet's rotation upon its axis. In this way Sir W. Herschel obtained for its duration 10^h 16^m. But the best determination yet made is by Professor Asaph Hall, from a white spot visible during nearly a month, beginning with December 7th, 1876, which gave a result 1^m 36^s less than Herschel's, viz., 10^h 14^m 24^s.

Next to the appearance of the planet's disc we feel tempted to proceed at once to discuss its wondrous rings,—so marvellous in themselves, so utterly unlike what we see in any other member of the Solar System. But we must first pause for awhile to mention, as briefly as possible, some of the accurate measurements of the size and shape, the weight and orbit, of Saturn, which, in spite of its vast distance, the perfection of modern instruments has secured.

The ball, or sphere, of this great planet measures (apart from the rings) about 74,000 miles in its equatoreal diameter. But its polar diameter is less by between $\frac{1}{9}$th and $\frac{1}{10}$th part; the proportionate compression at its poles being nearly twice as great as in Jupiter, but not so striking, owing to the smaller size of its apparent disc, and possibly from the effect upon the eye of the contour of the rings.

The surface of the globe of Saturn is fully 80 times, and its volume about 750 times, that of the Earth. It will be remembered that, in the case of Jupiter, the corresponding numbers are respectively about 120 and 1,300. Consequently the apparent volume of Saturn is not very much more than one-half of that of Jupiter. We must not, however, forget that our remarks, in Lecture XII., as to the difference which may exist between the size indicated by the visible disc and the real size of Jupiter, may also be applicable to Saturn. Our measurements are only of that which, after all, may be an outlying vaporous, or cloudy, envelope, within which a more solid and smaller globe may, or may not, exist. And in one respect this is all the more likely to be the case with Saturn, inasmuch as the average density of a globe extending to the boundary which we see, and of a weight such as that of the planet is found to be, would only be one-half as great as in the case of Jupiter; while a globe of the same size, if of the average density of the Earth, would weigh more than eight times as much as Saturn weighs.

The distance of the planet Saturn from the Sun varies from about 837 to about 937 millions of miles, its orbit being slightly more oval than that of Jupiter, but only about $\frac{1}{5}$rd as much so as that of the Earth.

Its mean distance from the Earth, when in Opposition to the Sun, is somewhat greater than 790,000,000 miles; Jupiter's being, under similar circumstances, about 390,000,000 miles. If its globe were of the same size, it would therefore appear, at such a time, to be of about one-half of the diameter of that of Jupiter, but, being smaller in itself, its apparent width is reduced to rather less than $\frac{5}{12}$ths of Jupiter's, while in area its disc barely exceeds $\frac{1}{8}$th of that of its larger neighbour. A magnifying power of 120 (*i.e.*, in linear dimensions), when applied to a telescope, therefore only enlarges the disc of Saturn to the same size as that to which the disc of Jupiter is enlarged by a power of 50; and of course the same proportion holds in the case of higher powers. But the apparent width of the image of Saturn, from the extreme boundary of the rings on one side across the planet's disc to their extreme boundary on the other side, is about the same as the apparent width of the orb of Jupiter.

Without attempting to recollect all the above statements, it may, perhaps, be well to commit to memory some such approximate and simple numbers as the following:—The Earth's diameter is about 8,000 miles; Jupiter's equatoreal diameter is about 88,000 miles, *i.e.*, about 11 times the Earth's; Saturn's equatoreal diameter is about 74,000 miles, *i.e.*, fully 9 times the Earth's; the diameter of the external contour of the rings is nearly 170,000 miles, or about 21 times that of the Earth.

We will now ask the attention of our readers to a beautiful drawing, in Fig. LXXXII., of Saturn and its rings, as seen in a powerful telescope. It was sketched by Mr. De La Rue, on March 27th and 29th, 1856, when the appearance of the rings closely resembled that which they will have in the year 1885. Apart from some minor divisions, it is clearly shown in this figure that there are two principal bright rings divided by a very decided dark gap. These are generally termed the rings A and B; A being the further of the two from the ball of the planet. Within these two bright rings a much fainter dusky-looking ring is seen, which is called the ring C.

We much regret that the enormous distance of Saturn from the Earth makes it impossible to study these interesting appendages with the minuteness which is desirable. Great assistance has, however, been derived from the ever-varying aspects under which they are seen at different times. Sometimes they appear as a very thin line of light, so that we have heard little children exclaim, when looking through a telescope, that the planet was like a ball with a knitting-needle stuck through it. At other times they are seen so widely opened out that their narrower diameter, as in Fig. LXXXII., exceeds that of the planet, when they consequently more or less surround it on all sides.*

Nor is it difficult to account for these changes of appearance, which are known as *the phases of Saturn's rings*. The rings lie very nearly in a plane, their thickness perpendicular to which is very small, and the planet is very nearly in their centre. If then their shape in the plane in which they lie be supposed to be circular (and it certainly is approximately so), it is clear that they would appear to us as circles surrounding the ball at a certain distance from it, provided their plane were at right angles to our line of sight. On the other hand, if at any time we should look at them edgewise, it follows, from the laws of perspective, that their apparent form would be reduced to that of a very thin straight line of light, which would extend to a certain distance upon each side of the planet, and probably be imperceptible in the portion to which Saturn's bright disc would form a background. At their great distance it might even be, that the whole of this line of light would appear so thin that its thickness would be inappreciable, and consequently invisible, to an observer looking at it *exactly* edgewise.

* At the present time, A.D. 1882, they are gradually approaching this appearance.

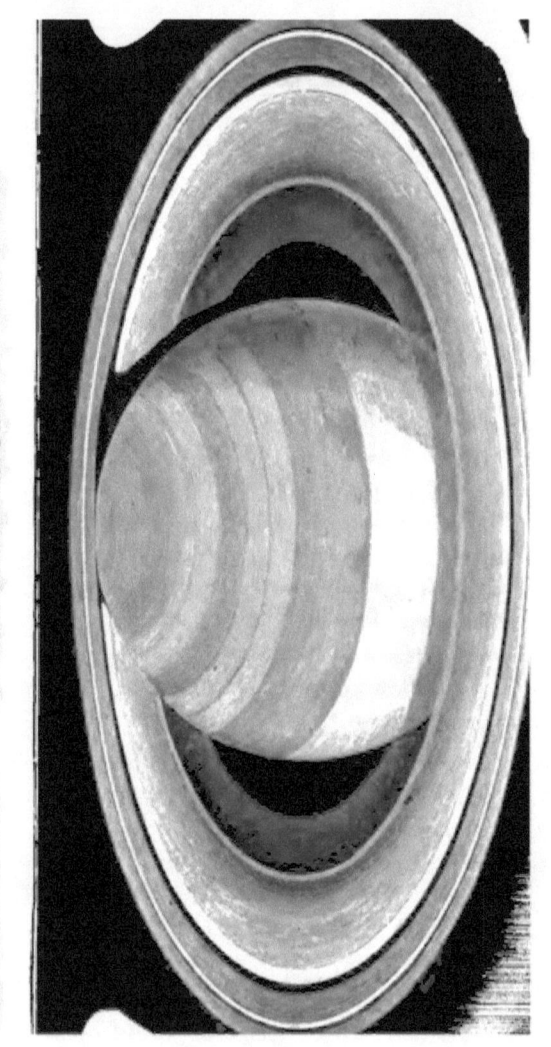

THE PLANET SATURN.

THE PLANET SATURN.

Under intermediate conditions the circular form of the rings would be seen, by the effect of perspective, as a more or less oval ellipse, according to the angle at which their plane might be inclined to the direction of the observer's gaze; the major, or larger, axis of the ellipse always remaining constant in magnitude; the minor axis gradually increasing, or diminishing.

Now observation shows that the plane of the rings is inclined at an angle of about 27° to the plane of Saturn's orbit, and

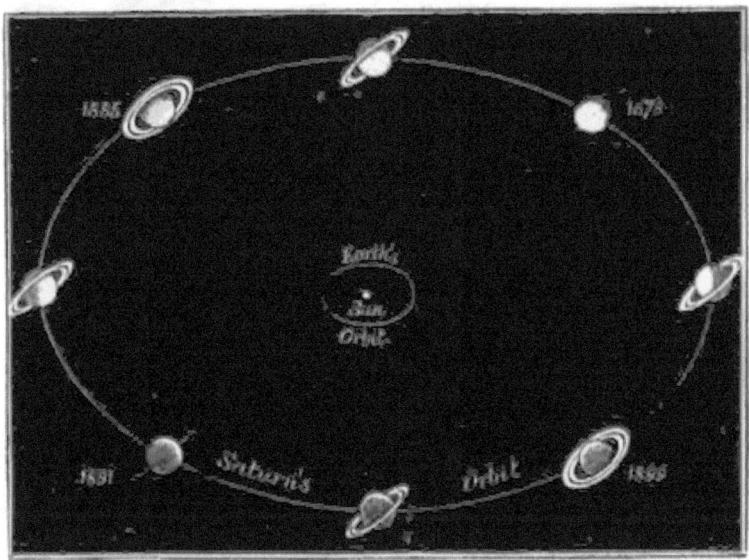

Fig. LXXXIII.*—The phases of Saturn's rings.

that it always retains a parallel position in space, while Saturn travels round the Sun with a speed barely one-third of that of the Earth in its long journey of nearly 29½ years' duration. The rings, if seen from the Sun, would therefore present in succes-

* This diagram is partly copied, by Mr. Bentley's kind permission, from one in the English edition of "The Heavens," by Guillemin. The irregularity in the intervals indicated in it, e.g., 1885–1891, 1891–1899, between each of which Saturn describes the same angle of 90° around the Sun, is partly due to the different months in each year (see page 360, lines 4-8) to which the positions of the planet correspond, but principally to its unequal velocity in its orbit; its maximum speed occurring when it is in perihelion, in the year 1885; its minimum when in aphelion, in the year 1900.

sion the appearances shown in Fig. LXXXIII. during the above-mentioned period.

Twice in the 29½ years, as in February 1878 and October 1891, they would be looked at edgewise, and would be reduced, as we have stated, to a very fine line, or for a few hours be invisible. Twice, as at the beginning of 1885 and in the middle of 1899, they would be seen in their most fully opened appearance, when they would, however, still fall short of a circular

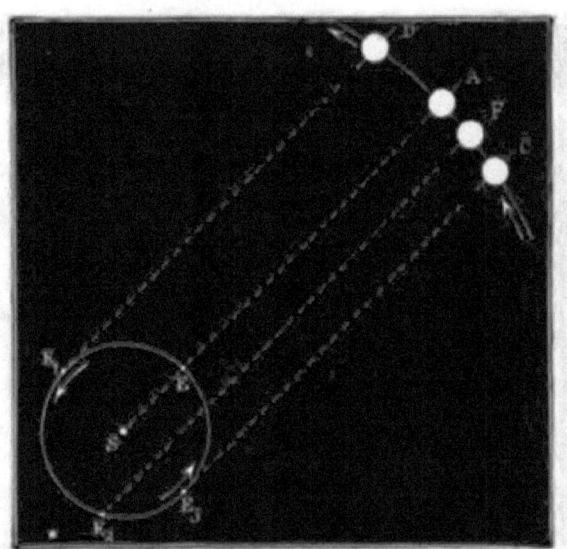

Fig. LXXXIV.—The plane of Saturn's rings sweeps across the Earth's orbit, $EE_1E_2E_3$, while Saturn is travelling from C to D. When Saturn is at A the plane of the rings passes through the Sun.

contour, the greatest ratio of the minor to the major axis of their apparent form being as 45 to 100. It is also to be noticed, as is shown in the figure, that during one-half of the time, or for about 14¾ years, the Sun would illuminate the northern surface of the rings; and during the other half of the time their southern side.

The view of the rings obtained by an observer upon the Earth does not in general differ much from that which would be seen from the Sun. Some slight allowance must be made for the fact that the Earth's plane of motion round the Sun (or, in other words, the ecliptic) is slightly different from the plane in which Saturn moves. And the alternate increase and decrease of the Earth's distance from Saturn, as well as the somewhat different directions in which the rings may be viewed from the Earth and from the Sun respectively, must be taken into consideration. Notwithstanding these slight differences, we are, however, justified in saying that the appearance is *in general* very much the same as it would be from the Sun.

But we must ask our readers carefully to notice that we specially emphasize the words "in general." The statement above made has one important exception. In fact, it ceases to be true near to those times when the rings would be seen *edgewise from the Sun*.

The diagram, in Fig. LXXXIV., shows that, at such times, the exact edgewise view from the Earth would only *concur* with that from the Sun, if the Earth should happen to be *exactly* in such a position, E, in its orbit, that it would be in the straight line, AES, joining the planet, A, and the Sun, S, at the very epoch, when the plane of the rings is passing through the latter, and an observer upon it would look edgewise at them. But such a coincidence in the relative positions of the Earth and of the Sun is very unlikely to occur; and it is consequently found to be necessary, in order to discover what may happen near to the date of a passage of the plane of the rings through the Sun, to allow both for the position and for the movement of the Earth in its orbit during several consecutive months, as well as to take into consideration the corresponding movement of Saturn.

In doing so, and in studying the meaning of Fig. LXXXIV., the first thing to be noticed is, that the interval, during which a passage of the plane of the rings through the Earth may occur near to the time of one of its periodical passages through the Sun, is limited by the length of time which that plane, while sweeping through space ever parallel to itself, occupies in passing across the Earth's orbit, $EE_1E_2E_3$.* During that

---

* We shall, throughout this explanation, suppose the plane of the Earth's orbit to be extended so as to reach as far as Saturn is from the Sun, and the plane of the rings to be extended so as to pass across the Earth's orbit. The intersection of the two planes will consequently be a straight line reaching from Saturn to the further side of the orbit of the Earth.

interval the plane of the rings, if extended far enough, would pass through the Earth (as the figure shows) whenever the direction of the straight line joining the centres of the Earth and Saturn might lie in it. This would be the case, for instance, if the Earth were at E_3 and Saturn at the same time at c; or if the Earth were at E_1 and Saturn at d; or if the Earth were in any intermediate position at E_2, and Saturn at the same time at f.

It is, moreover, evident that the time in which Saturn passes through this critical portion of its path, from c to d, is very nearly equal to that in which it travels through a distance equal to the diameter of the Earth's orbit (*i.e.*, through about 186,000,000 miles), since cd is, except for the curvature of the path, equal to E_1SE_3. It may be very easily calculated that this time is rather more than a year.

It is also an important fact, as a little further consideration will show, that it is only during such an interval that the Sun and the Earth can ever be on *opposite sides of the plane of the rings*. At all other times, whether it be the north side or the south side of that plane which is turned towards the Sun and is illuminated by it, that same side will be seen by the Earth. But if, for instance, the Earth were at E_3, and Saturn at the same time at f in Fig. LXXXIV., the bright side of the rings, *i.e.*, the side upon which the Sun would be shining, would be turned away from the Earth. Or, to take another example, let us suppose Saturn to have passed somewhat beyond a in the same figure, and the Earth to be approaching e. The Earth and the Sun would be both upon the same side of the rings, and we should see their bright side. But the Earth might soon afterwards have nearly overtaken their plane, and be about to pass through it. The shorter diameter of the rings would then become narrower and narrower, until at last they would be seen as an extremely fine straight line. Then, directly after the actual passage of the Earth through their plane, their dark side would be turned towards us, and they would become invisible. And this state of things, as we shall presently show, may sometimes last for several consecutive months, until the plane of the rings again passes, either through the Sun or the Earth, so that both are upon one and the same side of them again.

The explanation of the variations in the succession and in the number of the disappearances of the rings, which may at such times occur, is very instructive. And, if given in a somewhat rough and approximate form, it need not involve any mathematical calculations. We therefore hope that our readers will endeavour to follow it by the aid of Fig. LXXXV., in which $E_1 E_2, E_6 E_3$, etc., represents the Earth's orbit round the Sun.

Now we have already seen that, as Saturn revolves in his much larger orbit (a small portion only of which is drawn), the intersection of the plane of its rings with that of the Earth's orbit will sweep across the latter, at intervals of rather less than fifteen years, like an enormously long straight line, moving always parallel to itself. It will so pass in the direction indicated by the arrows in the diagram, while Saturn is moving from c to d; and in the opposite direction while Saturn is in a portion of its orbit which may be supposed to be drawn at an equal distance on the left of that of the Earth, although for convenience it is omitted in the figure. The time which Saturn takes to go from c to d (or through the corresponding and opposite portion of its path) is about thirteen months. During this time the Earth travels rather more than once round its own orbit in the direction shown by the arrows. We must also remember that, when the line in question sweeps across the Earth, or, in other words, when the Earth passes through it, we see the rings *edgewise*, and the Earth also passes from one side of their plane to the other. And when the line of which we are speaking sweeps across the Sun, the *Sun* shines *edgewise* upon the rings, after which it begins to illuminate the opposite side of them to that upon which it was previously shining. Moreover, whether their plane pass through the Sun or through the Earth, they are, in either case, apparently reduced for a while to the finest conceivable line of light, which may even be so fine as to be invisible.

We may investigate all possible varieties that can occur in the succession of such phenomena by supposing the Earth to occupy various different positions in its orbit, in Fig. LXXXV., at the time when, as is there shown, the intersection of the plane of the rings *begins* gradually to move across the circle E_1E_2, etc.

For instance, let us first of all suppose the Earth to be in such a position as E_1 when Saturn is at c. By about the time that the Earth has reached E_2 Saturn will be at S_2, and the Earth will pass through the plane of the rings, so that they will disappear. And not only so, but the Earth will now be upon the opposite side of them to the Sun, and will therefore look at their dark side. They will consequently *remain invisible* until Saturn having reached S_3, the Earth will have advanced to such a position as E_3, where it will cross their plane again, so as to see their bright side once more. But very soon after this,

Fig. LXXXV.—The disappearances and reappearances of Saturn's rings.

Saturn having reached S_4, their plane will pass through the Sun, and the Sun will immediately afterwards be on the opposite side of the rings to the Earth, the Earth having in the meanwhile sped on to such a position as E_4. They will consequently not only disappear for the moment as their plane passes through the Sun, but they will once more *remain invisible* for a while, because their dark side will be again turned towards the Earth. At some date before the completion of a year from the beginning of the time which we have been considering, Saturn having reached S_5, the Earth (when in such a position as E_5) will pass through the plane of the rings again, and having now attained the same side of them as the Sun, will see their bright surface once more.

In the above case, during the period considered, it is evident that the Earth would go through the plane of the rings *three times*, while that plane would also pass *once* through the Sun. There would be *two* disappearances of the rings, and *two* reappearances, with intervals of invisibility between. Such a series of events actually occurred in the few months succeeding November 1861.

But, if the Earth should happen to be in such a position as E_6, when Saturn arrives at c, then in half a year's time it would be at E_7, while the plane of the rings would have nearly reached the Sun, Saturn now being somewhere near to s_3. Very soon their plane would pass through the Sun; and they would not only momentarily disappear, but, the Sun and the Earth being now upon opposite sides of them, the Earth would look at their dark side, and they would continue for a while invisible. The Earth, however, would evidently soon afterwards pass through their plane, and they would then reappear, because it would once more see their bright side. In a year's time the Earth would get to E_6 again; and, in about another month, Saturn having reached D, the plane of the rings would have passed beyond the Earth's orbit, and no more passages, either of the Earth through the plane, or of the plane through the Sun, would occur for 14, or 15, years to come.

In such a case as this last, the Earth would only go *once* through the plane of the rings; that plane of course, as before, also passing *once* through the Sun. There would be only *one* disappearance and *one* reappearance. Such a passage of the plane of the rings through the Sun took place on February 6th, 1878, and of the Earth through their plane on March 1st of the same year. But the phenomena in question could not then be watched, because the Earth and Saturn were nearly in the same straight line upon *opposite* sides of the Sun at those dates; or in other words Saturn was so near to Conjunction with the Sun, that it was impossible to look at it in the glare of the Solar light. In 1891, the next epoch of a disappearance, the same hindrance will unfortunately again obstruct our view. But in 1907, a somewhat similar course of events to that of 1861 and 1862 will take place, so that *two* disappearances and *two* reappearances will occur. This will be the next occasion upon which it will be possible satisfactorily to observe the disappearance of the ring.

Certain further complications might arise in exceptional cases; as, for instance, if the plane of the rings should pass, or very nearly pass, through the Earth and the Sun at the same moment; but such instances are so rare that we need not discuss them.

It will, however, be a very useful exercise for the reader to work out what will happen if he suppose the Earth to start from various positions in the circle of its orbit, at the instant when Saturn passes through c. We hope that the two cases which we have discussed may suffice to show how any others may be treated, and how simple the explanation of so apparently complicated a series of phenomena really is.

Those who desire fuller information may consult Mr. Proctor's well-known work upon Saturn, in which he enters somewhat more elaborately into calculations than in many of his more recent astronomical treatises. He has there exhaustively discussed the whole question.

For a complete historical account of the successive *discoveries* of the rings, of their divisions and many interesting peculiarities, we refer our readers to Grant's "History of Physical Astronomy." We will only mention here a few of the principal episodes in the story. It is well known that Galileo Galilei wrote to Kepler in the year 1610, announcing a discovery connected with the planet Saturn in the form of a logogriphe, as follows:—

"smaismrmilmepoetaleumibunenugttauiras,"

which Kepler supposed to refer to Mars, and to be intended to represent the words:—

"Salve umbistineum geminatum Martia proles!"
"Hail, twin companionship, children of Mars!"

In fact, Kepler imagined that Galileo had discovered two satellites of Mars. But the real meaning intended was:—

"Altissimum planetam tergeminum observavi."
"I have observed the most distant of the planets to have a triple form." *

---

* See the "Sidereal Messenger of Galileo Galilei," translated by Rev. E. S. Carlos, pp. 88 and 90.

This proves that Galileo had actually detected the ring, although in his telescope it appeared like two satellites or companions, one upon either side of Saturn. But in the year 1612, when he looked at the planet, he no longer saw them; and it is said that he vowed that he would never look for them again, for fear that he might find out that there had been something wrong with his instrument.

Our readers will find in Newcomb's "Popular Astronomy" some interesting copies of ancient drawings from the *Systema Saturnium* of Huyghens, which show how the older observers were puzzled by the above-mentioned triple appearance of the planet, which seemed to them (as they said) sometimes as if it had arms or ears, and sometimes as if there were handles, or children, attached to it upon either side; while at other times they saw it without any such additions, as though, according to their quaint description, Saturn had devoured his offspring.

It was not until 1654 that Huyghens solved the riddle at a time when the rings disappeared; upon which he published his well-known logogriphe:—

"aaaaaaa ccccc d eeeee g h iiiiii llll mm nnnnnnnnn oooo pp q rr s ttttt uuuuu;"

from which the following sentence may be formed:—

"Annulo cingitur, tenui, plano, nusquam cohaerente, ad eclipticam inclinato;"

or, in other words, "The planet is surrounded by a slender flat ring, everywhere distinct from its surface, and inclined to the ecliptic."*

In 1676, Cassini perceived that the ring apparently consisted of two concentric rings.† And somewhat more than a

---

* See Grant's "History of Astronomy," p. 257.

† In Admiral Smyth's "Celestial Cycle," vol. i., p. 51, it is stated that "Mr. William Ball and his brother, Dr. Ball, of Minehead, in Devonshire, first saw Saturn's ring double." It seems, however, that Kitchener, in his work upon Telescopes published in 1818 (which concludes with the quaint remark that, if any one thinks the book has not a page in it worth a farthing, he should remember that less than a farthing was paid for it, as the whole 470 pages were offered for nine shillings) was the first to attribute the discovery to the Messrs. Ball. But an examination of the account of

century later, Sir W. Herschel, after more than ten years of careful observation, was convinced that, whether viewed from the one side or from the other, the division between the two rings was seen exactly in the same position; and that it was a real break through which the dark background of the heavens was visible, and not simply a dusky shading upon the ring-surface. The two rings, thus divided from one another, are those which, as we have already mentioned, are now generally named A and B; the outer being A, the inner B.

But in the year 1850 a wonderful extension of the system was discovered, inasmuch as Professor Bond, of the Cambridge Observatory, U.S., and Messrs. Dawes and Lassell, in England, detected a *third* ring, closer to the planet than the other two, and of a most remarkable character. It is that which is now called the ring C; its chief peculiarity being, that it has a dusky and obscure, although transparent, appearance, almost like that of gauze or crape, and such that the body of the planet can be seen through a considerable portion of it. Some time after its discovery in 1850, it was found that it had been noticed, so early as in 1838, to resemble a shadow where it

---

their observation, which is to be found in vol. i. of the "Philosophical Transactions," does not confirm their right to the credit of the discovery. They only noticed an apparent depression, or deformation, in the portions of the ansæ which appeared joined to the planet, which possibly suggested the idea that the two ansæ might be two independent appendages; but they saw no division in the breadth of the ring, the first indication of which is due to Cassini, whose drawing clearly indicates a shading in the ring where subsequent observations have conclusively proved that the principal division exists. We must therefore assign the discovery of the duplicity of the ring to Cassini. At the same time, we have no wish to detract from the reputation of William Ball as a skilful observer. Huyghens refers to his observations some years before the above date (having heard of them from Dr. Wallis) as confirming his own, in evidence of the appearance of a dark line seen across the planet at certain times (viz., when the rings are viewed edgewise), which proved the real form of the appendage, which had seemed so enigmatical, to be that of a wide and flat, but very thin, concentric ring surrounding the planet. Minehead (or, as it is in the "Philosophical Transactions," "Mainhead") is doubtless a misprint for Mamhead, a village a few miles south-west of Exeter.

PLATE IX.

SATURN IN 1874.

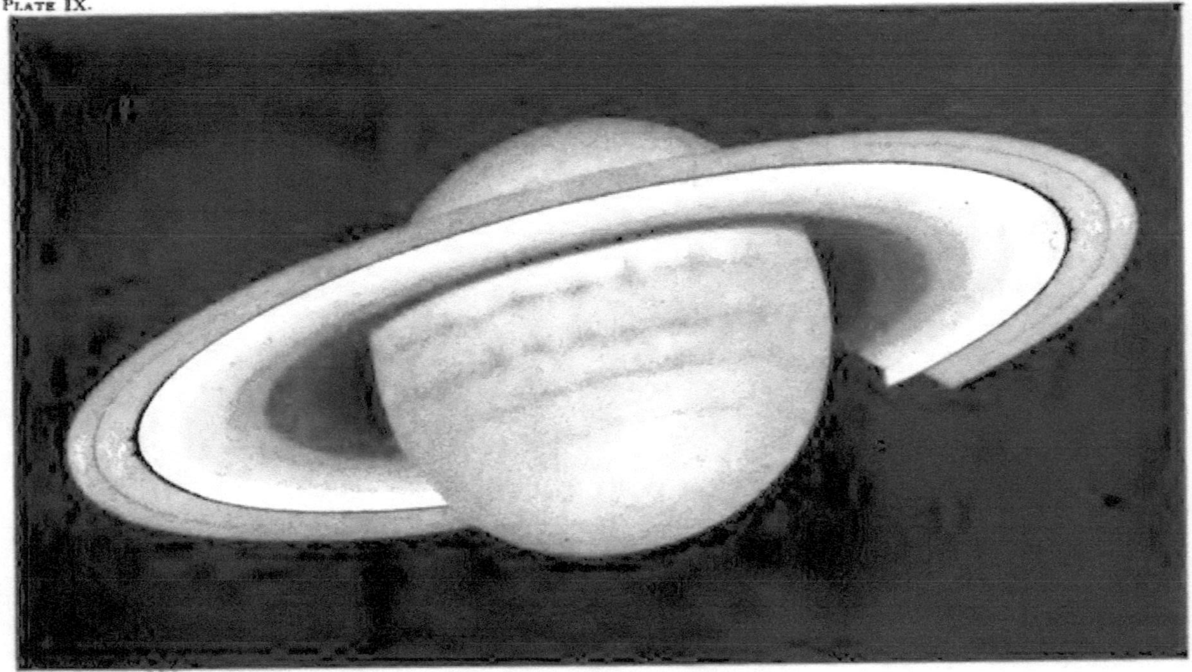

W. H. WESLEY, *Lith.*

From a Drawing by L. TROUVELOT,
In the "*Proceedings of the American Academy for 1875-76.*"

crossed the planet, by Dr. Galle of Berlin, who then published his observations and measurements, although he failed to detect its real character.* We hope that many important observations of it may be made during the next few years.

It may easily be understood that it is when the rings appear to be opened out to their widest (as will be the case in 1885), or near to such epochs, that the best opportunities are obtained for the observation of their various divisions and their comparative brightness, or for the study of their individual peculiarities of surface and condition. And in this connection it may be well to mention that in 1885 Saturn will have a large northern declination, which will raise it especially high above the horizon in latitudes such as our own; while in October of that same year it will pass through its Perihelion, or nearest approach to the Sun, two months after which it will be in Opposition to the Sun. From the joint effect of all these causes, it will consequently be unusually well situated for our observation.

It was under somewhat similar circumstances that Mr. De La Rue's drawing (see Fig. LXXXII., p. 358) was made. Our readers may therefore be glad to refer to it once more, and to notice in it several very interesting features of the ring-system, in addition to those hitherto mentioned, which have also been confirmed by various other eminent observers. For instance, the outer edge of the ring B, is decidedly the brightest part of all. There is also shown a distinct, but very delicate, division in the middle of the width of the ring A, which at times may be detected all round it. Indeed, we may remark that occasionally, in the portions seen furthest from the planet, or as it is technically expressed, near to the extremities of the *ansæ*, not only is a certain amount of shading often visible upon the ring-surface, as in the figure, but indications are seen of numerous divisions, so that it is impossible to decide into how many

---

* The Rev. T. W. Webb also mentions ("Celestial Objects," 4th Edition, p. 179) that an old assistant in the Observatory at Rome informed the late Father Secchi, that it had been noticed even so early as in 1848, although no further attention was paid to it.

rings, concentric, or nearly concentric, with one another, the whole may be divided.

When the rings appear widely opened-out, important observations may also be made of the shadow of the planet upon them. For instance, in a drawing by Mr. L. Trouvelot, in the year 1874, a copy of which is shown in Plate IX., as also in several published by the Messrs. Bond in Part I. of the beautiful and costly second volume of the "Annals of the Harvard College Observatory," various irregularities in the contour of the shadow are noticeable. These probably indicate differences of level in the surface of the rings, which thus distort the outline where it crosses them. In Plate IX. the edge of the shadow is certainly most remarkably notched as it passes from the ring A to the ring B. In some of the Harvard drawings a certain amount of shadow is also shown upon the opposite side of the ball to that upon which the principal shadow is seen, the cause of which extraordinary phenomenon we are quite unable to explain. All such peculiarities are, however, very important in relation to the physical condition of the rings, and the possible existence of an atmosphere belonging to them or to the planet, which may by its refraction produce such effects. Moreover, near to the ansæ certain remarkable irregularities, which are indicated by a series of small notches in Mr. Trouvelot's drawing (see Plate IX.), have been from time to time observed in the inner edge of the ring A, the occurrence of which is decidedly puzzling.

In addition to the shadow of the planet upon the rings, that of the rings upon the planet may be easily seen. Even when they altogether vanish from our view, it appears as a narrow black line across the disc. At other times it may be much wider, but then it is often to a great degree concealed by the rings themselves; while it is altogether hidden when they are opened out to their utmost extent. So much as we can see, therefore, in general appears as a narrow black line just above or below the rings.

Observations near to the time of the disappearance of the rings, or when they appear to be extremely thin, are, however, no less interesting than those made when they are widely opened-out. Many such, by the Messrs. Bond in 1848, are recorded in the second volume of the "Annals of the Harvard College Observatory," to which we have already referred. It is there stated, that at such times some of the satellites of the planet, of which we shall have more to say presently, may occasionally be seen to be apparently threaded, like beads on a needle, on the fine line of light to which the rings are reduced, while carried backwards and forwards along it by their orbital motions, which are nearly in the same plane with it.

It is believed that in this way an important proof may be

Fig. LXXXVI.—Saturn, December 26th, 1861, drawn by Mr. Wray.

obtained of the extreme thinness of the rings. It is said that even so small a satellite as the 7th, whose probable diameter is considerably less than 1,000 miles, has been seen, when thus situated, to extend upon either side beyond the thickness of the line of light of the rings. Other reasons also indicate that the thickness at any rate of the rings A and B, does not exceed at the most 100, or perhaps 200 miles, and may very possibly fall short of 50 miles.

At the times to which we have just referred, there is, however, an appearance of a greater thickness in the ring-system for about one-third of the distance to which their light extends upon each side of the disc. Thus far a fainter hazy light of somewhat greater breadth is seen, and was at one time thought to indicate the existence of some kind of atmosphere over that portion of the rings. It is, however, more probably produced by the dusky or crape-like ring, C; which, if so, may be considerably thicker than A and B. This is well shown—see Fig. LXXXVI.—in a drawing by Mr. Wray, dated December 26th, 1861, in which two of the satellites are also visible, apparently resting upon the line of the outer rings.

It is generally found that the above-mentioned fine line of light, before it altogether disappears, breaks up into several more or less isolated portions. An ingenious suggestion has been made by the Messrs. Bond,* that this may be simply the result of an optical effect produced by the inner and outer edges of the various rings being seen nearly edgewise, so that their light is united in certain parts, where it is consequently strong enough to be visible; while in other parts, where the bright edge of only one is turned towards an observer, the amount of light is too small to be perceived.

There are also some other classes of observations of great importance which may be made at such times. For instance, the line of light upon one side of the planet becomes sometimes so fine as to be invisible in telescopes of moderate power some days before this is the case upon the opposite side. This, of course, goes against the supposition of a rotation of the rings round Saturn; for, if they so revolved, we should expect that the more noticeable, or thicker, portion would go regularly round from one side to the other of the disc. It has no doubt been thought that certain observations of the vanishing rings by Sir W. Herschel, and by one or two other observers, have given indications of such a rotation, at least in one of them. We consider, however, and we shall presently endeavour to prove, that any such movement of any ring as a whole must have been apparent rather than real.

Very careful measurements indicate that the centre of the planet is not placed quite in the centre of the rings. If this

* A copy of the original diagram by which it was illustrated in vol. ii. of the "Annals of the Harvard College Observatory" may be seen in Chambers' "Handbook of Descriptive Astronomy," and in Guillemin's "The Heavens."

be so, any ring revolving round it somewhat eccentrically, would about every five hours alternately show a rather greater extension first on one side and then on the other. This, however, is not seen to be the case. Indeed, we do not think that it can be said that there is any certain and satisfactory evidence from *observation* of the rotation either of any one ring, or of a number of separate and concentric rings, each having its own independent period.

One thing we may very positively affirm, viz.,—that the rings *cannot* possibly rotate *as a whole* with any one and the same velocity. It will be remembered that Kepler's third law (see Lecture V., page 122) requires that a certain relation should exist between the distance and the speed of one body revolving round another in free space. Consequently the speed which would suit the position of the inner portion of the rings would be far too rapid for their outer parts; while that which would suit the outer parts would be far too slow for the inner portion. Such is the width of the rings, and so great the difference in the necessary speeds, that it is certain that any solid matter of which we can conceive would utterly break up, if, being composed of it, they were for a moment to be set rotating with one and the same speed throughout. At the inside of the ring B, the necessary velocity of rotation would be such as would carry a point round in somewhat over 7 hours; at the outside of the ring A, the rotation-period would need to be between 13 and 14. If a large grindstone when very rapidly revolved explodes with a terrific crash, surely the rings of Saturn, if solid, would in like manner be destroyed.

But it may be asked:—Would not the supposition previously referred to, viz., that the rings are split up into many narrow ones, each of which rotates in its own independent period, get over this difficulty? Unfortunately it would not; because it can be shown by the very refined investigations of the late Professors James Clerk Maxwell in England, and B. Pierce in America, that no such system could remain in stable equilibrium. Even if the rotations of its parts so acted, in conjunction with their mutual attractions and with the attraction of Saturn itself upon them, as for a moment to produce a balance of effect and a resulting equilibrium, the slightest disturbance would very soon bring about a general catastrophe. We cannot attempt to explain the mechanical and mathematical considerations involved in this statement. Our readers may, however, easily understand that the satellites of the planet, or the other planets of the Solar System, would by their attractions inevitably generate such disturbances. They may also be able to appreciate the extreme inherent weakness which would be involved in the construction of any body of so great a diameter as that of the rings, and of so excessive a thinness perpendicular to the plane of their extension. If their thickness be supposed to equal 100 miles, it would, in comparison with their diameter and their density, be only as though an arch of a hundred yards' span were cut out of a large sheet of some material of much less rigidity than iron, and only two inches in thickness. If such an arch were set up it certainly could not be expected to keep its shape. No more, therefore, could such a ring, balanced in space, support the attractions which those of Saturn undergo.

Nor is it in the least degree probable that the rings can be fluid; for, if so, it might also be shown that various external attractions would soon set up such waves in them as would continuously grow in intensity until their ruin would follow.

It seems, therefore, that by far the most reasonable hypothesis is, that the rings are composed of *myriads upon myriads of small satellites*, each revolving in its own independent orbit; but so thickly aggregated together as to produce the appearance of a continuous surface, with the exception that they are more scantily distributed where we see gaps, or divisions, in the system. This supposition also requires that they must be most thickly grouped in the neighbourhood of the inner edge of B, where the brightest appearance is found. And here we may state that it has been noticed by Professor Daniel Kirkwood,* of Bloomington, Indiana, U.S., that the attraction of some of the satellites of Saturn would produce comparatively vacant regions

---

* See also Lecture XI., page 289, as to the possible influence of Jupiter in producing some of the gaps which are met with in the region of the minor planets.

in a ring-system, thus constituted, just where they actually exist.

The explanation of the peculiar appearance of the dusky, or crape-like ring, C, upon the above hypothesis is, that it is composed of a collection of satellites much denser than in the divisions between the rings, but not dense enough to give the same bright appearance as in the other rings, A and B. It is also worthy of special notice, that there is some suspicion that this dusky ring, which at present reaches about half-way from the inner edge of B towards the planet, is becoming brighter than it was, and that its inner edge is also approaching the ball. If so, this is an intensely interesting fact. It would suggest that the orbits of some of these myriad satellites may be gradually contracting,—in other words, that a slow process may really be going on, by the continuance of which the rings would finally descend on to Saturn, and its chief beauty in the telescope be lost.

The principal dimensions of the rings are nearly as in the following table, with regard to which, however, it should be mentioned, that the best measurements do not exactly agree, and that we have only stated such approximate numbers as may most easily be remembered:—

Outer diameter of ring	A	about	166,000	miles.
Inner	,,	,,	A	,,	146,000	,,
Outer	,,	,,	B	,,	143,000	,,
Inner	,,	,,	B	,,	110,000	,,
Interval between A and B				,,	1,700	,,
Distance from the surface of the ball to the inner edge of . . . . . B				,,	18,000	,,
Equatoreal diameter of the ball				,,	74,000	,,

Although there seems to be very little, if any, reason for supposing that either the planet or the rings can be inhabited, it may be an useful and interesting geometrical exercise, and one which may help us to realize the actual relations existing between it and the rings, if we briefly consider what would be the experience of an observer upon either.

First one side and then the other of the rings being turned towards the Sun for nearly fifteen years, a resident upon them would alternately see the Sun for a long day, which would last for 14¾ of our years, and be deprived of its light for an equally long night. His long night of 14¾ years would, however, be partly compensated by the light that would be received from the glorious globe of Saturn itself, which, once in each revolution of the observer round it, would go through a series of phases corresponding to those which our Moon passes through in a month. Except, however, to any one situated close to, or within, the inner edge of the ring B, only one-half of any such phase would probably be visible, the other half being hidden by the flat expanse of the rings.

If the planet were in its full phase it would present, when seen from the inner edge of the ring B, a circular disc, whose area would be equal to about 25,600 times that of our Full Moon, and which would consequently afford, by the reflection of a sunlight of about ninety times less intensity, an illumination equal to about 280 times that which we receive from its full disc. But from a point on the rings further from the planet's globe, only one-half of a smaller area would be seen. Nevertheless, even from the outside edge of the ring A the semi-circular disc would still appear to have a diameter 100 times, an area fully 5,000 times, and a light about 55 times, as great as that of our Full Moon. Upon one-half of the disc the shadow of the rings would lie, at times obscuring a very wide zone, while bright and dark belts, and other physical features of the surface, would doubtless abound in interest.

Eclipses of the Sun would be very frequent occurrences to an observer on the rings, owing to the huge shadow of the globe of Saturn. There would, in fact, as a rule, be one in the course of each rotation, performed by the portion of the ring upon which he might be situated, around the globe of Saturn. The form of the shadow would, however, vary with the seasons of the Saturnian year, which would affect the occurrence, or the duration, of the eclipses. At an equinox it would be bounded by two straight lines, and be throughout of the same width as the planet. But as Saturn moved on towards a solstice, it would gradually become more and more elliptically curved in its periphery, so that its contour would bear some resemblance to that of the elongated projectiles which are usually fired from a large piece of modern artillery. For a certain time (viz., for about $1\frac{1}{2}$ years before, and $1\frac{1}{2}$ years after, a solstice) it would fall short of the outer boundary of the rings, just as, for a similar period, the rings, if seen from the Sun, would entirely surround the planet. At such a time, therefore, an observer upon the outer part of the ring-system would escape the Solar eclipses, while they would be of short duration for one only slightly overlapped by the range of the shadow. But at an equinox the eclipses would be of a maximum duration, the length of which, if we suppose our observer to go round the planet upon one of the myriad satellites which probably form the rings, may be easily calculated. It would, in fact, bear the same ratio to the whole period of his rotation (which would be between seven and fourteen hours) as the width of the shadow, which would then equal the diameter of Saturn's globe, would bear to the whole circumference of the orbit described by the observer round the planet.

On the other hand, at the time of an equinox, the nightly rising of the shadow upon the eastern part of the rings, as seen by an observer *on the planet* near to its equator, would almost coincide with the setting of the Sun in the west; and it would then sweep rapidly across them until, at midnight, by far the greater part of the rings would be hidden in its shade. At places in higher latitudes the shadow would not rise so soon, but at midnight it would probably overlie the whole of that portion of the rings which would otherwise be seen. It would of course pass away in the reverse direction before sunrise.

At midnight, in any latitude where it could be seen, the shadow of the planet would be always symmetrically placed in the centre of the visible portion of the rings.

Near to the time of the Summer Solstice, when the nights would be shortest, a large portion of the rings would be in shade, even before the setting of the Sun, as seen from places considerably removed from the equator.*

---

* The above statements are more fully explained, and are illustrated by an elaborate series of diagrams, in Mr. Proctor's "Saturn and its System," to whose valuable work upon the planet we are deeply indebted

As seen from the planet, the position and appearance of the *Rings* would greatly vary, both with the season of the year and with the locality of the observer. Since they lie in the plane of the planet's equator, they would only show their bright sides to a resident upon Saturn during the interval between a spring and an autumn equinox of the hemisphere upon which he might be; which interval would, however, last for about $14\frac{3}{4}$ consecutive years. Nor would they at any time appear to be of uniform breadth, for it is evident that, owing to Saturn's great diameter and their nearness, the portion furthest from the horizon would be much the closest to an observer, and would, therefore, appear to be so much the broadest. Moreover, they would seem to meet the horizon in points nearer to the east and west than would be the case with a circle of the celestial sphere described round the pole of the heavens visible to the observer, and drawn through their highest point above the horizon; *i.e.*, such a circle as the Sun, or a star, if it passed through that point when on the meridian, would describe in its daily course across the sky. It also follows that the arch formed by them would in general appear to be, not of a circular, but of an elliptical shape. At the equator only would it be a *vertical* arch of small width, but very considerably wider overhead than on the horizon; in which special case, although its outer boundaries would seem to be somewhat elliptical, its central section would appear to be circular.

That the above statements are a necessary result of the relative positions of the planet and of the rings, may be understood by a few moments' careful thought; or they may be illustrated by actually making a rough model of Saturn with a ball, or globe, and a ring; but in order to demonstrate them fully mathematical calculations would be necessary, into which we cannot enter here. The diagram in Fig. LXXXVII. may, however, to some extent illustrate the appearance which the rings would present, as seen from a place upon the planet, having a *North Latitude* of about 30°.

It shows, in accordance with our statements, that both the outer (and narrower) ring A and the inner ring B (which would apparently lie one above the other upon the sky), one-half of the visible portions of which are respectively marked A*a*bB, and B*b*rR, would seem to be wider upon the meridian from *a* to *b*, and *b* to *r*, than upon the horizon from A to B, and B to R respectively; and that a circle drawn round the pole and passing on the horizon through A, the outer edge of the outer ring, would cross the zenith at s^1, *i.e.*, above *a*, the corresponding point of the ring. While it is in like manner evident that a similar circumpolar circle through *a*, although not drawn in the figure, would meet the horizon to the south of A.

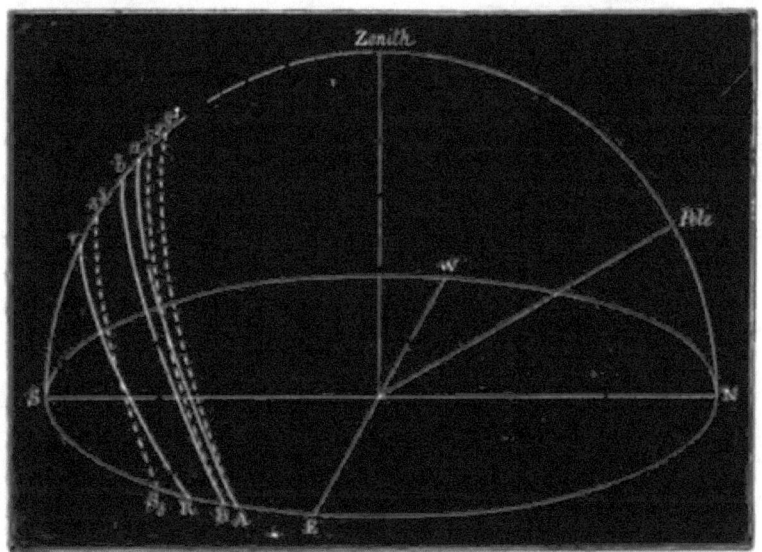

Fig. LXXXVII.—Saturn's rings, as seen from the Planet, and their effect in producing eclipses of the Sun.

This being the case, we may understand how erroneous are some of the statements that have occasionally been made with regard to the effects which the shadow of the rings would produce upon the half of the planet towards which their dark side would periodically be turned. It has even been said, that it would prevent the inhabitants of that hemisphere, for the most part, from seeing the Sun for nearly 15 years at a time.

We recommend those who wish to study this matter fully to consult the very instructive volume by Mr. Proctor, to which we have already referred. It must, however, be confessed that the explanation of it which he has given, although of the highest excellence, is, from the nature of the subject, by no means a very easy one to follow. Perhaps the simpler of the methods there shown, by which it may be treated, is to draw a series of diagrams representing for different seasons in the planet's year the size and position of the shadow upon it. Then it is not very difficult to calculate the way in which, day by day, the rotation of Saturn upon its axis will carry a place;—it may be, into the shadow in the morning, out of it at noon, and into it again in the afternoon;—or through it all day long;—or into it a certain time before noon, and out of it at an equal interval after noon.

But the reader may, if he choose, employ another method of procedure, and imagine himself to be upon the planet, while the rings in a great arch of wondrous beauty span his sky. He might be in such a northern latitude and in such a season of Saturn's year, that this arch would start from the horizon over the point at which the Sun would rise, and extend along the horizon some distance to the north of that point. In fact, we may suppose the Sun to rise at such a time, at some point between A and B in Fig. LXXXVII. Then the arch of the rings being broader as it ascends, its increase of width would tend to keep the Sun's path in it. On the other hand, since the central part of the arch, owing to the perspective effect of its comparative nearness, would (as we have stated) appear to be drawn down towards the horizon so as to cross the meridian below the position of a circle drawn round the pole through the point of sunrise, the Sun towards noon would from this cause tend to rise above it.

Whether the Sun would actually succeed in doing so, and thus become visible in the middle of the day, would depend on the distance measured along the horizon at which it might rise to the south of A. If it rose precisely at A, its diurnal path would be along As^1, s^1 being its place at noon upon the meridian. If it rose at B, it would pass along Bs^2, and it would evidently be visible at noon at s^2. In such a case, it would, of course, again be hidden by the arch of the rings before setting.

Or it might be that the Sun would rise uneclipsed at a point of the horizon to the south of the arch of the ring B, as at s_3 in Fig. LXXXVII., and set in a similar manner; but, as the dotted line s_3s^3 shows, during a certain interval before and after noon, be hidden by the arch. In this case there would be a mid-day eclipse, just as in that previously described there would be one in the morning and evening. Similarly, if the Sun rose in an intermediate position well-immersed in the arch, it might undergo eclipse throughout the day.

Such phenomena might continue to occur day after day for long consecutive periods. This may be proved, either by the method we have just described, or, according to that first mentioned, by calculating the size and position of the shadow of the rings upon Saturn. It may be shown that the shadow would grow wider and wider upon the hemisphere on which it would fall, while the planet would journey in its orbit from an equinox to a solstice; and that it would, at the same time, gradually travel from the neighbourhood of the equator towards the polar regions, for a while leaving a considerable extent of surface nearer to the equator free from all eclipses. But, after the summer solstice of the hemisphere in question, the shadow would begin to travel back from the polar regions, while the eclipses, in moderately high latitudes, would be still continuing.

For places situated upon the equator, the succession of eclipses would be different from that which would be experienced elsewhere, as may be understood by a consideration of the appearance which, as stated upon p. 378, the rings would there present. In such a locality it would be the *thickness* and not the *breadth* of the rings that would conceal the Sun; and Mr. Proctor has calculated* that, at each equinox, if the thickness of the inner ring B be supposed to be 100 miles, the Sun would be eclipsed all the day long, during a period equal to about 9·6 of our days, or 22 Saturnian days; both *before and after* which time there would be eclipses *in the middle of the day* during a period equal to 20·35 of our days, or about 47 Saturnian days.

* See "Saturn and its System," p. 181, and Table XI., p. 224.

On the other hand, in regions not far from the equator of Saturn, a series of eclipses caused by the width of the rings would take place, followed by an interval free from any; after after which they would again occur, until, the next equinox having been reached, the bright side of the rings would be visible for about 14¾ years to come. At a latitude of only 5° there would be 296 Saturnian days, after the autumnal equinox, free from eclipses, then 184 days of morning and evening eclipses, followed by 72 of all-day eclipses, and 684 of middle-day eclipses; or, if we measure in terrestrial time, nearly 1½ years of eclipses altogether, out of which there would, however, only be about one month during which the Sun would be hidden all the day. A similar series would, of course, precede the spring equinox. In latitude 10° each period of eclipses would last for nearly 3 terrestrial years.

But, in latitude 40° Mr. Proctor states that "the eclipses begin when nearly three years (*i.e.* terrestrial years) have elapsed from the time of the autumnal equinox. The morning and evening eclipses continue for more than a year, gradually extending until the Sun is eclipsed during the whole day. These total eclipses *continue to the winter solstice,* and for a corresponding period after it; in all, for 6 years 236·4 days, or 5,543 Saturnian days. This period is followed by more than a year of morning and evening eclipses. The total period during which eclipses of one kind or another take place is no less than 8 (terrestrial) years 292·8 days."

We must refrain from any further statement of the eclipses that would occur in other Saturnian latitudes. It may, however, be well to remark that nearly the whole of the portion of Saturn which lies within its arctic circles would have nothing to do with them. This may be shown as follows:—As we have already mentioned, an observer at the equator would only see the thin inner edge of the rings, which would come between himself and the sky. If, however, he were to travel either to the north or south upon the planet, the side of the rings which would be towards him would at first seem very narrow, owing to the foreshortening which would be produced by the direction in which he would look at it. It would, however, soon appear broader; but presently, when he might have reached such a latitude that a straight line from the inner edge of the ring c would touch the planet, that ring would graze his horizon. As he might journey to still higher latitudes, first the ring c, next the ring B, and finally the ring A would altogether disappear from his view.

This is indicated in Fig. LXXXVIII., in which it is shown that at c, in a latitude of rather less than 41°, c, the inner edge of the dusky ring, would touch the horizon, cc; while at b, in a latitude of about $51\frac{1}{4}$°, B, the inner edge of the next ring, would in like manner only just be seen. At a, in a

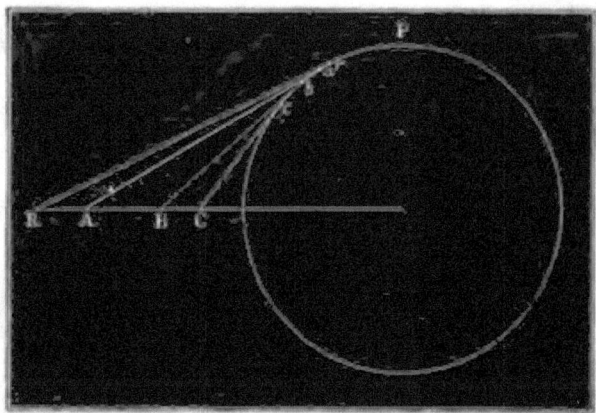

Fig. LXXXVIII.—Latitudes upon Saturn from which its various rings would be visible.

latitude of about $62\frac{1}{4}$°, A, the inner edge of the outermost ring, would disappear; and at r, in a latitude of not quite $66\frac{1}{2}$°, the whole system, whose breadth is shown by RABC, would be below the horizon, Rr, and consequently invisible; as would also be the case at all places nearer to either pole of the planet than r.

It may therefore be some satisfaction to any unfortunate inhabitants, if such there be within the arctic circles of Saturn, which are situated at about 27° from its poles (or in a latitude of 63°), that, except within some $3\frac{1}{2}$° of those circles, they have nothing to do with the shadow of the rings. The small amount of sunlight left to them during their winter is therefore not interfered with; since the shadows of the rings cannot fall upon those latitudes in winter, from which their bright side would not be seen in summer.

It is, however, a much more important fact that, at even so moderate a latitude as 36° from the equator of Saturn, the eclipses caused by the rings, when once they commence, continue through the winter solstice without any pause until they are done with, the shadow beginning to return again before its whole width has passed across that and any higher latitude. And it is, we think, upon the whole, very clear that this discussion strongly confirms our previous opinion against the habitability of Saturn. Apart from many other difficulties which the supposition would involve, it seems that, except in a comparatively very narrow zone of the equatoreal regions, the long duration of the eclipses caused by the rings would be sufficient to prevent it, unless the inhabitants were periodically to migrate north and south, and in that way avoid the winter seasons. Fascinating as the ring-system is when seen in our telescopes, we may be very thankful indeed that none such is possessed by the Earth. It appears to be replete with beauty to observers at a distance; it would inevitably be a very unpleasant obscurer of the Sun upon the planet itself.

We are then very glad that the Earth is a ringless planet. On the other hand, however, we might be very well content if our Moon, the fair queen of our nocturnal skies, had one or two companions, or even rivals in her domain. This is the case with Saturn; for of such larger satellites as circle round it, outside the minute myriads which we believe to constitute its rings, we know of *eight*.

We do not in any wise propose to discuss these Satellites with the same minuteness as those of Jupiter. Their much greater distance makes it impossible. Their eclipses, their mutual occultations and perturbations, would no doubt be very interesting if seen from the planet; but they are, if not altogether beyond our ken, certainly beyond the limits of this lecture. We must be glad if we can even get so much as a glimpse of the smaller ones amongst the eight of whose existence we are aware; while it is very likely that others may exist too small for us to see while some 750 millions of miles, or more, separate us from them.

The gradual discovery of the eight Satellites of Saturn has undoubtedly been one of the greatest triumphs that have rewarded the never-ceasing improvement of the telescope. The first was found by Huyghens in 1655; four more by Cassini between 1671 and 1684; two by Sir W. Herschel in 1789; and the 8th, Hyperion, in 1848, by Messrs. Bond and Lassell, the former noticing it in America on September 16th as a small star, the latter in England on September 18th quite independently, but both perceiving it to be a satellite on September 19th, 1848. The duplicate discovery in America and in England was perfectly independent. At that time, there was no Atlantic telegraph to flash the news of such a triumph across the ocean.

Titan, the first satellite detected, is much the largest, and is easily seen with a telescope of very moderate power; but some of the lesser ones—such as Mimas, which is also the nearest to the planet; and Hyperion, which, although farther away, is perhaps the smallest of all—are exceedingly difficult to observe except with the very largest telescopes. They may most easily be watched when the ring is nearly invisible, since its light does not then interfere with the observer's view.

Any estimate of their size is quite hypothetical. It has, however, been conjectured that Hyperion is only about 800 miles in diameter, and that the diameters of some of the others may perhaps measure 1,000 to 1,500 miles, while that of Titan may probably amount to 3,000, or even to 4,000, miles.

Their names, and the dates of their discovery, as also their periods of rotation round Saturn, and their distances from its centre, their primary, are approximately given below:—

Name.	Discoverer.	Date.	Distance from Saturn's Centre.	Period of Rotation round Saturn in Terrestrial Time.
Mimas.	Sir W. Herschel.	1789	about 117,000 miles.	about 0^d 22^h 37^m
Enceladus.	,,	,,	,, 151,000 ,,	,, 1 8 53
Tethys.	J. D. Cassini.	1684	,, 186,000 ,,	,, 1 21 18
Dione.	,,	,,	,, 238,000 ,,	,, 2 17 41
Rhea.	,,	1672	,, 332,000 ,,	,, 4 12 25
Titan.	Huyghens.	1655	,, 771,000 ,,	,, 15 22 41
Hyperion.	W. Bond and Lassell.	1848	,, 934,000 ,,	,, 21 7 28
Iapetus.	J. D. Cassini.	1671	,, 2,224,000 ,,	,, 79 7 54

It is especially worthy of notice, in the above table, that the period of Mimas is almost exactly one-half of that of Tethys; and that of Enceladus almost exactly one-half of that of Dione. Professor D. Kirkwood has also pointed out (see *The Observatory*, vol. i., p. 199) that the sum of five times the mean motion of the 1st (or innermost) satellite, added to that of the 3rd, and to four times that of the 4th, appears to be exactly equal to ten times that of the 2nd,—a relation which may be compared with that discovered by Laplace in the case of the three innermost of Jupiter's Moons (see Lecture XIII., p. 334).

It may also be interesting to notice that Mimas, the nearest of all the eight satellites, revolves at a distance of only about 34,000 miles from the outer boundary of the rings; and that the above table shows, that the distances of the first four, or five, satellites are apparently arranged upon a different system from that of the others. There is a certain regularity in the increase of distance for the first four, but a much larger interval between the 4th and 5th; while the gap between Hyperion and Iapetus is again exceedingly wide in comparison with that between Hyperion and Titan. Certainly no such regularity is apparent in the distances of these satellites as in those of the four moons of Jupiter, which it will be remembered are nearly as the numbers 6, 9, 15, 27, in which series each successive increase is about double of the preceding.*

---

* The following are the only attempts with which we are acquainted to find any series at all resembling Bode's so-called Law of Planetary Distances, in the case of the distances of the Saturnian satellites. The former is quoted from Mr. Proctor's "Saturn," p. 63; the latter from the *English Mechanic*, vol. xxiii., p. 198.

The distances of the satellites from the centre of the planet are approximately as the numbers—

		4	5·13	6·36	8·14	11·37	26·36	31·88	76·60
Now if we take		4	4	4	4	4	4	4	4
and add		0	1	2	4	8	16	32	64
we obtain		4	5	6	8	12	20	36	68

most of which are in very fair accordance with the real values.

Or we may take		1	2	3	4	5	6	7	8
add		1	2	4	8	16	32	64	128
add		6	6	6	6	6	6	6	6
and we obtain		8	10	13	18	27	44	77	142
the halves of which		4	5	6½	9	13½	22	38½	71

also resemble the true ratios.

We are almost tempted to wonder whether it can be that, from some cause related to that which originated the rings of the Saturnian system, several satellites were also formed in a somewhat regular order of distance near to the outer boundary of the rings, while the same cause permitted others further away to be differently arranged. In any case, we cannot but be struck with the analogy of the position occupied in the Saturnian system by the largest satellite, Titan, and that of the planet Jupiter in the Solar System. As four planets, or perhaps five, or even more (if we admit the existence of one, or several, intra-mercurial planets), are found around the Sun, succeeded by a gap, in which only the minute forms of the Minor Planets are seen, after which the huge globe of Jupiter appears, with other large planets located beyond its orbit at more irregular distances, —so there are five satellites round Saturn, and then a gap, after which we come to Titan, which may well be termed huge, and then again to one so much smaller that we wonder whether it has any companions in the wide region between Titan and Iapetus. Indeed, we are almost inclined to ask :—May there not, perhaps, be a multitude of little moons scattered through this vast and apparently vacant space, as there is of Minor Planets between Mars and Jupiter?

We see, in each case, a great central aggregation of matter in the Sun or Saturn, then several smaller aggregations, and, at a certain distance, one much smaller than the central one, but much larger than all the others. We also find that the Minor Planets and Hyperion, the most minute of all, are respectively found next to the largest, Jupiter and Titan. Can it be that there is some law, or some effective process, of which we are at present ignorant, connected with the condensation of a nebula into a central body with others revolving round it, whether it be into a Sun and its planets, or into a planet and its moons, which would explain all this?

Thin as the ring-system is, it would when seen edgewise from places upon, or very near to, the planet's equator, almost always conceal the seven innermost satellites, since they revolve in, or very nearly in, its plane. From other parts of Saturn they would never be hidden by the rings. But just as we have shown, in Fig. LXXXVIII., that it is impossible to see the rings from any place situated at more than a certain distance north or south of the planet's equator; so it may be calculated that, in order to see Mimas, the nearest satellite, it would be necessary to descend about $17\frac{1}{2}°$ from either pole; and $13\frac{1}{2}°$, and $10\frac{1}{2}°$, respectively, in order to see Enceladus and Tethys.

The satellites would, of course, go through their various phases in a manner similar to our Moon, only much more rapidly. When in the phase of Full Moon they would also, at times, pass through the shadow of the planet, and be eclipsed. Indeed, for a considerable period on each side of an equinox an eclipse of a full moon would be the rule, except in the case of Iapetus, which, owing to its greater distance from Saturn, would almost always escape. But the eclipses would not continue to be so frequent during the whole of Saturn's year as in the case of Jupiter's satellites, because the equator of Saturn, and the orbits of its moons, are so much more inclined to the plane of its path. As a solstice approaches it will be remembered that we have already stated, that the outer portion of the rings escapes the shadow of the planet, owing to the tilt of their path. Consequently the satellites, whose orbits are equally tilted, so that they would appear, as seen from the Sun, to describe curves similar to the outer boundary of the rings, only farther away from Saturn, would still more easily escape the shadow; the outer almost invariably, the innermost for about two years both before and after, as well as during, the time in which the shadow falls short of the more distant portion of the ring.

We have explained (see Lecture IV., p. 91) that the Earth's Moon really looks $\frac{1}{16}$th wider in diameter when seen in the zenith than when seen in the horizon; and we have shown that a similar, but more striking, effect of the same kind takes place in the case of those of Mars (see Lecture X., p. 257). Those of Saturn, owing to their proximity to the planet, would be so especially near to an observer when seen close to the zenith, that the alteration of their apparent size would be very important. The nearest, for instance, would look about twice as large in area, while even the third would appear half as large again, when seen in the zenith, as when seen in the horizon.

Nevertheless the total amount of light received from all, even under the most favourable circumstances, would be very small. The Solar light which they reflect to Saturn, being of about $\frac{1}{90}$th of the intensity of that which we receive, it is not difficult,—if we assign a possible diameter of about 1,000 miles for each of the first two satellites; 1,500 miles for each of the two next; 2,000 miles for Rhea; 4,000 miles for Titan; about 3,000 miles for Iapetus, and perhaps 800 for Hyperion;*—to calculate that the total light which they would reflect, even if it were possible for them all to be seen at the same time in their full phase, would not amount to $\frac{1}{15}$th of that which we receive from the moon when full; while the light of the outer ones would be exceedingly faint. For instance, with such a diameter as we have stated, Titan, at the distance belonging to its orbit, would only appear to have a disc of about $\frac{9}{25}$ths of the area of that of our Moon, if seen in the zenith of an observer, and would therefore give about $\frac{9}{25}$ of $\frac{1}{90}$th, or only about $\frac{1}{250}$th of its light. The nearest, Mimas, might, in a similar position, have an apparent disc about double that of our Moon, and give $\frac{1}{45}$th of its light; but, when seen upon the horizon, it would only give about $\frac{1}{90}$th.

Another point which deserves attention is the remarkable smallness of the distances of the Satellites, in comparison with the size of Saturn itself. Three of them are actually nearer to Saturn's centre than the Moon is to that of the Earth, and the furthest is only about ten times as far away. But Saturn's equatoreal diameter is more than nine times that of the Earth. The outermost is therefore proportionally at a not very different distance from that of our Moon, while the innermost is, comparatively speaking, at such a distance as the Earth's Moon would have, if it were removed less than 13,000 miles, or about $3\frac{1}{4}$ radii of the Earth from its centre; in which case, when overhead, it would only be 9,000 miles distant from us. The inner of the two satellites of Mars is no doubt proportionally nearer still, its distance being only about $2\frac{3}{4}$ radii of the planet. But,

* See Mr. Proctor's "Saturn," p. 184.

in comparison with the diameter of their primary, the nearest of Jupiter's moons is nearly twice as far away as the nearest of Saturn's.

From the nearest satellite, Saturn's disc would present a magnificent spectacle, its diameter subtending an angle of about $37°$. Its disc would, therefore, appear to be about 5,000 times as large as that of our Moon, or nearly 400 times as large as that which the Earth would present if seen from the Moon. From the furthest satellite Saturn would appear of about $3\frac{5}{8}$ times the width, and of about 13 times the area, of our Moon. From the 6th its area would be more than 100 times, and from the 7th somewhat less than 80 times, that of the Moon. It would be seen to pass through phases similar to those observed from the rings, except that instead of the one-half (see p. 376), which would in general be seen from them, the whole of the illuminated phase would be visible, a comparatively narrow zone of the planet being all that would be hidden by the ring-system.

An observer upon the satellites would look at the rings nearly edgewise, and see them as a line of light,* widest in its middle portion where it would be nearest to him, but everywhere very narrow. At the times when Saturn's disc would be seen in its full phase, the full extent of the line of rings upon each side would also be visible; at other times only a portion of their circumference corresponding to the extent of the planet's phase, although a faint additional effect might be produced by any part which might rise above the general level of the rest. Were it not for the thinness of the rings their appearance would be very beautiful, as they would, when viewed from the innermost satellite, extend about half-way across the whole width of the heavens, the nearest portion of which, if they be supposed to be 100 miles in thickness, would be of about $\frac{1}{3}$rd of the apparent width of the Moon.

It would not be profitable, nor can we afford the requisite space, to discuss the eclipses of the Sun by these satellites, as

---

* The outermost satellite, owing to the greater tilt of its orbit to that of Saturn, might occasionally see a little more of the flat surface of the rings.

we did in the case of those of Jupiter. Of course such eclipses occur upon portions of the surface of the planet from time to time, but the shadows of the satellites when they are in the phase of new moon will much more often than not miss its disc, owing to the considerable tilt of their orbits to the plane of Saturn's path. If such an eclipse is occurring, sufficient telescopic power ought no doubt to show the shadow to us, as a small dark spot, travelling across the disc, similar to those which we so often see upon Jupiter. And it is very interesting to know that, in the case of Titan, this phenomenon may occasionally, although rarely, be observed. It was once seen

Fig. LXXXIX.—Saturn, December 9th, 1877, 6h. 45m. p.m. Transit of the shadow of Titan.

by Sir W. Herschel in 1789, by Gruithuisen in 1833, in 1862 by Dawes and others, and again on November 7th and December 9th, 1877. Also on Christmas Day, 1877, by Lord Lindsay at Dun Echt (see *Monthly Notices*, Royal Astronomical Society, vol. xxxviii., p. 100), and by the writer of this lecture, who made a drawing at the time which agreed very closely with the more accurate observations taken at Dun Echt.* The transit of December 9th, 1877, as seen with an 8¼ in. Browning reflector, by Mr. J. Rand Capron, is shown in Fig. LXXXIX.†

* It is also worthy of mention that upon one occasion, viz., in 1692, a star was seen by Cassini to be occulted by Titan; and that in 1862 Dawes saw an eclipse of Titan by its immersion in shadow of Saturn.

† This interesting sketch is extracted, by the kind permission of the Astronomer Royal, from *The Observatory*, vol. i., p. 289.

Iapetus, the fifth satellite in order of discovery, but the furthest from Saturn, shows a considerable variability in its light. This was especially remarked by J. D. Cassini, who discovered it, and subsequently by Sir Wm. Herschel, and may be supposed to indicate an axial rotation, or a variation in the reflective power of different portions of its surface, similar to that which has been suspected in the case of some of the satellites of Jupiter.

And here we must leave this most fascinating system. It is sad that we know so little of its beauties and its mysteries. We look forward, however, with the utmost interest and hopefulness to the observations of the rings that may be made during the next few years, while they will appear to be very widely opened out. Since they were last so seen, telescopes of greatly improved power and precision have been constructed in the observatories of Washington, Vienna, Ealing, and elsewhere. We know not how soon we may be startled by some surprising discovery, some new revelation, which shall show us why Saturn is girdled with rings such as no other planet possesses. We do not in the least anticipate the occurrence of any violent catastrophe by which the rings will be suddenly destroyed, but we may ere long be taught whether they are to be considered absolutely permanent, or whether they are undergoing some slow process of change, which, in course of time, may aggregate their constituent parts to form additional satellites of the planet, or gradually draw them down upon it, to add a slight increase to its bulk.

The Saturnian system seems to be unique; let us hope that its teaching with regard to itself, and the Solar System as a whole, may also, in course of time, be unique in interest and instruction.

LECTURE XV.

THE PLANETS URANUS AND NEPTUNE.

> "When the planets,
> In evil mixture to disorder wander,
> What plagues, and what portents! what mutiny!
> What raging of the sea! shaking of earth!
> Commotion in the winds! frights, changes, horrors!"
>
> *Troilus and Cressida.*

WE must now take another gigantic stride, more than twice as great as that which carried us from the orbit of Jupiter to that of Saturn ; for our next step onwards, over the space that intervenes between those of Saturn and Uranus, measures almost exactly 900,000,000 miles. It brings us to a planet whose distance from the Earth is therefore always so enormous, that it is with difficulty that we can discover any points of interest connected with it. Nor are we surprised that until March 13th in the year 1781 its existence was altogether unknown; or that, even then, when Sir W. Herschel unexpectedly detected it, he thought that what he saw was a distant comet moving amongst the stars visible in the field of view of his telescope.

And here we may draw special attention to one very interesting and instructive fact involved in the history of the discovery of the planet Uranus, viz., that a certain peculiarity in its appearance indicated to the experienced eye of that great astronomer, within (as he himself has stated) a minute of his first seeing it, that it was no ordinary star. There was a haziness, and a comparative faintness about its light, which he at once noticed. This led him to apply a higher magnifying power to his instrument, in order to see what effect would be thereby produced. Sir W. Herschel was well aware that such a higher power would not diminish the brightness of the image of a star, in which all the light received is as nearly as possible condensed by the lenses of the telescope to a point. He knew that, whatever magnifying power may be employed, the stars are so far distant that their images afford no true disc capable of being enlarged; while, on the contrary, the apparent disc of a planet, or of a comet, increases, and its apparent brightness in a like degree decreases, with every increase in the magnifying power used. When Sir W. Herschel tried the effect of a higher power upon the object which he saw, his suspicion that it was not stellar was immediately confirmed. He therefore continued to watch it until he was assured that it had a proper motion of its own amongst the stars, and he then announced his discovery to the scientific world as that of a comet.

But Dr. Maskelyne and other English astronomers, the President de Saron in France, and especially Lexell, the astronomer of Finland (who is celebrated for his investigations connected with a remarkable comet discovered in 1770, and who happened to be visiting England at the time of the discovery of Uranus), soon perceived that a nearly circular orbit would probably best satisfy the observations of its path. Nor was it long before an approximate value of the radius of its orbit was calculated. It was also found to possess an ordinary planetary disc.

We need not recount the various discussions that took place as to the name to be given to this new brother amongst the planets, for which at first the unsuitable appellation of *Georgium Sidus* was proposed, in honour of King George III. It was also suggested that it should be named after Herschel himself. Finally, however, the more appropriate name of Uranus, suggested by Bode, was chosen, in agreement with the system of mythological names already belonging to the other planets; while the astronomical symbol ⛢ was also adopted for it, the principal portion of which worthily commemorates the initial of its discoverer's name.

We may remark in passing that, although it may in one sense be said, that Sir W. Herschel discovered the planet by chance, its discovery was, in another sense, the just reward of many years of most arduous labour. When we remember how vastly he improved by his own inventive skill and manual dexterity the powers of the telescope; and that, year after year, he investigated, in the most regular and persevering manner, vast regions of the sky,—or, to use his own forcible expression, *swept the heavens*,—moving his mighty telescope from time to time, so that each successive strip of space which it fathomed might begin exactly where the preceding strip ended, and no object of interest be missed ; when we note that, at the very time at which he discovered the planet, he was engaged in a most careful series of such observations of the particular region in which it was found, with a view to select certain special stars whose distances from the Earth he might be able to determine; and when we read how skilfully he at once noted and tested the peculiarity of its appearance,—we must acknowledge that the honour of achieving so important a discovery was never more fully merited. We rejoice to think that the recent occurrence of its centenary, in the year 1881, has led Professor Holden to publish a most interesting account of the great astronomer's life, which, if possible, has caused his worth and talents to be even more highly appreciated than was previously the case.

Indeed, the credit due to Sir W. Herschel becomes still more apparent when we notice, as subsequent investigations have proved, that various astronomers before his time, who were by no means devoid of skill and ability, had frequently observed this very planet without detecting its planetary character, and had noted its positions as those of a star. The earliest of such records is one by Flamsteed in December 1690 ; while from that date to the year 1771, the number of observations amounted altogether to six by Flamsteed, three by Bradley, one by Mayer, and twelve by Lemonnier; six of which last were in one month, viz., January 1769. It has been well remarked that it was only a want of order and method that prevented Lemonnier from securing the discovery as his own.*
It is also surprising that Flamsteed did not make it.

* Chambers' "Handbook of Descriptive Astronomy." p. 159.

These previous positions of the planet were of the utmost use. They at once gave astronomers much information as to its orbit and movements, in order to obtain which it would otherwise have been necessary to wait for a long series of subsequent observations; and, as we shall presently see, they happened to have been made in an especially important portion of the planet's orbit.

As soon as that orbit was even roughly calculated, it was observed that its mean distance from the Sun was rather more than double that of the orbit of Saturn, and in very remarkable accordance with the distance that would be given by Bode's so-called law.

We may remind our readers* that Bode's law suggests for Saturn's distance the number 100, if the Earth's distance be represented by 10, while the next term in the series is 196. Now it is found that the actual distances of Saturn and Uranus are respectively very nearly as 95 and 192. That of Uranus is therefore somewhat less than Bode's series would require, but only by about $\frac{1}{30}$th part; the actual difference, however, owing to the enormous scale of its orbit, being about 36,000,000 miles.

The eccentricity, or degree of ovalness, of the orbit of Uranus is found to be about three times that of the Earth's orbit, although less than those of all the other planets except Venus and Neptune. Its mean, or average distance, which is about 1,785,000,000 miles, consequently increases, by rather more than a $\frac{1}{20}$th part, to about 1,868,000,000 miles when Uranus is at its furthest from the Sun, or in *Aphelion;* and in like manner decreases to about 1,702,000,000 miles when the planet is in *Perihelion,* or at its nearest to the Sun.

It may be observed that the difference between these two last-named distances, viz., 166,000,000 of miles, is between four and five times as great as the amount by which the mean distance falls short of that which Bode's law suggests. The agreement with that law may therefore be considered remarkably close, although it certainly requires a little familiarity with the huge numbers involved before we can bring ourselves

* See Lecture V., page 123.

THE PLANETS URANUS AND NEPTUNE.

to speak of a difference of 36,000,000 miles as a comparatively trifling matter.

In an orbit at such a distance from the Sun, the period of Uranus is necessarily very long; in fact, it occupies rather more than eighty-four years in performing its circuit once round it.

Its onward velocity in its orbit is not much more than $\frac{1}{5}$th of that of the Earth, but is nevertheless sufficient to carry it forward at the rate of some 15,000 miles per hour; *i.e.*, at rather more than four miles per second.

When due allowance is made for its enormous distance, it is found that the apparent diameter of its globe, which is equal to about four seconds of angular measure, indicates a real diameter of about 32,000 miles. Its surface is therefore between sixteen and seventeen times as great as that of the Earth, and its volume about sixty-six times. Its volume is nevertheless only about $\frac{1}{25}$th of that of the apparent globe of Jupiter, and about $\frac{1}{11}$th of that of Saturn.

When its weight is calculated from the observed movements of its satellites, it is found that it amounts to about fourteen times that of the Earth. The above-mentioned volume consequently involves a value for the mean density of the planet equal to about $1\frac{1}{4}$ times that of water, or between $\frac{1}{4}$th and $\frac{1}{5}$th of that of the Earth. In other words, if the materials of which Uranus is made were of the same average density as those of the Earth, the planet would weigh between four and five times as much as it does.

The attraction of gravity upon its surface, as the result of its lighter density but larger size, is found to be somewhat less than upon the Earth's surface, but only to such an extent, that a weight of one pound transferred from the Earth to Uranus would be reduced to about $14\frac{1}{2}$ ounces. It is, perhaps, rather a curious coincidence that, upon a planet so different in size and constitution from the Earth, weights would be so little altered.

Owing to its great distance from us it is difficult to decide as to the existence, or otherwise, of any ellipticity in the shape of the globe of Uranus. In 1842 and 1843 Mädler, however, made measurements which indicated that it was compressed at its poles in about the same degree as Saturn, but subsequent observations have not supported his conclusion, nor have they revealed any other decided physical features. Some traces of a belt were at one time supposed to be seen upon it, but this was exceedingly doubtful; while the suspicion of the existence of a ring surrounding it, similar to those of Saturn, has been quite given up.

In 1870 and 1872 Mr. Buffham, observing in Bonner's Road, Victoria Park, with a 9-inch mirror by With, thought that he detected a rotation of the planet upon its axis in about 12 hours, from the motion of certain spots, or portions of a belt, upon its surface; but since that date, so far as we are aware, no other observer has confirmed his statement.

It is useless to discuss the habitability of Uranus, and very hard to conceive its possibility. We may, however, remark that, if any inhabitants could exist upon it, they might, at intervals of nearly fifteen years, see Saturn at its greatest elongation from the Sun as a morning or evening star. But, even when so seen, its light would only amount to about $\frac{1}{12}$th part* of that which it exhibits to the Earth when most favourably situated in Opposition.

On the other hand, Neptune would be far better seen when near to Opposition than from the Earth, and might shine with seven times as much light. It is just possible that Jupiter might be detected by a very careful scrutiny, even as Mercury is with difficulty seen by us; but Mars and the Earth, and all the other planets, would always be much too near to the Sun to be visible.

So remote is the planet Uranus that the intensity of light and heat which it receives from the Sun is only equal to about $\frac{1}{368}$th of that which the Earth enjoys. That same remoteness

---

* Saturn, in the two cases referred to, would be at distances from Uranus and the Earth respectively, of which the former would be fully 2¼ times as great as the latter. In the former case its phase would also be only that of a half-moon, in place of the fully illuminated disc which it shows to the Earth when in Opposition. The increase of distance would diminish its light in the ratio of the square of that distance, or between five and six times; and the half-moon phase would further reduce it by one-half; or to about $\frac{1}{12}$th of that which we see under the most favourable circumstances.

also involves that its distance from the Earth is never less than about 1,610 millions of miles. And yet, at so great a distance, four Satellites of Uranus have been discovered which have been so carefully followed in their orbits that their ever-changing places can be regularly predicted.

Of these Satellites Sir W. Herschel discovered two in January 1787. He also imagined that he obtained glimpses of four others, but his observations were not, in any wise, so precise and accurate as were those of the first two which he detected, nor have they received subsequent confirmation. It is now generally considered that only four satellites of Uranus are visible in our largest telescopes: two of which were found by Sir W. Herschel, and two subsequently by Mr. Lassell, in 1851,* both of which latter are nearer to the planet than either of Herschel's, so that their periods in their orbits are very short; viz., about $2\frac{1}{2}$ and 4 days respectively. The periods of the outer two have, by subsequent observation, been found to accord very closely with the values of about $8\frac{3}{4}$ days, and $13\frac{1}{2}$ days, first announced by Herschel.

To these four Satellites the names stated in the following table have been assigned.

Satellites of Uranus.	Periodic Times.	Distances from Uranus.
Ariel	2 days $12\frac{1}{2}$ hours.	about 120,000 miles.
Umbriel	4 ,, $3\frac{1}{2}$,,	,, 170,000 ,,
Titania	8 ,, 17 ,,	,, 280,000 ,,
Oberon	13 ,, 11 ,,	,, 370,000 ,,

In some text-books of astronomy, of not very recent date, a list is given of the *six* Satellites which Sir W. Herschel

---

* Some glimpses of one, or both, of these were, perhaps, obtained by Mr. Lassell, and possibly of one of them by M. Otto Struve, somewhat before the above date, but their undoubted discovery by Mr. Lassell dates from 24th October, 1851, at Starfield, near Liverpool. He announced to the Royal Astronomical Society in November of that year very approximate periods for the two new Satellites, which were only slightly corrected by the observations subsequently made in the clearer atmosphere of Malta, to which island he removed his telescope of 2 feet in diameter, in the autumn of 1852.

believed that he had seen, and of their *distances and periodic times*. A reference to his original memoir in the "Philosophical Transactions" for the year 1798 will, however, show, that he simply calculated those *periodic times* by Kepler's 2nd law, from the *supposed* distances from the planet of *four* out of the six ; and that he did not obtain them, except in the case of those now called Titania and Oberon, by watching the satellites as they performed successive revolutions. He also only fixed the *distance* of the innermost of the six by an observation of its place, when he thought (although he could not be very positive about it) that it was most probably not far from its greatest elongation. Then he somewhat vaguely suggested that the orbit of one, which he believed to revolve between Titania and Oberon, might probably be just *half-way* between their orbits; and that, of two others outside them, one might move in an orbit of *double*, and the other of *four* times the radius of that of Oberon. The periodic times calculated to suit these hypothetical distances are, of course, no more certain than the distances themselves.

We do not, in mentioning these facts, intend for an instant to derogate from the skill and accuracy of Herschel's work as an observer. We only desire to caution our readers not to attribute, as we think some writers have, an accuracy to the above statements far greater than he ever assigned to them. There is no doubt that, with an instrument of his own construction, he discovered Titania and Oberon nearly fifty years before the year 1834, when they were next seen again by Sir John Herschel [*] with a telescope of 20 feet in focal length similar to that which his illustrious father used, and nearly sixty-five years before Mr. Lassell discovered Ariel and Umbriel. Whether Sir W. Herschel, in every case, mistook some small stars for additional Satellites, or in some of his supposed observations (as Professor Holden, after a careful consideration of them, thinks to be possible, or even probable) really perceived one of Mr. Lassell's two, it matters not. The only observations of any importance made by Herschel were confined to Titania and Oberon. He himself says as plainly

---

[*] Professor Holden's "Life and Works of W. Herschel," p. 143.

as possible in his memoir, that "further observations," for which we presume that he had neither time nor opportunity, must "furnish us with proper data for more accurate determinations;" and that "the accuracy of the periods stated depended entirely upon the truth of the assumed distances."

But much the most remarkable fact relating to these satellites and their movements still remains to be mentioned. It is this, that they revolve in orbits which are so tilted (if we may be allowed the expression) that they are nearly perpendicular to the plane of the orbit of Uranus. This latter plane only differs by less than 1° of inclination from the plane of the ecliptic in which the Earth moves. In fact, the two are more nearly coincident than in the case of the orbit of any other planet; and yet we find this very anomalous position of the planes in which the satellites revolve.

It is well-known that the orbits of comets may be inclined at all possible inclinations to the ecliptic; but we have hitherto, in our description of the other planets and satellites of the Solar System, met with nothing like the case of these satellites. If the inclination of their orbits were equal to a right angle, it would technically be said to be 90°; as it is, it is believed to amount to about 82°.

It moreover follows, that, if these satellites rotate in, or nearly in, the plane of the planet's equator, as do those of Jupiter and Saturn, that equator must also be tilted in a like manner.

It must, however, be remembered that this last supposition, although it may be probable, is in no wise proved to be true. If it be so, the state of things is very nearly that which would result in the case of such a planet as Jupiter (whose equator is only inclined very slightly to the plane of the ecliptic), if we could arrange as follows. First, that an axis of its figure should project for some distance from each of its poles, and secondly that the plane of its equator should be extended, beyond the boundary of its globe, to a distance equal to that of its furthest satellite. Then, if we could take hold of the two ends of the projecting axis, and turn it through an angle of 90 degrees, so as to bring it very nearly into the plane of the ecliptic, as is indicated further on in Figs. XC. and XCI., the planet's equator and the orbits of its satellites would become nearly perpendicular to that plane.

Let us, however, for a moment or two suppose the coincidence of the planet's axis with the ecliptic to be exact, and consider the exceedingly remarkable results which would follow. They would, of course, be such as would exist upon the Earth, if its polar axis were, in like manner, in the plane of the ecliptic, instead of being inclined to it at an angle of $66\frac{1}{2}°$. It is easy to understand that the peculiarities of the polar regions would under such circumstances extend over the whole of the Earth, as would also those of the tropics; the arctic circles would be brought down to the equator, and the tropics of Cancer and Capricorn would be moved up to the poles.

At the poles the sun would remain visible as at present for six months at a time, but in the height of summer it would be vertically overhead at noon; while for three months after the spring equinox it would describe a spiral curve in the sky, first going round close to the horizon, then gradually ascending higher, and travelling day by day in smaller and smaller circles, until it would be nearly fixed in the zenith on the day of the summer solstice and for some days before and after that date. Upon one-half of the globe the days would increase from 12 hours in length at an equinox, until, at the summer solstice, the whole hemisphere would see the Sun during a day of 24 hours. Upon the other half the nights would increase in like manner, until at the winter solstice they would be 24 hours long, and the Sun would be invisible for the whole of the day throughout the hemisphere in question. At places near to the equator, when the days would be longest, the Sun would all through the day keep very close to the horizon, since it would go round the pole of the heavens, which would itself be so situated, in a very small circle; just as at places near to the poles, it would, at the same time, appear to go round the celestial pole, in a similar small circle, near to the zenith.

All this may be easily worked out, by imagining the globe of such a planet to circuit round the Sun while keeping its axis of rotation always parallel to itself and *in the plane of the ecliptic;* so that the axis in question would pass through the Sun at the solstices, while the Sun's distance from a pole of the heavens would change from zero at a solstice to 90° at an equinox. Upon this supposition, figures may easily be drawn, similar to those used for the Earth in the earlier part of Lecture VIII., which would indicate the Sun's daily path for various seasons and localities.

In such a case it can also be shown, if the satellites moved in the plane of the equator of Uranus, that they would appear, when seen from any place upon the planet, to describe a daily path in the sky which would always intersect the horizon in its east and west points. If an observer were upon the equator, they would pass through his zenith at mid-day; if he were near to a pole, they would all the day long keep close to his horizon; while they would cross his meridian at an intermediate elevation for intermediate places of observation. All this can be readily concluded from the fact that they would constantly be 90° from either pole of the heavens, the altitude of which above the horizon of any place, as in the case of the Earth (see Lecture VIII., p. 180), would be equal to the observer's latitude.

Let us, however, next consider what appearance they would present to an observer upon the Earth, if they thus rotated round Uranus in a plane at right angles to the ecliptic. It is evident that, if at any time that plane were turned edgewise to our view, they would seem simply to travel up and down in straight lines perpendicular to the ecliptic, passing alternately before and behind the planet. On the other hand, if that plane should at any time be at right angles to our view, they would seem to go round the planet in circles. At intermediate times their paths would appear as ellipses of a greater or less degree of ovalness. But in every case their motion would be in a plane whose direction would be to the north and south of the ecliptic, and always perpendicular to it. They would not have any motion at all round Uranus from east to west, or from west to east, relatively to the ecliptic.

The actual state of things in the case of the Satellites is not, however, *exactly* thus. But it really is such as would happen, if, in our imaginary procedure, by which we have suggested that such an axial position as is supposed to exist in Uranus might be brought about, we had tilted a planet's axis, NS, supposed to be originally perpendicular to the ecliptic (see Fig. XC.), through a little more than 90°; in fact, through about 98°, or to the position N_1S_1, so as to bring the end that was above, or *north* of the ecliptic, a little *below* it to N_1; and the other, or *south* end of the axis, a little *above* it to S_1.

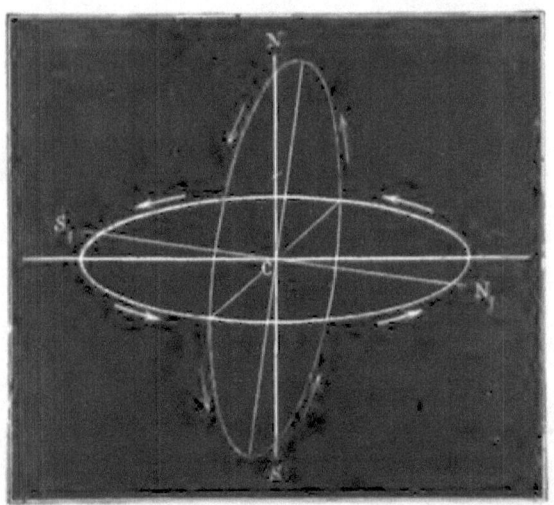

Fig. XC.—NS is a planet's axis perpendicular to the plane of the ecliptic; N_1S_1 the same axis turned through 98°. The corresponding positions of the plane of the planet's equator, and the directions of the motion of a satellite revolving in it from west to east, are also shown.

It is easy to see that, in this case, the satellites of the planet would have a small amount of motion from east to west with regard to, or (as it is technically termed) resolved upon, the ecliptic, because their orbits would not be *quite* perpendicular to it. And it would result, as a careful consideration of the arrows in Fig. XC. shows, that any such part of their motion would appear to us to be from *east* to *west*, and not from *west* to *east*; *i.e.*, its direction would be *contrary* to the general direction of the motion of other planets and satellites, owing to the *south* pole of the planet having been transferred to a position slightly to the *north* of the ecliptic. And the planet itself, if we could see spots upon its surface, and thus observe its rotation, would also appear to go round on its axis in the opposite direction to that in which the other planets turn. Indeed, we think that this may very probably prove to be actually true in the case of Uranus, if we ever detect its rotation.

Under the above-mentioned circumstances, the motion of the satellites in their orbits would of course appear to an observer *upon the planet* to be in the *same* direction as that of the rotation of Uranus upon its axis. It would matter not whether he might choose to describe it as being from west to east, or from east to west; east and west being in any such case merely relative terms. But what is more important is, that such an observer would see the Sun and the other planets and their satellites, if he could watch their axial and orbital rotations, all apparently turning and travelling in an *opposite* direction, with the single exception of the motion of the satellite of Neptune in its orbit, and possibly of Neptune's own movement upon its axis.

It should, however, be noticed that we might have equally well supposed the axis of the planet to have been brought nearly into the ecliptic, by turning it from the position NS, in which it is perpendicular to it, through a little less than 90°, or, to take the actual case of Uranus, through about 82°, in the *opposite* direction, so that it would take up the position $N_2 S_2$ in Fig. XCI. But if so, the north pole, N_1, originally above the ecliptic, would *remain above it* in the position N_2, and it would be necessary that the satellites should have revolved originally, not from *west* to *east*, according to our use of these terms, but from *east* to *west*, in order that after the change of the direction of the planet's axis they might travel as those of Uranus are seen to do. We therefore think that it is, upon the whole, better to describe what takes place as if it had occurred upon the other supposition which is shown in Fig. XC.

The above statements, in which we have endeavoured, as far as possible, to explain what is undoubtedly a rather complicated matter, may perhaps help some of our readers to understand what is meant in ordinary text-books of astronomy, when it is said that, for *direct* motion of the satellites of Uranus, we must consider the plane of their orbits to be inclined at about 98° to the ecliptic, that is at rather *more* than a right angle; and for *retrograde* motion at about 82°, that is at rather *less* than a right angle.

After all it does not, however, very much matter whether we take it, that the actual amount of retrograde (or east to west) motion relatively to the ecliptic, which we observe the

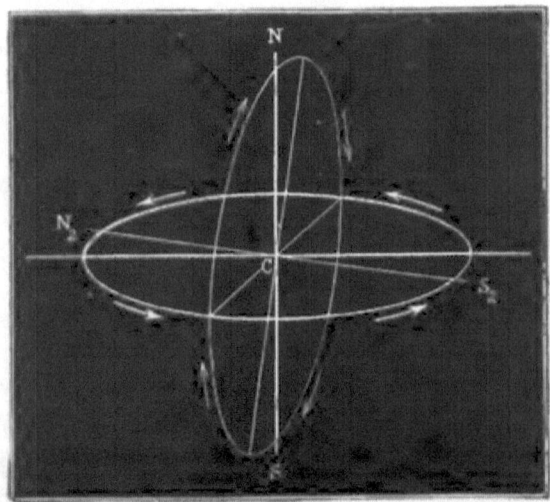

Fig. XCI.—NS is a planet's axis perpendicular to the plane of the ecliptic; $N_2 S_2$ the same axis turned through 82°. The corresponding positions of the plane of the planet's equator, and the directions of the motion of a satellite revolving in it from west to east, are also shown.

satellites to possess, is such (in agreement with the former of our two suppositions) as would be produced, if the *south* pole of Uranus had been brought up a little *above* the plane of the ecliptic, while the satellites *originally* travelled round it, as others in general do, from west to east; or whether it be said, (in agreement with the other supposition) that they move as they would, if the *north* pole of the planet had been brought *down* nearly to the ecliptic, while they may have *originally* gone round in the reverse direction. For, in either case, the great peculiarity really is, that the orbits of the satellites are so tilted up, that there is at all times hardly any motion in them either from *west* to *east*, or from *east* to *west*, round a perpendicular to the plane of the ecliptic. In the same way it would probably not be so much a matter of astonishment to an observer upon the planet, if other neighbouring planets, such as Jupiter and Saturn and their satellites, were seen to have a different direction of rotation, as it would be to find, that their axes, instead of being nearly in the plane of the ecliptic, were nearly perpendicular to it, and the orbits of their Satellites nearly in it.

And here we may anticipate one portion of what we should otherwise state somewhat further on in this lecture with regard to the satellite of the still more distant planet Neptune, viz., that its motion relatively to the ecliptic is also apparently retrograde; and not only so, but in a much more decided degree, for the inclination of the plane of its orbit to that of the ecliptic is only about 35°. The orbit is therefore far less nearly perpendicular to that plane than those of the moons of Uranus; and the satellite of Neptune consequently possesses a very considerable amount of apparent motion, as it journeys round the planet, which is in a direction the reverse of that of the general movements of the Solar System. It therefore follows from our previous explanation that;—if we are to suppose the planet Neptune to have originally revolved upon its axis from west to east, and its satellite to have revolved around it in the same direction, as is certainly the case for all the other planets and their satellites with the exception of Uranus, and if the satellite also moves nearly in the plane of Neptune's equator; —what was originally the north pole of the planet *above* the ecliptic must have been depressed not only to a small angle, but to something like 55°, *below* it.

The probable position of the axes of these planets is consequently anomalous. It seems very difficult to conceive how it could have been brought about. And yet, to imagine, that in some way the end of each axis primarily above the ecliptic has been brought below it, as in Fig. XC., and the plane of the planet's equator and of the orbits of its moons turned through a corresponding angle, appears to be the best way to get over the difficulty (in which we should otherwise be involved) of supposing the orbital motion of the satellites to have been originally reverse to that obtaining in the rest of the Solar System, and possibly reverse to that of the axial rotation of the planets themselves. We certainly hesitate to believe that any such reverse motion can have existed in the early history of our System, not only upon the grounds of symmetry and analogy, but, if for no other reason, because it would involve in the greatest possible difficulties the important Nebular Theory of Laplace, which necessarily assumes all the original motions to have been in one and the same direction. It is, however, almost equally hard to conceive by what possible cause so vast a change in the supposed direction of the axes of the planets, as we have suggested, could have been brought about.*

So little is known about Uranus, that there is only one other fact relating to it to which we wish to draw attention. We refer to an interesting comparison between its light and that of one of the satellites of Jupiter, which is to be found in Flammarion's "Popular Astronomy" (page 573). He mentions

* It must, however, be remembered, as we have already mentioned, that we have no positive proof as to the position of the plane of the equator of Uranus, inasmuch as we cannot detect its axial rotation. It is only upon the ground of a probable analogy with Jupiter and Saturn that it is suggested, that the plane of its equator may nearly coincide with the orbits of its satellites. If, however, this is not the case, we believe that it would make a very great difference to observers upon Uranus, whether the axis of the planet may have a greater, or less, inclination than that of the orbits of the satellites to the ecliptic, which is most probably about 82°. Those of our readers who are acquainted with mathematics may be able to see that, if its inclination be less, and there be a precession of the equinoxes upon Uranus, then, during alternate halves of the period of that precession, the apparent orbital motion of the satellites, as regards rotation from east to west, or *vice versâ*, round the axis of Uranus would be reversed. A similar change would also occur during alternate halves of the period of the revolution of the nodes of the orbit of any one of the satellites. We do not notice any such effects in our own Moon's apparent orbital path among the stars, because the inclination of the Earth's axis to the plane of the ecliptic, which is 66½°, is *greater* than the inclination of the plane of the Moon's path (which is in fact only about 5°).

that, on June 5th, 1872, the two planets might have been seen at an apparent distance only equal to 1½ times the diameter of Jupiter, had not daylight prevented the observation. But, at 9 p.m., Jupiter and its four satellites, together with Uranus, were beautifully grouped in the field of view of the telescope; Uranus being situated almost exactly above the 3rd or largest satellite. The two were apparently of the same size, but the planet was rather more brilliant in its light. It therefore follows that the light of Uranus (at that time) slightly exceeded that of a 6th magnitude star, and that it is, as observation has often proved to be the case, under such circumstances just perceptible by the naked eye.

The Planet Neptune.

We have already more than once mentioned the planet Neptune as the most remote from the Sun of all those with which we are acquainted. We have spoken of its orbit as lying very far beyond that of Uranus; but we have not as yet referred to the very remarkable character of the investigations by which it was discovered in the year 1846.

That discovery, which we now proceed to explain, is undoubtedly the greatest modern triumph of the mathematics of astronomy; a triumph which makes us feel how much more appropriate the word "*now*" would be, if substituted for the twice recurring "*nor*," in the lines of Prior:—

> "Each planet shining in his proper sphere,
> Doth with just speed his radiant voyage steer;
> And in his passage through the liquid space,
> *Nor* hastens, *nor* retards, his neighbour's race."

Most of our readers may already be aware, that this last addition to the known members of the Sun's family was detected in its far distant path by no mere chance or accident. There was nothing unexpected in the first observation of it, as in the case of Uranus. On the contrary, two great mathematicians who, in common with the rest of the scientific world, had good reasons for believing that such a planet must exist, calculated, by long and arduous processes, in what direction to point the telescope by means of which it should be seen.

They were Mr. Adams (now Lowndean Professor of Astronomy in the University of Cambridge), and M. Le Verrier* (late Director of the Observatory of Paris), whose comparatively recent death is mourned as that of one who, by his untiring assiduity and his splendid ability, apart from the discovery of the planet Neptune, has conferred benefits almost beyond all estimation upon the science he so much loved. The honour of the discovery is so great that it can well afford to be divided between these two distinguished astronomers.

The planet was practically found, and its place was pointed out, within a degree of longitude, by Le Verrier, and within less than $2\frac{1}{2}$ degrees by Mr. Adams, simply and solely as the result of theoretical calculations of the most elaborate character. But the difficulty of the problem thus solved was immense. Some idea of it may perhaps be gathered from the following statement.

When the observations made of the planet Uranus during some fifty years after its discovery in 1781 were compared with the previous observations of it, to which we have referred in the earlier part of this lecture, *i.e.*, with the records of its place, which were found to have been made at various times without its real character being known, it was noticed, that there was apparently a considerable amount of irregularity in its past movements; and that this irregularity showed a tendency to increase again. For a considerable number of years before the year 1822, and especially after the year 1800, the planet seemed to have gained speed in addition to that which would properly correspond with its distance from the Sun; but, as the year 1822 drew nigh, the rate of increase in its speed diminished, and after that date it seemed that some retarding influence began to act upon it, the effect of which, in a few years' time, became decidedly vigorous.

Fig. XCII. illustrates the hypothesis by which it was suggested that this peculiarity in the movement of Uranus might be explained. It was thought, that an additional planet might possibly be situated in an orbit exterior to that of Uranus, the attraction of which might have acted (although the real effect

* It may be noticed that we have put the two names in alphabetical order.

is considerably more complicated) somewhat after the manner indicated by the arrows in the diagram, and thus have alternately accelerated and retarded the velocity which Uranus would otherwise have possessed.*

The problem of investigating where that planet might be was altogether an indeterminate one, and the number of equations of condition and unknown quantities involved, very numerous. It was evident that a larger planet further off, or a smaller one at a nearer distance, might equally well produce

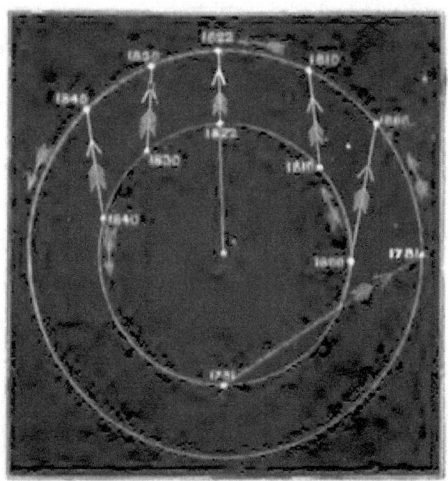

Fig. XCII.—Illustrating the perturbation of motion of Uranus by Neptune.

the observed result. Nor was it possible to make a preliminary

* In Herschel's "Outlines of Astronomy," sec. 760 *et seq.*, a full explanation of the theory of the action of Neptune upon Uranus is given. It is there shown, by a beautiful and lucid geometrical process, that the direct, or tangential, effect of Neptune upon the velocity of Uranus, which is all that Fig. XCII. indicates, would be of comparatively little importance except for a moderate number of years on each side of the date of its opposition in 1822, near to which time its effect would first experience a rapid increase, followed by an equally rapid decrease. It is, moreover, proved that Neptune would also exert another effect normal to the path of Uranus, by which it would draw it from time to time nearer to, or farther from, the Sun, and thus *indirectly* alter its velocity, by changing the power of the Sun's attraction upon it. This latter effect is, however, much slower in action, less important, and more difficult of explanation than the other.

calculation of the disturbing effects of the various other planets upon the path of Uranus (or rather of those of Saturn and Jupiter, which would be by far the most important), and then to say, that any *outstanding* irregularity must be caused by an exterior planet.

If this could have been done, it might have been a comparatively easy matter to calculate at any given moment, from the extent of that irregularity, where, and how large, such a planet must be. But the only way in which mathematical processes at present enable us to investigate the positions and movements of the planets, is by a method of successive approximation, which involves the effect of any one upon all the rest, and of all the rest upon it, in a sort of inextricable intimacy, such that it is impossible to separate any one of these effects from the remainder. The planets must all be supposed in our calculations to be acting and reacting upon the movements of each and all, at one and the same time, and even to be mutually acting and reacting upon one another's disturbing effects.

It was, however, only natural that both Le Verrier and Adams should imagine, from the analogy of the known planets, that, if another large one existed still more remote from the Sun, the size of its orbit would approximately follow Bode's law, and that it would, like the rest, most probably move nearly in the plane of the ecliptic. Having made these suppositions, they next determined, by very complicated calculations, the probable position of such a planet at the time at which their investigations were made, *i.e.*, in the year 1846. And so nearly did their results indicate its true place, that it was found, by Dr. Galle, of Berlin, the first night that he looked for it, in the locality assigned by Le Verrier; while it was afterwards shown, that the calculations of Adams were amply sufficient to have detected it with almost equal facility. In fact, even before Dr. Galle's announcement of Le Verrier's discovery, it had been twice seen at the Observatory at Cambridge, in the neighbourhood in which Mr. Adams had requested that search might be made for it. For want, however, of sufficiently accurate charts, it was for the time being recorded as a star (and not as a planet) in two different positions, which, as soon as time permitted, would by their comparison have demonstrated its onward movement, and consequently have proved its real character.*

We think we are indeed justified in calling this a triumph. Perhaps it may be some day repeated, and irregularities in Neptune's orbit be used to deduce the place of a yet more distant planet. The movements of Neptune have indeed been already watched with this object in view, but any attempt to solve the problem is at present premature. The investigation would, of course, be much more difficult than it was in the case of Uranus. Nor are we, at present, able to calculate the orbit of Neptune itself with such accuracy that we can be perfectly certain of the extent to which irregularities occur in its motion; although we are aware that the attractions of some of the other planets do undoubtedly, and very considerably, perturb its path.

One striking illustration of such perturbations of its movement, first brought under our notice by the kindness of the Astronomer Royal (Mr. Christie), may be both interesting and instructive. It is, that, at the end of 1881, Neptune was about 500,000 miles further from the Sun than it was in 1876, or than it probably will be in 1887, or near to that date.

Now the year 1881 is that in which, if Neptune's motion were undisturbed, it would pass through its Perihelion, or make its nearest approach to the Sun; but the result of the above perturbation (which is chiefly due to Jupiter) is, that its distance is really 500,000 miles less, about five (or six) years, both before, and after, that date. That is to say, it practically passes through two Perihelia, one of which occurred in 1876, while the other will be in, or about, 1887. This may, at any rate,

* It should also be mentioned that, just as it was found after the discovery of Uranus that it had been previously observed, and its place recorded as that of a star; so in like manner two observations of Neptune were found to have been made by Lalande, on May 8th and 10th, 1795, fifty-one years before its discovery in 1846. Indeed Lalande noticed that the places observed on these two days did not agree, but, instead of imagining that the object might be a planet whose proper motion caused the discordance, he simply rejected that of May 8th as erroneous.

make it clear, that some of these planetary perturbations are of no small moment, and help to indicate how difficult it was to determine the probable position of Neptune from the disturbance of the movements of Uranus, when it is seen how Neptune itself is perturbed.*

Far away from the influence of the Sun's genial warmth, or perhaps we should rather say, almost beyond any possibility of benefit from it, although not beyond the control of its mighty gravitating attraction, Neptune pursues its tedious journey. The observations made since its discovery prove its mean distance from the Sun to measure very nearly 2,800,000,000 miles. It is therefore about thirty times as far away as the Earth from the centre of the Solar System.

It may be well to notice that such a distance is very different from that which Bode's so-called law would involve. To correspond with this series, it should be 800,000,000 miles greater than it is ; which would also involve a periodic time for the planet's rotation round the Sun about sixty years longer than its real period, which is nearly 165 years.

Neptune's velocity as it travels round the Sun is about $3\frac{1}{3}$ miles per second. Its speed is therefore only about $\frac{1}{8}$th of that of the Earth, and $\frac{1}{3}$th of that of Mercury, but nevertheless 200 times as great as that of an express train. Its probable diameter—as to which there is, however, some doubt, since its immense distance makes it difficult accurately to measure the apparent width of its disc—is about 35,000 miles, *i.e.*, it is about $\frac{1}{11}$th greater than that of Uranus.

* It afterwards proved that the hypothetical planet employed in the calculations of Le Verrier had been supposed to be located in an orbit, the mean distance of which from the Sun was about 570,000,000 miles greater than that of Neptune, while its mass was about twice as great. In like manner, the mean distance, assumed by Mr. Adams for his planet, exceeded that of Neptune by about 675,000,000 miles, and its mass was three times the real value. Nevertheless, the effect that each would have produced upon the movement of Uranus, in the portion of its orbit which it described in about twenty years on either side of the date of the year 1822, was so nearly the same as that of Neptune itself, that in each case the result of the calculations made almost exactly indicated the true *direction* in which the unknown planet was to be found in 1846.

Its surface is not quite 20 times, its volume somewhat more than 80 times, that of the Earth.

Its density, which is slightly less than that of Uranus, and somewhat greater than that of water, is about equal to $\frac{1}{5}$th of that of the Earth. Notwithstanding its much greater volume, its weight is therefore only about 16 times the Earth's weight. The attraction of gravity upon its surface is almost exactly the same as upon that of Uranus, or about $\frac{9}{10}$ths of what it is upon the Earth.

The intensity of the Sun's light and heat received by Neptune, and the apparent area which its disc would present to an observer upon the planet, are only about $\frac{1}{900}$th of what they are for an observer upon the Earth. The Sun seen from it would therefore offer no appreciable disc to the naked eye, but would simply look like a very brilliant star.

Neptune possesses, as we have already mentioned in our discussion of the moons of Uranus, one Satellite, which was discovered by Mr. Lassell in October, 1846. Its distance from the planet's centre is about 220,000 miles; its period in its orbit about 5^d 21^h 8^m. As far as we can judge from its visibility at its enormous distance from the Earth, this satellite is probably much larger than any other with which we are acquainted.

It may be interesting to notice that, in the case of Neptune (see also our remarks with regard to the planet Mars, p. 262), it is because it possesses a Satellite that we are able to weigh it, as we have above stated, with very considerable accuracy. If it were not for this we should have to use indirect methods which would render the problem vastly more difficult.

We suppose that our readers must by this time have perceived that we really know but very little about either of these two far-distant planets, Uranus and Neptune. Nevertheless it may not be amiss if we endeavour to extend our gaze somewhat further still by asking:—Do any other planets, even more distant than these, exist beyond them?

At present, as we have mentioned, the study of the perturbations of Neptune's movement has not afforded any satisfactory indications in favour of an affirmative reply to this query, although Mr. Todd has even gone so far as to make a careful search with the great Washington telescope, in a region of the heavens where he had some reasons for thinking that such an outlying planet might possibly be found.

Nor does another ingenious method by which the same difficult problem has recently been attacked by Professor Forbes of Glasgow, depending upon a well-known peculiarity of certain cometary orbits, seem likely to achieve any greater success. It is found, that the maximum distance from the Sun of several comets, whose paths have been accurately determined, is attained when it is nearly equal that of the orbit of Jupiter; while there are others which (speaking approximately) attain a maximum distance which is nearly the same as that of the orbit of Saturn, or of Uranus, or of Neptune. Having attained such a position, they (so to say) turn round and begin to approach the Sun again. But this is just the sort of path which would now belong to any comet, which at some past date might have been previously moving in an altogether different orbit, but while so moving might have happened, on a special occasion, to have passed very close to one of the above-named planets. In such a case the attraction of the planet would very likely have altogether changed the comet's orbit, so that it would be for the future such an one as we have described.

What we have stated, then, comes to this :—A certain number of comets are found, whose maximum distances from the Sun (or, to speak more technically, the distances of whose *Aphelia*) are nearly equal to that of Jupiter. This can be shown to be a probable result of Jupiter's attraction upon them, and this series of comets may consequently be described as having orbits regulated, or *dominated*, by Jupiter. Also certain other comets are found whose orbits have in like manner been affected by Saturn, or by Uranus, or by Neptune.

Professor Forbes has, therefore, suggested that, if a planet really exist beyond the orbit of Neptune, we may obtain some evidence of its presence by its effect upon the orbits of such comets as may, at some time or other, have passed very near to it. If we can find some of these bodies, whose paths show that they have undergone a certain effect *common to several of them*, it may arise from the attraction of such a planet. If others, travelling to a greater distance, show a repetition of another and a different effect, they may indicate the power of another and still more distant planet. Suitable calculations may point out the position in which to search for such planets; but any such investigation must by its very nature be vastly less accurate than one which follows the method of Adams and Le Verrier. It is, however, so ingenious that it well deserves some consideration.

We will now draw our remarks upon the planets to a conclusion by a brief reference to a hypothesis which we some time since discussed more fully in another course of lectures delivered at GRESHAM COLLEGE. We refer to the supposition that important results may arise from the passages of the planets, and especially of those of the larger planets, through their *Perihelia*, or nearest positions to the Sun.

It is, of course, well known that the Conjunctions and the apparent near approaches of the planets to one another (which only involve their being seen very nearly in the same direction, although really many millions of miles apart), have always been believed by astrologers to have a baneful or beneficent influence; a belief which is illustrated by the quotation from Shakespeare which forms a heading to this lecture. And any one may assure himself that such opinions are still maintained, if he will take the trouble to read "Zadkiel's Almanac," or some more elaborate astrological work. He will there find that these planetary positions are supposed to act in combination with the relation of certain planets and their houses in the heavens to individuals and countries; nor need he be surprised if a considerable proportion of the vague prophecies made in connection with them sometimes come true, while probably a much larger proportion, to the failure of which little attention is subsequently paid, prove to be false.

We might quote many another poet in support of this belief, to which Byron refers in the well-known lines:—

> "Ye stars! which are the poetry of heaven,
>   If in your bright leaves we would read the fate
>   Of men and empires!"

But it is not only among poets or astrologers that such notions prevail. They maintain a remarkable hold upon the popular mind. Any oracular statement connected with them spreads with marvellous rapidity when once publicity is given to it. And, of late, certain prophets, or seers, seem to have taken up the theory that, in addition to any results dependent upon their Conjunctions or Oppositions, the passages of the larger planets, *i.e.* of Jupiter, Saturn, Uranus, and Neptune, through their *Perihelia*, may be expected to have most disastrous and disturbing effects upon the Earth, and may bring about plagues and pestilences of the most awful and destructive character. An attempt has also been made to show, from the history of plagues, that, on one or two previous occasions, viz., in the sixth and sixteenth centuries, the effect of a near coincidence in the times of their passing through their Perihelia has been actually such.

It has been stated that about every twelve years, when Jupiter thus passes, evil consequences result ; and that, once in every fifty-nine years, when the Perihelion-passages of Jupiter and Saturn nearly coincide in date, they are especially injurious.

We have not now space to discuss this subject at length. We can only say that, although it is impossible to deny that the approach of a large planet may in some way act upon the Sun, or upon the Earth, we have, after considerable search, met with no sufficient record of any apparent connection between times of pestilence and such near approaches. On the contrary, if any actual effect takes place, we should think that it might, by disturbing the Sun and increasing its heat, rather bless than curse the planets in general. And we should expect that the approach of a planet to the Sun when passing through its Perihelion might affect that planet itself (whether for good or evil) much more than it would, through its influence upon the Sun, affect any other planet ; so that some special commotion in the spots, or belts, of such a planet as Jupiter might possibly have some connection with its own Perihelion-passage.

It must, of course, be carefully noticed that the near coincidence of the Perihelion-passages of the planets does not mean

THE PLANETS URANUS AND NEPTUNE. 419

that, when they so pass, they are nearly in the same direction as seen from the Sun, or all nearly in a straight line drawn from the Earth. On the contrary, it is only a near coincidence *in time* that is referred to, while the Perihelia in question may be in very different longitudes in the heavens. This is shown in Fig. XCIII., in which the various dates and positions of the

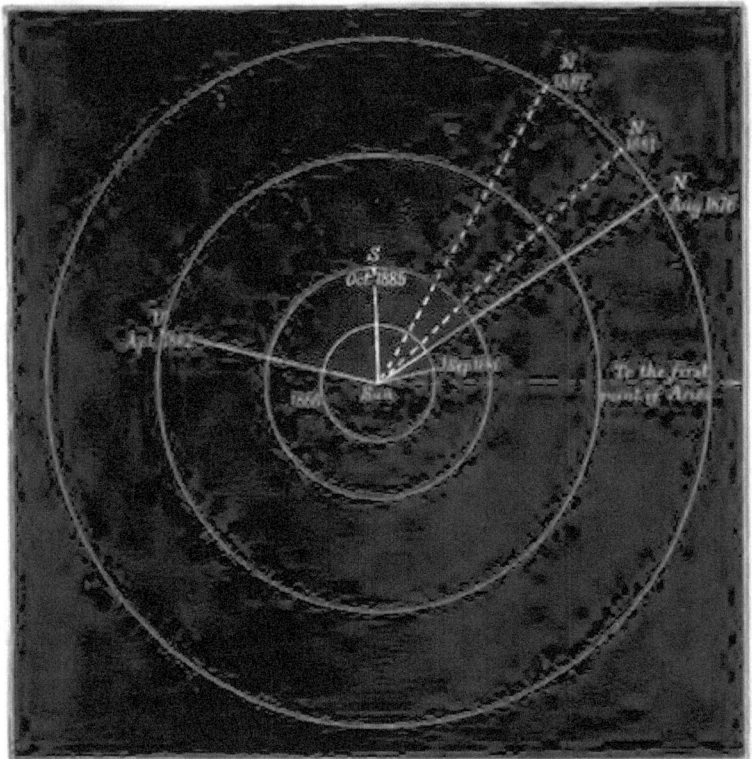

Fig. XCIII.—Positions J, S, U, N, of the Perihelia of the orbits of Jupiter, Saturn, Uranus, and Neptune, between A.D. 1876 and A.D. 1887; and of the Aphelion of Jupiter in 1886.

Perihelia of the orbits of Jupiter, Saturn, Uranus, and Neptune, which occur during several consecutive years, beginning with 1876, are indicated. It is, however, impossible, with the scale upon which the diagram is drawn, to represent the orbits otherwise than as circles.

It is quite true, as the above figure indicates, that Jupiter passed through Perihelion on September 25th, 1880; that

Saturn will so pass on October 19th, 1885; that Uranus so passed on April 9th, 1882; and that Neptune, as we have previously explained, would have done so in 1881, if its movements had not been perturbed; but that it actually was at one nearest position to the Sun in March 1876, and will be so again in, or about, 1887, or 1888.

In connection with any such supposed effects as those to which we just now referred, it may also be remarked, that it cannot matter much whether, when any one of the above planets is in Perihelion, the Earth is very nearly *between it and the Sun*, or not. For such is their comparatively slow motion, that, if the Earth were, at the date in question, exactly upon the opposite side of the Sun, Neptune would only have moved through $\frac{1}{330}$th of its orbit away from its Perihelion, Uranus through about $\frac{1}{170}$th, Saturn through about $\frac{1}{57}$th, and Jupiter through about $\frac{1}{22}$nd, before the Earth would catch them up and come between them and the Sun. In fact, even Jupiter would still be comparatively close, and all the others would be very close indeed, to Perihelion, when the Earth, between six and seven months afterwards, would make its next especially near approach to them by passing between them and the Sun.

It is rather to be noticed, and is much more to the point, that, if any such effect be produced, the Earth may be under its influence for a considerable time, since, as it goes round *its whole orbit* it will be decidedly nearer to any one of these planets which may be about to pass, or may have just passed, through its Perihelion, than if the planet were in, or near to, its Aphelion.

For instance, Jupiter would, in the former case, be something like 46,000,000 miles nearer to the Sun *during the whole of the Earth's year* than in the latter case; so that, instead of the Earth's distance from it varying, as it would near to its Aphelion passage, from about 414,000,000 miles to 600,000,000 miles during the year, its greatest distance would be about 554,000,000 miles, while its least distance would be reduced to about 368,000,000 miles. In other words, its *average distance* all the year through would be diminished by 46,000,000 miles.

We are willing to allow all this, and not only so, but in addition, that, in the year 1880, the Earth did also happen to pass between Jupiter and the Sun most remarkably near to the time of Jupiter's passage through its Perihelion, viz., within eleven days; and yet we see no necessity for believing that any special influence must, under such circumstances, arise,—certainly not that any harmful results must ensue.

We must, moreover, recollect that we have no idea, if any such action should take place, by what law it would be governed. It might be proportional to the mass of the body so acting divided by its distance; or divided by that distance squared, or cubed. It might follow laws similar to those of electricity, or of magnetism; or it might simply act according to the ordinary law of gravitation. If the latter be the case (a possible, although not a very probable, supposition, which we may use for the purpose of illustration as a mean between various imaginable laws), we may calculate, that the maximum attractive effect of Jupiter upon the Earth would be nearly fourteen times that of Saturn, Saturn's about thirty-three times that of Uranus, and that of Uranus more than twice that of Neptune; so that (speaking approximately) Jupiter's maximum influence would approach 500 times that of Uranus, and exceed 1,000 times that of Neptune.

We may also remark, that not only do the great distances of Neptune and Uranus from the Earth almost put them out of the question altogether, but the very near approach of Neptune's orbit to a circular form makes its Perihelion-passage utterly unimportant. If any Perihelion effects exist, they must practically be limited to those of Jupiter and of Saturn; and Jupiter is so much the more important, both in mass and proximity, that all we should think it necessary to do to settle the question (and this may be worth the while), is carefully to compare meteorological records, and to see whether any periodic change can be detected which corresponds with the dates of *Jupiter's* Perihelion-passages.

At the same time, it must be remembered, that the period of Sun-spot maxima and minima (see Lecture II., p. 37) has an average duration of about 11·1 years, and might therefore correspond, for a considerable length of time, with that of any effect produced by Jupiter, and cause an apparent, although not real, coincidence, with the period of 11⅔ years, which elapses between its Perihelion-passages. A sufficiently long-continued series of observations would, however, distinguish between the two classes of phenomena.

As to prophecies of impending trouble connected with the near coincidence of the dates of the Perihelia which occur between 1880 and 1885, we most confidently believe that no one need regard them, although it is certainly somewhat remarkable that the dates should just now follow one another quite so rapidly. It is also an unfortunate fact for the prophets in question, that, Jupiter's period being barely twelve years, that huge planet will be very nearly in its *Aphelion*, or at its greatest distance from the Sun, the position of which is marked 1886 in Fig. XCIII., before Saturn will have reached its Perihelion in 1885. In 1885 we might therefore expect the influence of Jupiter, and that by far the most important of all such possible influences, to be nearly reversed, and to oppose that of Saturn.

It may also be well to notice another important fact; viz., that the *change* in the attractive effect of the Sun upon the Earth, in the course of *each single year*, is vastly greater than the *change* in the attractive effect, even of Jupiter, when its Perihelion and Aphelion passages respectively draw it especially near to the Earth, or remove it to an unusual distance from it. This is, of course, upon our previous supposition, which we have adopted for purposes of illustration, that the action referred to takes place according to the law by which gravity attracts. So huge is the Sun that a change of 3,000,000 miles in its distance, at different times in the course of each year, alters its attraction upon the Earth about 10,000 times as much, as a change of 46,000,000 miles of distance alters the average value of that of Jupiter. It may also be calculated that the annual change in the Sun's attraction upon the Earth is about 1,800 times as great as the whole average attraction of Jupiter. Hence we may easily conclude that, if the Perihelion-passage of Jupiter affects the Earth, the Earth's own Perihelion passage might, year by year, be expected to do so vastly more.

It is, of course, just possible that Jupiter might, when near to Perihelion, affect the Sun, and that such an effect *transmitted through the Sun* to the Earth might be much greater than its direct action. And we are aware, that it has at times been suspected that Jupiter's position relatively to the Sun actually influences its spots, or other phenomena, to some slight extent. Nevertheless, no decided proof has yet been obtained of any influence exerted by Jupiter (huge as is its bulk) upon its much greater ruler, which deserves to be termed important.

We may, therefore, say that, up to the present time, no effect, *direct or indirect*, of the passage of Jupiter, much less of any other planet, through its Perihelion, has been proved to act upon the Earth, either for good or evil. On the other hand, we cannot absolutely affirm its impossibility. And we may further allow that Jupiter is so important a body, that it may be well to watch for any synchronism of other phenomena with its periodic time, which, if detected, would deserve the most careful attention.

Here we must for the present close our discussion of the Solar System;—its Sun, its Planets, and its Moons. Not that we can pretend that our treatment of either Satellite or Planet has been in any wise exhaustive, or as complete as we could have wished to make it; much less were we able to give due space in our first two short lectures to the description of the marvels of the Sun, or to follow out the many suggestive lines of thought which they involve, and the important issues to which they lead.

Nor do we forget, not only that there may be many bodies in the Solar System yet unknown to us—Planets circling beyond the path of Neptune, or within that of Mercury—Minor Planets, the succession of whose discovery has hitherto never ceased,—Comets, whose known numbers increase as each year passes, and of which many more would probably be seen if sufficient time could be devoted to a systematic scouring of the heavens in order thereto;—but that, even amongst the known members of the Solar System, we are regretfully leaving many unnoticed.

To it belong some Comets, whose orbits are so precisely determined, that we can calculate their daily places almost as accurately as those of the Planets; some, whose periods round the Sun are so short * that they come within the limits of those of the Minor Planets, while their orbits are not very much more oval, or excentric, than those of some of these last-named little bodies. There are also, as we have previously mentioned, several well-known Comets which, in journeying round the Sun, never go much farther away than the orbit of Jupiter; others whose range is approximately bounded by that of Saturn, or of Uranus; others which sweep outwards a little beyond that of Neptune, amongst which last is included the great comet of Halley;† while there are others, still under the control of our Sun, which, in their elongated paths, rush away to distances far more remote, before they turn back to a nearer proximity again. All these are members of our Solar System.

And many a swarm of meteorites, mostly minute, but mingled occasionally with some of larger bulk, revolve in regular orbits round the Sun, in which in some cases (if not in most) they lag behind comets to which they have once belonged, or of the dissipation of which they are the remnants. They are ever falling upon the Sun as fuel; or being exploded into dust by the furious heat which the friction of our own, or of some other, atmosphere generates as they hurry through it; or, by collision with Satellites and Planets, or with one another, losing their individuality and becoming aggregated into some larger mass. While they thus revolve, they too are members of our Solar System.

And it has its visitors as well—Comets and Meteorites which rushing onwards from distant regions of space travel once, and

* As, for instance, $3\frac{1}{3}$ years in the case of Encke's comet, and $5\frac{1}{2}$ years in the case of Winnecke's.

† To its return in 1910, in what will probably be a most effective position for observation in our latitudes, those who are now young may look forward with pleasing anticipations.

only once, through its boundaries, in a hyperbolic path from which they never return—visitors that are its links to the universe beyond.

As to all these we must at present say nothing. Perhaps our best consolation may be, that, as yet, we know so little that is positive and certain with regard to them, that, if our space had allowed us to discuss them, we might very likely have been led away from the realms of positive and well-ascertained truth, to which we have endeavoured as far as possible to confine the lectures, to others rich in fascination, but full of uncertainty, and vaguely hypothetical.

Let us hope, that the information which, to the best of our ability, we have put before our readers may help them, if they should be disposed in their further studies to venture upon the stormy and doubtful sea of hypothesis, to start as from a safe and well-surveyed harbour, to which they may ever and anon return, and, casting anchor, pause awhile to meditate upon the results of their voyage. May the facts here recorded be as ballast to their vessel, to keep it from drifting to shipwreck on the rocks of error, before some gale of false deduction, or hasty conclusion; and as warning beacons point to a careful and persevering study of accurate and positive truth, and to the mathematical investigation of the laws that govern facts definitely observed, as the means by which Astronomy has, in the past, gained its marvellous triumphs and achieved its wondrous successes, and shall, in the future, gain triumphs more marvellous and achieve a success more wondrous by far.

It may be an appropriate reminder of the subject matter which this course of lectures has embraced, and a not unsuitable conclusion to the whole, to put before our readers the following simple diagram, which shows how vastly the mighty globe of our great central luminary surpasses that of the largest Planet; and how small and unimportant is the Earth compared with several of its brethren in the family of the Sun. It has been drawn as nearly as possible upon a correct scale, and is intended to represent, according to the statements made from time to time in the preceding pages :—

A diameter for the Sun			of about	865,000	miles
an equatoreal	,,	,, Jupiter	,,	88,000	,,
,,	,,	,, Saturn	,,	74,000	,,
an exterior	,,	,, Saturn's Rings	,,	166,000	,,
a	,,	,, Neptune	,,	35,000	,,
,,	,,	,, Uranus	,,	32,000	,,
an equatoreal	,,	,, the Earth	,,	7,926	,,
a	,,	,, Venus	,,	7,700	,,
,,	,,	,, Mars	,,	4,200	,,
,,	,,	,, Mercury	,,	3,000	,,

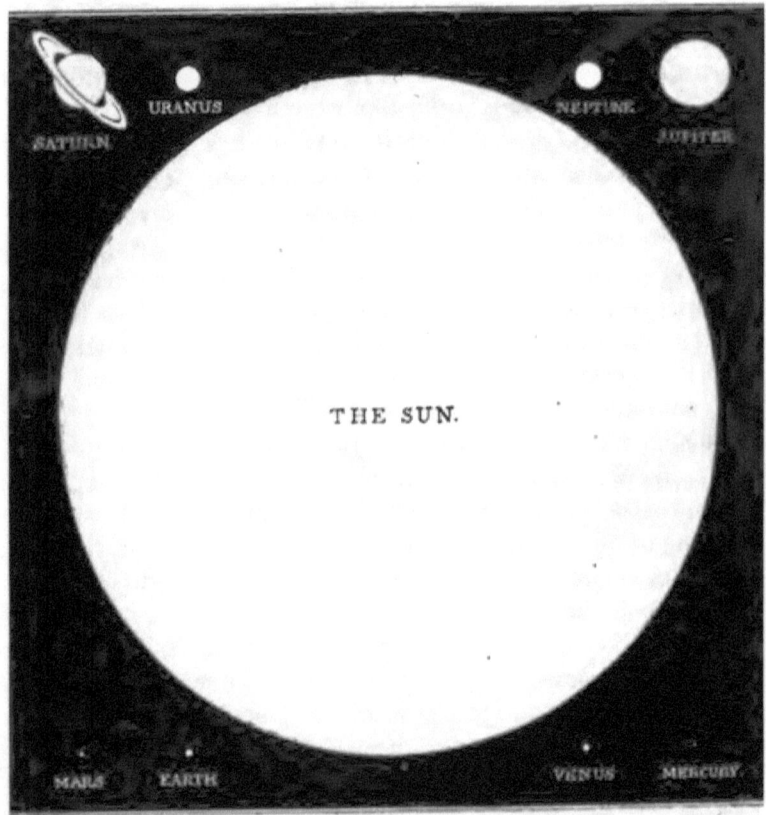

Fig. XCIV.—Comparative sizes of the Sun and the planets.

INDEX.

Aberration of light, 93.
Adams, Prof., discovery of Neptune, 410, 414.
Æthra, minor planet, orbit of, 267, 273.
Airy, Sir G. B., Harton colliery experiment, 217; rotation of Jupiter, 310; solar prominences, 32.
Apogee, the Sun when in, 228.
Arctic latitudes, attained by explorers, 190.
Artillery experiments, and the Earth's rotation, 206.
Asteroids, 266 (see *Planets, Minor*).
Astronomy, nautical, and Gresham College, 93.
Atmosphere, lunar, 79; of Venus, 165.
Aurora, the, and Sun-spots, 39.
Baily, Mr., the Cavendish experiment, 214.
Ball, the Messrs., and Saturn's ring, 367.
Beck and Co., Messrs., stereographs of the Moon, 84.
Bianchini, observations of Venus, 163.
Bode's so-called law of planetary distances, 128; its application to Uranus, 396; to Neptune, 414; to Jupiter's satellites, 329; to Saturn's, 386.
Bond, Prof., discovery of Saturn's ring C, 368; disappearances of Saturn's rings, 372.
Calendar, the, and the Moon, 94.
Carlos, Rev. E. S., his translation of the *Sidereal Messenger*, 321.
Carrington, observations of Sun-spots, 33.
Cassini (I), J. D., apparent paths of the planets, 101; observations of Venus, 163; rotation of Jupiter, 311; duplicity of Saturn's ring, 367.
Cavendish experiment, the, 208—216.
Centrifugal force, 203.
Ceres, minor planet, colour of light of, 278; discovered, 269; distance of, 267; size of, 278.
Chart of Mars (see *frontispiece*).
Chemical elements, dissociation of, 26, 45.
Christie, Mr. (Astronomer Royal), transit of Mercury in May 1878, 151.
Colbert, Prof., sizes of Jupiter's satellites, 326.
Comets, affected by planets, 416; belonging to the solar system, 424; comets and ultra-Neptunean planets, 416.
Commercial panics and Sun-spots, 40.
Conic sections, 281.
Conjunction, inferior and superior, 125.
Copernican theory, the, 117, 118; tested, 127, 324; and Jupiter's satellites, 324.

Corder, Mr., drawings of Jupiter, 309.
Corona, the, 43; Mercury seen on the, 153.
Corporation of London, the, library of, 115.
Crabtree, William, 4.
Craters, lunar (see *Moon*).
Creswick, Mr., transit of Mercury, 151.
Croll, Dr., "Climate and Time," 187.
Crystal spheres, 116.
Cycles and epicycles, of Ptolemy, 107; of Mars, Jupiter, and Saturn, 110; of Venus and Mercury, 107.
Cyclones, upon the Sun, 35.
Darwin, Mr., on prehistoric tides, 92.
Day, the, its length, 175; secular change in, 225; sidereal and solar, 222, 228.
Daylight observations of Venus, 157, 162.
Dawes, Mr., discovers Saturn's ring C, 368.
Dearborn Observatory, observations of Jupiter at, 308, 312, 317.
Declination, 134.
Deimos, satellite of Mars, 257.
De La Rue, drawing of Jupiter, 304; of Saturn, 358, 369.
Denning, Mr., rotation of spots on Jupiter, 312; announces the *Great Red Spot*, 306.
Diameter, the, of the Earth, 228.
Distances, of the planets, 117; Bode's so-called law of, 128; method of lunar, 93.
Duponchel, M., the Sun-spot period, 37.
Earth, the, axis of, its inclination, 228; centre of gravity of, and of the Moon, 54; centrifugal effect of rotation of, 205, 220; density of, 216, 228; its day, 222, 225, 228; its diameters, 228; eccentricity of its orbit, secular change in the, 188; flattened at the poles, 195; fluid in the past, 207; gravity upon, diminished at the equator, 205; heat, internal, of, 218; interior of, 220; magnetism of, and Sun-spots, 39; nearest to the Sun, when, 189, 228; rotation of, proofs of, 197; seasons of, 178, 228; seen from the Moon, 80; shape of, 193; solidity of, 219; tides upon, 55; volume of, 228; weight of, methods of determining, 207, 228.
Easter and the Moon, 94.
Eccentricities, of the orbits of the planets, 120; of those of the minor planets, 273.
Eclipses, lunar, 82; frequency of, 89; number in any year, 87.
Eclipses, solar, 87; of August 17th, 1868, 92; of July 29th, 1878, 28; of

May 17th, 1882, 28; duration of, 88, 90; frequency of, 89; next in England, 88; upon Jupiter, 341; number in any year, 89; phenomena visible in, 43.
Ecliptic, the plane of the, 124.
Electrical phenomena upon the Sun, 36.
Elements, chemical, dissociation of, 26.
Ellipse, description of an, 11
Ellis, Mr., the Earth's magnetism, and Sun-spots, 38.
Encke, perturbations of minor planets, 285; the solar parallax, 13.
Engelmann, Prof., observations of Jupiter's satellites, 346; their sizes, 326.
Equator, diminution of gravity at the, 205.
Faculæ, solar, 21, 35.
Feasts, fixed and movable, 95.
Fisher, Rev. O., physics of Earth's crust, 219.
Flammarion, M., the satellites of Jupiter, 346.
Forbes, Prof., on comets and ultra-Neptunean planets, 416.
Foucault's pendulum, 199.
Frigga, variability in light of, 279.
Galileo, Galilei, name of, 302; Jupiter's belts, 302; his "Sidereal Messenger," 321; perceives Saturn's ring, 366; his telescopes, 325; discovers the satellites of Jupiter, 321; his diagrams of them, 322; their light, 345; their periods, 323; the phases of Venus, 170, 324.
Galle, Dr., sees Saturn's ring C, 368.
Gill, Mr., the Sun's distance by observations of Mars, 17.
Glacial epochs, 187.
Gledhill Mr., observations of Jupiter, 309.
Granules, solar, 42.
Gravity, diminution of, at the equator, 205.
Green, Mr. N. E., chart and views of Mars, 239.
Greenwich observations of the Earth's magnetism, 38.
Gresham College, Foster, Samuel, Professor in, 4; Hooke, Dr., Professor in, 93; the aberration of light, 93; experiments with falling balls, 198; observations of Mars, 236; rotation of Jupiter, 311; minor planets, lecture upon at, 267; nautical astronomy in, 93; observatory of, 93.
Gwilliam, Mr., transit of satellite and shadow over the Great Red Spot on Jupiter, 317.
Gyroscope, the, 202.
Hall, Professor Asaph, discovers the satellites of Mars, 249; rotation of Saturn, 355.
Harmonia, minor planet, 275.
Harmony, the celestial, 116.
Heat of the Earth, internal, increase of, 218.
Helium, 34.
Herschel, Sir Wm., life of by Prof. Holden, 395; rotation of Jupiter, 312; size of Jupiter's satellites, 326; observations of their light, 345; of Saturn, 355; of its rings, 368, 372; his theory of Sun-spots, 21; discovers Uranus, 393; and two of its satellites, 399.
Herschel, Sir J., perturbations of Uranus, 411.
Hilda, minor planet, orbit of, 267; long period of, 276.
Hirst, Mr. G. D., drawings of Jupiter, 304; observation of satellite's shadow, 344.
Holden, Prof., life of Sir W. Herschel, 395.
Homer, references to Venus, 106; names for moons of Mars from, 251.
Hooke, Dr., Professor in Gresham College, experiments with falling balls, 198; observations of γ Draconis, 93; of Mars, 236; of Jupiter, 311.
Horizon, rational and sensible, 179.
Hornstein, Herr, sizes of minor planets, 280.
Horrox, Jeremiah, 4.
Hough, Prof., observations of Jupiter, 308, 312, 317.
Huggins, Dr., observation of a solar prominence, 30; spectra of Mars and of the Moon, 237; transit of Mercury, 151.
Huyghens, on Saturn's ring, 367; his "Systema Saturninum," 367.
Hyginus, the lunar crater, 78.
Inclination of the plane of an orbit, 126, 274; of orbits of minor planets, 274.
Inferior planets, greatest elongation of, 125.
Intra-Mercurial planets, when likely to be seen, 139; supposed observations of, 139.
Irene, minor planet, long period of, 276.
Iron, lines in spectrum of, 27.
Janssen, M., daylight observations of solar prominences, 29; photographs of the Sun, 41.
Juno, the minor planet, colour of its light, 278; discovery of, 269; distance of, 267; found by Olbers' hypothesis, 284; observation of by Mr. Lassell, 278; size of, 278; the Sun's distance found by, 292.
Jupiter, attraction of, compared with the Sun's, 422; belts of, 302; their colours, 308; not seen by Galileo, 302; centre of gravity of, and of the Sun, 298; cycle and epicycle of, 110; Dearborn Observatory, observations of, 308, 312, 317; density of, 295; diameter of, 297; distance of, 298; drawings of, by Conder, 309; De la Rue, 304; by Gledhill, 309; Hirst, G. D., 304; Pritchett, 306; Terby, 310; Trouvelot, 307; Various, 305; early observation of, 302; gravity on, 296; gravity diminished on equator, of 313; habitability of, 318, 319; heat of, 295, 353; oblateness of, 297; orbit apparent of, 102, 112; inclination of, 299; oval markings on, 305; period of, in orbit, 299; perturbations of minor planets by, 289; physical condition of, 295, 313; polar flattening of, 297;

rotation period of, 299, 310; satellites of (see *Satellites*); apparently without satellites, 347; seasons of, 318; shadow of, 332; size of, 294, 308; size of inhabitants of, 319; solar eclipses on, 341; solar light and heat on, 318; solar system, a miniature, 352; spots upon, 303; compared with Sun-spots, 311, 316; equatoreal white spot of, 312, 313; small black spots of, 313; Sun, the, its similarity to, 316, 353; surface of, 297; synodic period of, 129; velocity of, 299; of point on equator of, 300; weight of, 292, 295.

    The *Great Red Spot* upon, 305; first announcement of, 306; colour of, 307; first discovery of, 307; earlier drawings of, 309; present condition of, 314; rotation period of, 312; self-luminosity of, 317; transits of satellites and shadows across, 317.

Kepler, the three laws of, 121.
Kirkwood, Prof., on minor planets, 289; on Saturn's rings, 374; on the periods of satellites of Saturn, 386.
Labrador, the cold of, 186.
Langley, Prof., atmospheric absorption of solar radiation (*Preface*); Sun-spots, 22; the Sun's temperature, 45.
Laplace, the movements of Jupiter's satellites, 333; his nebular hypothesis, 287; its relation to the satellites of Uranus, 408.
Lassell, Mr., discovery of Saturn's ring C, 368; of 2 satellites of Uranus, 399; of the satellites of Neptune, 415.
Latitude, celestial, 134; highest Arctic attained, 190; length of a degree of, 195.
Ledger, Rev. E., observation of transit of Mercury, 152; naked-eye views of Mercury, 144; daylight observations of Venus, 162.
Lescarbault, his intra-Mercurial planet, 141.
Le Verrier, discovers Neptune, 410, 414; the Sun's distance, 13; on intra-Mercurial planets, 138; on matter between Mars and Jupiter, 279.
Light, aberration of, 15, 93; bright lines in spectrum of, 25; dark lines, 24; refrangibility of, 23; spectrum of, 23; velocity of, 14, 348; the zodiacal, 44.
Linné, the lunar crater, 78.
Little, Dr., transit of Mercury, 152.
Lockyer, Mr., daylight view of prominences, 29; dissociation of chemical elements, 26; solar eclipse of May 17th, 1882, 28.
Lohse, Dr., rotation of Jupiter, 312.
Lomia, the minor planet, 274.
Longitude, celestial, 134; by eclipses of Jupiter's satellites, 348; by lunar distances, 93.
Lyman, Prof., the atmosphere of Venus, 165.
Madan, Mr., names of satellites of Mars, 251.

Mädler, diameters of minor planets, 277
Mamhead, a village near Exeter, 368.
Mars, chart of (see *frontispiece*); atmosphere of, 246; axis of, 243; canals on, 240; cycle and epicycle of, 110, 113; diameter of. 248; and the Sun's distance, 16; Mr. Gill's observations of, 17; habitability of, 245; oppositions of, 231; orbit of, 230; path of, seen from the Earth, 105; phases of, 242; phenomena seen from, 253; physical features of, 234; polar snows of, 238; red colour of, 242; rotation of, 236; satellites of (see *Satellites*); seasons of, 244; size of, 248; solar heat and light on, 244; southern hemisphere of, 237; spectrum of, 237; synodic period of, 129; weight of, 262.
Marth, Mr., phenomena seen from Mars, 253.
Maskelyne, Dr., the Schehallien experiment, 208.
Maunder, Mr., articles on Mars, 246.
Maxwell, Prof. J. C., on Saturn's rings, 373.
Medusa, minor planet, short period of, 276.
Mercury, axis of, 145; brilliancy of, compared with Venus and Jupiter, 149; corona, seen on the, 153; cycle and epicycle of, 107; elongation, greatest, of, 129; habitability of, 149; mountains on, 145; naked-eye views of, 144; names, early, of, 106; observations, early, of, 143; orbit of, 143; path of, seen from the Earth, 100, 103; solar light and heat on, 146; spectroscopic observations of, 153; synodic period of, 129; transits of, in May and November, 126; transits of, list of, 150; phenomena seen in, 151.
Meteorites, orbits of, 424; swarms of, near the Sun, 44.
Meudon, observatory of, solar photography at, 41.
Michell, Rev. Jn., inventor of the torsion-balance, 209.
Milton, references in, to cycles and epicycles. 98; to the Earth, 197; to Venus, 106, 155.
Millosevich, diameters of minor planets, 278.
Minor planets (see *Planets, Minor*).
Moon, the, acceleration of, secular, 225; albedo of, 163; the Alps and Apennines of, 72; atmosphere of, 79; attraction of Sun and Earth upon, 57; the calendar, its relation to, 95; celestial scenery of, 80; centre of gravity of and of the Earth, 54; its concave orbit, 58, 349; cones upon, 76; Copernicus, the mountain, 71; craters of the Moon, 75; its distance from the Earth, 50; Easter, its relation to, 95; Earth, the, seen from, 80; eclipses of, 82; eclipses caused by, 87; their frequency, 89; health, its relation to, 96; Hyginus,

the crater, 78; lunar librations, 67; light streaks, 74; Linné, the crater, 78; longitude by, 93; maps of, 73; its mountains, 70; nodes of orbit of, 85; orbit of, 58, 62; phases of, 63; a planet, 62; photographs of, 81; rills upon, 73; rotation of, 66; Mr Rutherfurd's photograph of, 68, 72; seas, so-called, of, 72; secular acceleration of, 225; size of, 52; stereographs of, 81; surface, how much seen, 68; temperature of surface of, 80; tidal power of, 55; tides, benefits of, 92; variations in distance of, 51; velocity of, 57; volcanic action on, 77; the weather, its relation to, 96; weight of, methods of determining, 53, etc.; zenith, seen in the, 90.

Nasmyth, Mr., model of the lunar mountain Copernicus, 71.
Newton, Mr. F. M., occultation of the shadow of a satellite of Jupiter, 344.
Newton, Sir I., the spectrum of light, 22; the Earth's density, 208.
Nisten, M., on minor planets, 267; their orbits, 290.
Nodes, of the Moon's orbit, 85.
Neptune, Bode's law, and, 414; density of, 415; discovery of, 409; distance of, 414; gravity upon, 415; perihelia of its orbit, 413; perturbations of, 413; perturbations of Uranus by, 410; seen by Dr. Galle, and at Cambridge, 412; observations of, in 1795, 413; satellite of, 415; size of, 414, 415; solar heat and light on, 415; speed of, 414; ultra-Neptunean planets, 415, 416.
Olbers, the hypothesis of, 282.
Orbits, of the planets, 117, 118; of the minor planets, 290, etc. (see also *Mercury*, *Venus*, etc.)
Palisa, Dr. J., discoverer of many minor planets, 270; rediscovers Hilda, 273.
Pallas, the minor planet, colour of light of, 278; discovery of, 269; distance of, 267; size of, 278.
Parallax, solar, 4, etc.
Pendulum, experiments in Horton Colliery, 217; Foucault's, 199.
Perigee, the Sun when in, 228.
Periods, orbital, of the planets, 117; synodic, 129.
Peters, Prof., numerous discoveries of minor planets by, 270; on intra-Mercurial planets, 141.
Phases, of an inferior planet, 127; of the Moon, 63; of Venus, 59; half-moon phase of Venus, 164; phases of Mars, 242.
Phobos, satellite of Mars, 251.
Photographs, of the Sun, 41; of the Moon, 68, 81.
Photosphere, of the Sun, 22.
Pickering, Prof., diameters of minor planets, 278; photometry of satellites of Mars, 251.
Pierce, Prof. B., on Saturn's rings. 373.
Planets, attractions of, upon the Earth, 421, 422; Bode's law of distances of, 123; conjunctions of, with the Sun, 125, 417; distances of, 117; eccentricities of orbits of, 119; elongations, greatest, of, 129; inclinations of orbits of, 126, 274; intra-Mercurial. 139; orbits of, 118; paths, apparent, of, 100; perihelion passages of, 417; their possible effects, 417, etc.; their positions and dates, 419; periods of in orbits, 117; phases of, 127, 159 164, 242; retrogression and stations of, 102, 130; sizes of, 425, 426; synodic periods of, 129; ultra-Neptunean, 415, 416; velocities of, 117
Planets, minor, discoverers of, 269, 270, 271; distances of, 267, 275, 289, 290; eccentricities and inclinations of orbits of, 273, 274; Jupiter's perturbations of, 289; Jupiter weighed by, 292; light, comparative, of, 271; meteoric matter and origin of, 288; names, curious, of, 271; the nearest to the Earth, 275; orbits of, 12, 290; near coincidences of, 277; origin of, possibly by a collision, 286; origin of and the nebular hypothesis. 287; Olbers' theory, 281, 283; periods of, 276; possible beyond Jupiter, 288; rediscovery of, 272; uses of, 291; and the satellites of Mars, 252, 293; sizes of, 277; Sun's distance found by, 291; variability of light of, 279; weight of, aggregate. 270; zone of, 268.
Planetoid, the name, 266 (see *Planets, Minor*).
Pole, celestial, altitude of, 180; definition of, 177; position, apparent, of, 178.
Polyhymnia, the minor planet, 273.
Pritchett, Prof., drawing of Jupiter, 306.
Primum Mobile, the, 116.
Proctor, Mr., map of Mars, 239; rotation of Mars, 236; his work upon Saturn, 366.
Prominences, solar (see *The Sun*).
Repulsive force in the Sun, 44.
Retrogression of planets, 102, 130, 136.
Right ascension, 134.
Rills, lunar, 73.
Römer, the velocity of light, 348.
Rosse, Lord, the temperature of the Moon's surface. 80.
Rotation of the Earth, proofs of, 197.
Russell. Mr. H. C., observations of Jupiter, 309.
Rutherfurd, Mr., lunar photographs by, 68, 72.
Sandwich Isles, great crater in, 76.
Sappho, the minor planet, 292.
Saturn, belts upon, 354; cycle and epicycles of, 110; distance of, 256; drawings of, 358, 369, 370, 371; observations of, by the Balls, 367; by Bond, 368; Cassini, 367; Dawes, 368; Galle, 368; Hall, 355; Herschel, 355, 368; Huyghens, 367; Lassell, 368; orbit of, 356; path of, seen from the Earth, 102; phases of rings of, 357, 359, etc.; polar compression of, 355; position of, fa-

vourable in 1885, 369; Proctor, Mr., his work upon, 366; satellites of (see *Satellites*); seen from its rings, 376; from its satellites, 390; from Uranus, 398; its shadow on the rings, 370, 376, 377; size of, 355; compared with Jupiter, 355; spots upon, 355; synodic period of, 129.

Rings, the, of Saturn—possibly approaching Saturn, 375; brightest part of, 369; bright side of, when visible, 362; constitution of, 373, 374; description of, 357, 367, etc.; dimensions of, 375; disappearances of, 363, 364; when next to occur, 365; discovery of, history of successive, 366, 369, etc.; divisions in, 291, 367, 374; eclipses of Sun by, 380, 384, eclipses of Sun on, 376; eccentrical, 372; gaps in, 291, 374; latitudes, whence invisible, 383; named A, B, C, 357; passage of their plane through the Earth, 360; their phases, 357, 359, 360, etc.; rotation of, 372, 373; satellites, composed of, 374; satellites threaded on, 371; Saturn seen from, 376; Saturn, shadow of on, 370, 376, 377; seen from Saturn, 378; seen from satellites, 390; thickness of, 371; when next most widely opened out, 357; ratio of their axes at such times, 360.

Satellites of Jupiter, atmospheres of, 346; Bode's law, and, 329; Copernican theory, and, 324; densities of, 330; discovery of, 321; distances of, 327, 329; eclipses of, 14, 332, 333, 348; longitude found by, 348; eclipses and occultations of, order of, 337; Jupiter seen without satellites, 347; Jupiter as a sun to, 353; light of, variations in, 345, 346; longitude found by, 348; movements of, Laplace's relation between, 333; observations of, by Englemann, 326, 346; Flammarion, 346; Galileo, 321, 345; Herschel, 345; Laplace, 333; occultations of, 335, 337; orbits of, when convex to the Sun, 349; the diameters of, 327; planes of, 331; paths in space of, 349, 351, 352; periods of, sidereal, 329; synodic, 323; phases of, 331; rotation of, 346; seen by the naked eye, 325; seen in horizon and in zenith, 331; seen through the edge of Jupiter, 347; shadows of, size of upon Jupiter, 341; transits of, 339; shadow preceding or following satellite, 340; occulted by satellite, 352; sizes of, 326, 327; solar eclipses caused by, their duration, 341; their frequency, 343; solar system, a miniature, 352; transits of, and of their shadows, 339, 345; variations in light of, 345, 346; velocity of light determined by, 14, 348; weights of, 330.

Satellites of Mars, 249, etc.; discovery of, 249; inhabitants of, 261; light of, 257; Mars seen from, 261; Mars weighed by, 262; may be minor planets, 252, 298; movements, apparent, of, 253; seen in horizon and zenith, 256; seen from Mars, 261; sizes of, 251; tides caused by, 258; transits of across Sun, 262.

Satellite of Neptune, discovery of, 415; distance, size, and orbit of, 415; inclination of its orbit, 407; retrograde motion of, 407.

Satellites of Saturn, Bode's law and, 386; compared with the solar system, 387; discovery of, 385; distances of, 385, 389; eclipses of, 388; Iapetus, variability in light of, 392; latitudes in which visible, 388; light of, 389; periods of, 385, 386; rings composed of, 374; rings, where hidden by, 387; rings seen from, 390; seen in horizon and zenith, 388; Saturn seen from, 390; sizes of, 385; threaded on the ring-line, 371; Titan, shadow of, 391.

Satellites of Uranus, discovery of, 399; distances of, 399; orbits, inclination of, 401; retrograde motion of, apparent, 404, 406; its relation to the nebular hypothesis, 408; periods of, 399.

Satellite of Venus, 166.

Schehallien experiment, the, 208.

Schiaparelli, Prof., the axis of Mars, 243; canals on Mars, 240.

Schröter, observations of Mercury, 145; of Venus, 164, 168.

Scott, Mr. Benjamin, supposed intra-Mercurial planet, 141.

Seasons of the Earth, 173; of the Tropics, 185.

Secchi, Padre, observations of Sun-spots, 33; of prominences, 31.

Sidereal and solar days, 222, 228; years, 227.

Siemens, Dr., regeneration of solar energy, 46.

Smyth, Admiral, observation of a satellite of Jupiter, 347; Saturn's ring, division of, 367.

Solstices, the, 175.

Specular reflection, of Venus, 165.

Spectrum, continuous, 24; diffraction, 26; lines in, 25; pure, 23; solar, 24.

Spectroscope, the, and Mars, 237; and Mercury, 153; and minor planets, 278; nature of, 23.

Spectroscopic Society, memoirs of Italian, drawings of prominences in, 31; of Sun-spots, 35; solar photographs in, 42.

Spheres, crystal, 116.

Spheroids, how generated, 194.

Stations of the planets, 102, 130, 136; of Venus, 135.

Stone, Mr., diameters of minor planets, 277; transits of Venus in 1769 and 1874, 11, 13.

Sun, the, in apogee, 228; attraction of, upon the Moon, 57; attraction of,

432　　　　　　　INDEX.

compared with Jupiter's, 422; centre of gravity of, and of Jupiter, 316; corona of, 43; daylight view of prominences of, 29; distance of, 4, 6, 13, 14, 16, 17, 291; most probable value of, 18; diurnal path of, 175, 183; eclipses of, 87; next, visible in England, 88; frequency of, 89; duration of, 90; phenomena of, 43; total of, August 1868, 92; July, 1878, 28; May, 1882, 28; granules, solar, 42; heat of, how maintained, 46; light, velocity of, and the Sun's distance, 14; magnetism of the Earth and Sun-spots, 36, 39; meteors revolving round, 44, 424; parallax of, 4; in perigee, 228; prominences of, connected with Sun-spots, 34; drawings of, 31; eruptive and quiescent, 34; very lofty, 31; seen in daylight, 29; repulsive force of, 44; spots on (see *Sun-spots*); substances existing in, 27; temperature of, 45; tidal power of, 55; upon Mars, 260; transits of Mercury across, 18, 150; of Venus, 2, 6, 9, 10, 11, 12, 126; Young's layer upon, 28; zodiacal light of, 44; zones of spots and prominences, 32.

Sun-spots, 21; cyclones in, 35; drawings of, 35; compared with those of Jupiter, 311, 316; large, 39; maxima and minima of, 37; origin of, 34; affected by the planets, 41, 423; show solar disturbance, 36; spectra of, 33; their relation to the weather, 40.

Swift, Mr., supposed intra-Mercurial planet, 142.

Synodic periods of the planets, 129; minor planets, sizes of, 278.

Tacchini, Prof., prominences, drawings of, 31; Sun-spots and faculæ, drawings of, 35.

Tebbutt, Mr., transit of Mercury, 152.

Telescope, the, power of the, 328.

Temperature, the, of the Moon's surface, 80; of the Sun, 45.

Teneriffe, the great crater of, 75.

Terby, M., drawings of Jupiter, 310.

Thomson, Sir Wm., solidity of the Earth, 219.

Tides, benefits of, 92; prehistoric, 92; on Mars, 258; and the Earth's rotation, 225; and the solidity of the Earth, 219; tidal power of the Sun and Moon, 55.

Tisserand, M., on intra-Mercurial planets, 141.

Titan, shadow of, upon Saturn, 391.

Titius, law of the planetary distances, 123.

Todd, Mr., observations of the satellites of Jupiter, 347.

Transits, of Venus (see *Venus*); of Mercury (see *Mercury*).

Tropics, seasons of the, 185.

Trouvelot, Mr. L., drawing of Jupiter, 307; of Saturn, 370.

Ultra-Neptunean planets, 415, 416.

Uranus, axis of, its inclination, 402, 408; Bode's law and, 393; discovery of, as a comet, 393; distance of, 396; gravity upon, 397; light of, compared with satellite of Jupiter, 408; past observations of, 395; orbit of, 396; perturbations of, by Neptune, 410; proved to be a planet, 394; planets, the, seen from, 398; rotation of, supposed, 398; satellites of (see *Satellites*); shape of, 397; size of, 397; solar heat and light upon, 398; velocity of, 397; weight of, 397.

Velocity of light, 14, 348.

Velocities of the planets, 117

Venus, albedo of, 163; atmosphere of, 165; axis of, 168; brilliancy of, 149; when greatest, 157; cycle and epicycle of, 107; daylight observations of, 162; drawings of transit of, in 1874, 11; elongation, greatest, of, 129; habitability of, 169; *lumière cendrée* of, 166; mountains on, 164; names early, of, 116; orbit of, 156; path of, in 1884, 133; path of, seen from the Earth, 104; phases of, 159, 164, 324; rotation of, 163, 170; satellite of, 166; size of, 155; specular reflection of, 169; when stationary, 135; synodic period of, 129.

Venus, transits of, 2-12. Black Drop, the 10; contact phenomena of, 10; dates of, 2; Delisle, method of, 9; duration of, 6; Halley, method of, 9; in June and December, 126; photographs of, 12; results of, in 1874, 11.

Vesta, minor planet, colour of light of, 278; discovery of, 269; distance of, 267; Olbers' hypothesis and, 284; size of, 278; spectrum of, 278; visible to the naked eye, 270.

Vico, De, observations of Venus, 164.

Victoria, minor planet, 292.

Vincent, Mr., supposed intra-Mercurial planet, 142.

Vogel, Dr., the rotation of Venus, 170; spectroscopic observations of minor planets, 278.

Watson, Prof., supposed intra-Mercurial planet, 142.

Weather, the, and the Moon, 96; and Sun-spots, 40.

Winter, moonlight in, 82.

Wolf, Prof., the Sun-spot period, 37.

Wolfius, on inhabitants of Jupiter, 319.

Wray, Mr., drawing of Saturn, 371; supposed intra-Mercurial planet, 141.

Year, the sidereal, length of, 227.

Young, Dr., reversal of the solar spectrum, 27; on Sun-spots and prominences, 33, 34.

Zodiac, the, 105.

Zodiacal light, the, 44, 139.

Zöllner, the albedo of the Moon, 163.

Zones of Sun-spots and prominences, 32.